Data Analysis and Data Mining

Data Analysis and Data Mining

An Introduction

ADELCHI AZZALINI

AND

BRUNO SCARPA

OXFORD
UNIVERSITY PRESS

OXFORD
UNIVERSITY PRESS

Oxford University Press, Inc., publishes works that further
Oxford University's objective of excellence
in research, scholarship, and education.

Oxford New York
Auckland Cape Town Dar es Salaam Hong Kong Karachi
Kuala Lumpur Madrid Melbourne Mexico City Nairobi
New Delhi Shanghai Taipei Toronto

With offices in
Argentina Austria Brazil Chile Czech Republic France Greece
Guatemala Hungary Italy Japan Poland Portugal Singapore
South Korea Switzerland Thailand Turkey Ukraine Vietnam

Published by Oxford University Press, Inc.
198 Madison Avenue, New York, New York 10016
www.oup.com

Library of Congress Cataloging-in-Publication Data

Azzalini, Adelchi.
[Analisi dei dati e "data mining". English]
Data analysis and data mining : an Introduction /
Adelchi Azzalini, Bruno Scarpa; [text revised by Gabriel Walton].
p. cm.
Includes bibliographical references and index.
ISBN 978-0-19-976710-6
1. Data mining. I. Scarpa, Bruno.
II. Walton, Gabriel. III. Title.
QA76.9.D343A9913 2012
006.3'12—dc23 2011026997

9780199767106

English translation by Adelchi Azzalini, Bruno Scarpa and Anne Coghlan.
Text revised by Gabriel Walton.

First published in Italian as Analisi dei dati e "data mining", 2004, Springer-Verlag
Italia (ITALY)

---

9 8 7 6 5 4 3 2 1

Printed in the United States of America
on acid-free paper

CONTENTS

PREFACE

When well-meaning university professors start out with the laudable aim of writing up their lecture notes for their students, they run the risk of embarking on a whole volume.

We followed this classic pattern when we started jointly to teach a course entitled 'Data analysis and data mining' at the School of Statistical Sciences, University of Padua, Italy.

Our interest in this field had started long before the course was launched, while both of us were following different professional paths: academia for one of us (A. A.) and the business and professional fields for the other (B. S.). In these two environments, we faced the rapid development of a field connected with data analysis according to at least two features: the size of available data sets, as both number of units and number of variables recorded; and the problem that data are often collected without respect for the procedures required by statistical science. Thanks to the growing popularity of large databases with low marginal costs for additional data, one of the most common areas in which this situation is encountered is that of data analysis as a decision-support tool for business management. At the same time, the two problems call for a somewhat different methodology with respect to more classical statistical applications, thus giving this area its own specific nature. This is the setting usually called *data mining*.

Located at the point where statistics, computer science, and machine learning intersect, this broad field is attracting increasing interest from scientists and practitioners eager to apply the new methods to real-life problems. This interest is emerging even in areas such as business management, which are traditionally less directly connected to scientific developments.

Within this context, there are few works available if the methodology for data analysis must be inspired by and not simply illustrated with the aid of real-life problems. This limited availability of suitable teaching materials was an important reason for writing this work. Following this primary idea, methodological tools are illustrated with the aid of real data, accompanied wherever possible by some motivating background.

Because many of the topics presented here only appeared relatively recently, many professionals who gained university qualifications some years ago did not have the opportunity to study them. We therefore hope this work will be useful for these readers as well.

Although not directly linked to a specific computer package, the approach adopted here moves naturally toward a flexible computational environment, in which data analysis is not driven by an "intelligent" program but lies in the hands of a human being. The specific tool for actual computation is the R environment.

All that remains is to thank our colleagues Antonella Capitanio, Gianfranco Galmacci, Elena Stanghellini, and Nicola Torelli, for their comments on the manuscript. We also thank our students, some for their stimulating remarks and discussions and others for having led us to make an extra effort for clarity and simplicity of exposition.

Padua, April 2004 Adelchi Azzalini and Bruno Scarpa

PREFACE TO THE ENGLISH EDITION

This work, now translated into English, is the updated version of the first edition, which appeared in Italian (Azzalini & Scarpa 2004).

The new material is of two types. First, we present some new concepts and methods aimed at improving the coverage of the field, without attempting to be exhaustive in an area that is becoming increasingly vast. Second, we add more case studies. The work maintains its character as a first course in data analysis, and we assume standard knowledge of statistics at graduate level.

Complementary materials (data sets, R scripts) are available at: `http://azzalini.stat.unipd.it/Book-DM/`.

A major effort in this project was its translation into English, and we are very grateful to Gabriel Walton for her invaluable help in the revision stage.

Padua, April 2011 — Adelchi Azzalini and Bruno Scarpa

1 Introduction

He who loves practice without theory
is like the sailor who boards ship without a rudder and compass
and never knows where he may cast.

—Leonardo da Vinci

1.1 New Problems and New Opportunities

1.1.1 Data, More Data, and Data Mines

An important phase of technological innovation associated with the rise and rapid development of computer technology came into existence only a few decades ago. It brought about a revolution in the way people work, first in the field of science and then in many others, from technology to business, as well as in day-to-day life. For several years another aspect of technological innovation also developed, and, although not independent of the development of computers, it was given its own autonomy: large, sometimes enormous, masses of information on a whole range of subjects suddenly became available simply and cheaply. This was due first to the development of automatic methods for collecting data and then to improvements in electronic systems of information storage and major reductions in their costs.

This evolution was not specifically related to one invention but was the consequence of many innovative elements which have jointly contributed to the creation of what is sometimes called the *information society*. In this context, new avenues of opportunity and ways of working have been opened up that are very different from those used in the past. To illustrate the nature of this phenomenon, we list a few typical examples.

- Every month, a supermarket chain issues millions of receipts, one for every shopping cart that arrives at the checkout. The contents of one of these carts reflect the demand for goods, an individual's preferences and, in general, the economic behavior of the customer who filled that cart. Clearly, the set of all shopping lists gives us an important information base on which to direct policies of purchases and sales on the part of the supermarket. This operation becomes even more interesting when individual shopping lists are combined with customers' "loyalty cards," because we can then follow their behavior through a sequence of purchases.
- A similar situation arises with credit cards, with the important difference that all customers can be precisely identified; there is no need to introduce anything like loyalty cards. Another point is that credit card companies do not sell anything directly to their customers, although they may offer other businesses the opportunity of making special offers to selected customers, at least in conditions that allow them to do so legally.
- Every day, telephone companies generate data from millions of telephone calls and other services they provide. The collection of these services becomes more highly structured as advanced technology, such as UMTS (Universal Mobile Telecommunications System), becomes established. Telephone companies are interested in analyzing customer behavior, both to identify opportunities for increasing the services customers use and to ascertain as soon as possible when customers are likely to terminate their contracts and change companies. The danger of a customer terminating a contract is a problem in all service-providing sectors, but it is especially critical in subsectors characterized by rapid transfers of the customer base, for example, telecommunications. Study of this danger is complicated by the fact that, for instance, for prepaid telephone cards, there can be no formal termination of service (except for `number portability`), but merely the fact that the credit on the card is exhausted, is not recharged after its expiration date, and the card itself can no longer be used.
- Service companies, such as telecommunications operators, credit card companies, and banks, are obviously interested in identifying cases of fraud, for example, customers who use services without paying for them. Physical intrusion, subscriptions with the intention of selling services at low cost, and subverting regulatory restrictions are only some examples of fraud-implemented methods. There is a need for tools to design accurate systems capable of predicting fraud, and they must work in an adaptive way according to the changing behavior of both legitimate customers and fraudsters. The problem is particularly challenging because only a very small percentage of the customer base will actually be fraudulently inclined, which makes this problem more difficult than finding a needle

in a haystack. Fraudulent behavior may be rare, and behavior that looks like an attempt at fraud in one account may appear normal and indeed expected in another.

- The Worldwide Web is an enormous store of information, a tiny fraction of which responds to a specific query posted to a search engine. Selecting the relevant documents, the operation that must be carried out by the search engine, is complicated by various factors: (a) the size of the overall set of documents is immense; (b) compared with the examples quoted previously, the set of documents is not in a structured form, as in a well-ordered database; (c) within a single document, the aspects that determine its pertinence, or lack thereof, with respect to the given query, are not placed in a predetermined position, either with respect to the overall document or compared with others.
- Also, in scientific research, there are many areas of expertise in which modern methods produce impressive quantities of data. One of the most recent active fields of research is microbiology, with particular reference to the structure of DNA. Analyses of sequences of portions of DNA allow the construction of huge tables, called DNA microarrays, in which every column is a sequence of thousands of numerical values corresponding to the genetic code of an individual, and one of these sequences can be constructed for every individual. The aim—in the case of microbiology—is to establish a connection between the patterns of these sequences and, for instance, the occurrence of certain pathologies.
- The biological context is certainly not the only one in science where massive amounts of data are generated: geophysics, astronomy, and climatology are only a few of the possible examples. The basic organization of the resulting data in a structured way poses significant problems, and the analysis required to extract meaningful information from them poses even greater ones.

Clearly, the contexts in which data proliferation manifests itself are numerous and made up of greatly differing elements. One of the most important, to which we often refer, is the business sector, which has recently invested significantly in this process with often substantial effects on the organization of marketing. Related to this phenomenon is the use of the phrase *Customer Relationship Management* (CRM), which refers to the structuring of "customer-oriented" marketing behavior. CRM aims at differentiating the promotional actions of a company in a way that distinguishes one customer from another, searching for specific offers suited to each individual according to his or her interests and habits, and at the same time avoiding waste in promotional initiatives aimed at customers who are not interested in certain offers. The focus is therefore on identifying those customer characteristics that are relevant to specific commercial goals, and then drawing information from data about them and what is relevant to other customers with similar profiles. Crucially, the whole CRM system clearly rests on both the availability of reliable quantitative information and the capacity to process it usefully, transforming raw data into knowledge.

1.1.2 Problems in Mining

Data mining, this new technological reality, requires proper tools to exploit the mass elements of information, that is, *data*. At first glance, this may seem paradoxical, but in fact, more often than not, it tells us that we cannot obtain significant information from such an abundance of data.

In practical terms, examining the data of two characteristics of 100 individuals is very different from examining the results of 10^2 characteristics of 10^6 individuals. In the first case, simple data-analytical tools may result in important information at the end of the process: often an elementary scatterplot can offer useful indications, although formal analysis may be much more sophisticated. In the second case, the picture changes dramatically: many of the simple tools used in the previous case lose their effectiveness. For example, the scatterplot of 10^6 points may become a single formless ink spot, and 10^2 characteristics may produce $100 \times 99/2$ of these forms, which are both too many and at the same time useless.

This simple example highlights two aspects that complicate data analysis of the type mentioned. One regards the *size* of the data, that is, the number of *cases* or *statistical units* from which information is drawn; the other regards the *dimensionality* of the data, that is, the number of features or *variables* of the data collected on a certain unit.

The effects of these components on the complexity of the problem are very different from each other, but they are not completely independent. With simplification that might be considered coarse but does help understand the problem, we may say that *size* brings about an increase primarily in computational aspects, whereas *dimensionality* has a complex effect, which involves both a computational increase similar to that of size and a rapid increase in the conceptual complexity of the models used, and consequently of their interpretation and operative usage.

Not all problems emerging from the context described can be ascribed to a structure in which it is easy to define a concept of size and, to an even lesser extent, of dimensionality. A typical counterexample of this kind is extracting those pages of the Web that are relevant to a query posted to a specific search engine: not only is it difficult to define the size of the set of cases of interest, but the concept of dimensionality itself is vague. Otherwise, the most classic and common situation is that in which statistical units are identified, each characterized by a certain predetermined number of variables: we focus on this family of situations in this volume. However, this is the structure in which each of the tables composing a database is conceptually organized; physical organization is not important here.

We must also consider the possibility that the data has 'infinite' size, in the sense that we sometimes have a *continuous stream* of data. A good example is the stream of financial transactions of a large stock exchange.

In the past few years, exploration and data analysis of the type mentioned in section 1.1.1 has come to be called *data mining*. We can therefore say that:

> data mining represents the work of processing, graphically or numerically, large amounts or continuous streams of data, with the aim of extracting information useful to those who possess them.

The expression "useful information" is deliberately general: in many cases, the point of interest is not specified a priori at all and we often search for it by mining the data. This aspect distinguishes between data mining and other searches related to data analysis. In particular, the approach is diametrically opposed, for example, to clinical studies, in which it is essential to specify very precisely a priori the aims for which data are collected and analyzed.

What might constitute useful information varies considerably and depends on the context in which we operate and on the objectives we set. This observation is clearly also true in many other contexts, but in the area of data mining it has additional value. We can make a distinction between two situations: (a) in one, the interesting aspect is the global behavior of the phenomenon examined, and the aim is the construction of its *global model*, taken from the available data; (b) in the other, it is characterization of detail or the *pattern structures* of the data, as we are only interested in cases outside standard behavior. In the example of telephone company customers, we can examine phone traffic data to identify trends that allow us to forecast customers' behavior according to their price plans, geographical position, and other known elements. However, we can also examine the data with the aim of identifying behavioral anomalies in telephone usage with respect to the behavior of the same customer in the past—perhaps to detect a fraudulent situation created by a third party to a customer's detriment.

Data mining is a recent discipline, lying at the intersection of various scientific sectors, especially statistics, *machine learning*, and *database* management.

The connection with database management is implicit in that the operations of data cleaning, the selection of portions of data, and so on, also drawn from distributed databases, require competences and contributions from that sector. The link with artificial intelligence reflects the intense activity in that field to make machines "learn" how to calculate general rules originating from a series of specific examples: this is very like the aim of extracting the laws that regulate a phenomenon from sampled observations. This, among the methods that are presented later, explains why some of them originate from the field of artificial intelligence or similar ones.

In light of the foregoing, the statements of Hand et al. (2001) become clear:

> Data mining is fundamentally an applied discipline ... data mining requires an understanding of both statistical and computational issues. (p. xxviii)
>
> The most fundamental difference between classical statistical applications and data mining is the size of the data. (p. 19)

The computational cost connected with large data sizes and dimensions obviously has repercussions on the method of working with these data: as they increase, methods with high computational cost become less feasible. Clearly, in such cases, we cannot identify an exact rule, because various factors other than those already mentioned come into play, such as available resources for calculation and the time needed for results. However, the effect unquestionably exists, and it prevents the use of some tools, or at least renders them less practical, while favoring others of lower computational cost.

It is also true that there are situations in which these aspects are of only marginal importance, because the amount of data is not enough to influence the computing element; this is partly thanks to the enormous increase in the power of computers. We often see this situation with a large-scale problem, if it can be broken down into subproblems, which make portions of the data more manageable. More traditional methods of venerable age have not yet been put to rest. On the contrary, many of them, which developed in a period of limited computing resources, are much less demanding in terms of computational effort and are still valid if suitably applied.

1.1.3 SQL, OLTP, OLAP, DWH, and KDD

We have repeatedly mentioned the great availability of data, now collected in an increasingly systematic and thorough way, as the starting point for processing. However, the conversion of raw data to "clean" data is time-consuming and sometimes very demanding.

We cannot presume that all the data of a complex organization can fit into a single database on which we can simply draw and develop. In the business world, even medium-sized companies are equipped with complex IT systems made up of various databases designed for various aims (customers and their invoices, employees' careers and wages, suppliers, etc.). These databases are used by various operators, both to insert data (e.g., from outlying sales offices) and to answer *queries* about single entries, necessary for daily activities—for example, to know whether and when customer *X* has paid invoice *Y* issued on day Z. The phrase referring to methods of querying specific information in various databases, called *operational*, is *OnLine Transaction Processing* (OLTP). Typically, these tools are based on *Structured Query Language* (SQL), the standard tool for database queries.

For *decision support*, in particular analysis of data for CRM, these operational databases are not the proper sources on which to work. They were all designed for different goals, both in the sense that they were usually created for administrative and accounting purposes and not for data analysis, and that those goals differ. This means that their structures are heterogeneous and very often contain inconsistent data, sometimes even structurally, because the definitions of the recorded variables may be similar but are not identical. Nor is it appropriate for the strategic activities of decision support to interfere with daily work on systems designed to work on operational databases.

For these reasons, it is appropriate to develop focused databases and tools. We thus construct a *strategic* database or Data WareHouse (DWH), in which data from different database systems merge, are "cleaned" as much as possible, and are organized round the postprocessing phase.

The development of a DWH is complex, and it must be carefully designed for its future aims. From a functional point of view, the most common method of construction is progressive aggregation of various data marts—that is, of finalized *databases*. For example, a data mart may contain all the relevant information for a certain marketing division. After the DWH has been constructed, the later aggregation must achieve a coherent, homogenous structure, and the DWH must be periodically updated with new data from various operational databases.

After completing all these programming processes (which can then progress by means of continual maintenance), a DWH can be used in at least two ways, which are not mutually exclusive. The first recomposes data from the various original data marts to create new ones: for example, if we have created a DWH by aggregating data mart for several lines of products, we can create a new one for selling all those products in a certain geographical area. A new data mart is therefore created for every problem for which we want to develop quantitative analysis.

A second way of using a DWH, which flanks the first, directly generates processing (albeit simplified) to extract certain information about the data summary. This is called *OnLine Analytical Processing* (OLAP) and, as indicated by its name, is made up of querying and processing designed in a certain way to be a form of data analysis, although it is still raw and primarily descriptive.

For OLAP, the general support is a structure of intermediate processing, called a *hypercube*. In statistical terms, this is a *multiway table*, in which every dimension corresponds to a variable, and every cell at the intersection of different levels contains a synthetic indicator, often a frequency. To give an example of this, let us presume that the statistical units are university students. One variable could be constructed by place of residence, another by department or university membership, gender, and so on, and the individual cells of the cross-table (hypercube) contain the frequencies for the various intersecting levels. This table can be used for several forms of processing: marginalization or conditioning with respect to one or more variables, level aggregation, and so on. They are described in introductory statistical texts and need no mention here. Note that in the field of computer science, the foregoing operations have different names.

As already noted, OLAP is an initial form of the extraction of information from the data—relatively simple, at least from a conceptual point of view—operating from a table with predefined variables and a scope of operations limited to them. Therefore, strictly speaking, OLAP returns to data mining as defined in section 1.1.2, but limited to a form that is conceptually a very simple way of processing. Instead, "data mining" commonly refers to the inspection of a strategic database and is characteristically more investigative in nature, typically involving the identification of relations in certain significant ways among variables or making specific and interesting patterns of the data. The distinction between OLAP and data mining is therefore not completely clear, but essentially—as already noted—the former involves inspecting a small number of prespecified variables and has a limited number of operations, and the latter refers to a more open and more clearly focused study on extracting knowledge from the data. For the latter type of processing, much more computational than simple management, it is not convenient to use SQL, because SQL does not provide simple commands for intensive statistical processing. Alternatives are discussed later.

We can now think of a chain of phases, starting as follows:

- one or more operational databases to construct a strategic database (DWH): this also involves an operation in which we homogenize the definition of variables and data cleaning operations;
- we apply OLAP tools to this new database, to highlight points of interest on variables singled out previously;
- data mining is the most specific phase of data analysis, and aims at finding interesting elements in specific data marts extracted from the DWH.

The term *Knowledge Discovery in Databases* (KDD) is used to refer to this complex chain, but this terminology is not unanimously accepted and *data mining* is sometimes used as a synonym. In this work, data mining is intended in the more restricted sense, which regards only the final phases of those described.

1.1.4 Complications

We have already touched on some aspects that differentiate data mining from other areas of data analysis. We now elaborate this point.

In many cases, data were collected for reasons other than statistical analysis. In particular, in the business sector, data are compiled primarily for accounting purposes. This administrative requirement led to ways of organizing these data becoming more complex; the realization that they could be used for other purposes, that is, marketing analysis and CRM, came later.

Data, therefore, do not correspond to any sampling plan or experimental design: they simply 'exist'. The lack of canonical conditions for proper data collection initially kept many statisticians away from the field of data mining, whereas information technology (IT) experts were more prompt in exploiting this challenge.

Even without these problems, we must also consider data collected in spurious forms. This naturally entails greater difficulties and corresponding attention to other applicative contexts.

The first extremely simple but useful observation in this sense has to do with the validity of our conclusions. Because a company's customer database does not represent a random sample of the total population, the conclusions we may draw from it cover at most already acquired customers, not prospective ones.

Another reason for the initial reluctance of statisticians to enter the field of data mining was a second element, already mentioned in section 1.1.2—that is, research sometimes focuses on an objective that was not declared a priori. When we research 'anything', we end up finding 'something' . . . even if it is not there. To illustrate this idea intuitively, assume that we are examining a sequence of random numbers: ultimately, it seems that there is some regularity, at least if we examine a sequence that is not too long. At this point, we must recall an aphorism coined by an economist, which is very fashionable among applied statisticians: "If you torture the data long enough, Nature will always confess" (Ronald H. Coase, 1991 Nobel Prize for Economics).

This practice of "looking for something" (when we do not know exactly what it is) is therefore misleading, and thus the associated terms *data snooping* or *data dredging* have negative connotations. When confronted with a considerable amount of data, the danger of false findings decreases but is not eliminated altogether. There are, however, techniques to counter this, as we shall see in chapter 3.

One particularity, which seems trivial, regards the so-called leaker variables, which are essentially surrogates of the variables of interest. For example, if the variable of interest is the amount of money spent on telephone traffic by one customer in one month, a leaker variable is given by the number of phone calls made in that same month, as the first variable is recorded at the same moment as the second variable. Conceptually, the situation is trivial, but when hundreds of variables, often of different origin, are manipulated, this eventuality is not as remote as it may appear. It at least signals the danger of using technology blindly, inserting whole lists of variables without worrying about what they represent. We return to this point in section 1.3.1.

Bibliographical notes

Hand et al. (2001) depict a broad picture of data mining, its connections with other disciplines, and its general principles, although they do not enter into detailed technical aspects. In particular, their chapter 12 contains a more highly developed explanation of our section 1.1.3 about relationships between data management and some techniques, like OLAP, closer to that context.

For descriptive statistics regarding tables of frequency and their handling, there is a vast amount of literature, which started in the early stages of statistics and is still developing. Some classical texts are Kendall & Stuart (1969, sections 1.30–1.34), Bishop et al. (1975), and Agresti (2002).

For a more detailed description of the role of data mining in the corporate context, in particular its connections with business promotion, see the first chapters of Berry & Linoff (1997).

1.2 All Models are Wrong

All models are wrong but some are useful.

—George E. P. Box

1.2.1 What is a Model?

The term *model* is very fashionable in many contexts, mainly in the fields of science and technology and also business management. Because the important attributes of this term (which are often implicit) are so varied and often blurred, let us clarify at once what we mean by it:

> A model is a simplified representation of the phenomenon of interest, functional for a specific objective.

In addition, certain aspects of this definition must be noted:

- We must deal with a *simplified representation*: an identical or almost identical copy would not be of use, because it would maintain all the complexity of the initial phenomenon. What we need is to reduce it and eliminate aspects that are not essential to the aim and still maintain important aspects.
- If the model is to be *functional for a specific objective*, we may easily have different models for the same phenomenon according to our aims. For example, the design of a new car may include the development of a mechanical or mathematical model, as the construction of a physical model (a real object) is required to study aerodynamic characteristics in a wind tunnel. Each of these models—obviously very different from each other—has a specific function and is not completely replaceable by the other.
- Once the aspect of the phenomenon we want to describe is established, there are still wide margins of choice for the way we explain relationships between components.
- Therefore, this construction of a "simplified representation" may occupy various dimensions: level of simplification, choice of real-life elements to be reproduced, and the nature of the relationships between the components. It therefore follows that a "true model" does not exist.
- Inevitably, the model will be "wrong"—but it must be "wrong" to be useful.

We can apply these comments to the idea of a model defined in general terms, and therefore also to the specific case of mathematical models. This term refers to any conceptual representation in which relations between the entities involved are explained by mathematical relationships, both written in mathematical notation and translated into a computer program.

In some fields, generally those connected with the exact sciences, we can think of the concept of a "true" model as describing the precise mechanics that regulate the phenomenon of interest. In this sense, a classical example is that of the kinematic laws regulating the fall of a mass in a vacuum; here, it is justifiable to think of these laws as quite faithfully describing mechanisms that regulate reality.

It is not our purpose to enter into a detailed discussion arguing that in reality, even in this case, we are effectively completing an operation of simplification. However, it is obvious that outside the so-called exact sciences, the picture changes radically, and the construction of a "true" model describing the exact mechanisms that regulate the phenomenon of interest is impossible.

There are extensive areas—mainly but not only in scientific research—in which, although there is no available theory that is complete and acquired from the phenomenon, we can use an at least partially accredited theoretical formulation by means of controlled experimentation of important factors.

In other fields, mostly outside the sciences, models have purely operative functions, often regulated only by the criterion "all it has to do is work," that is, without the pretext of reproducing even partially the mechanism that regulates the functioning of the phenomenon in question. This approach to formulation is often associated with the phrase "black-box model," borrowed from the field of control engineering.

1.2.2 From Data to Model

Since we are working in empirical contexts and not solely speculatively, the data collected from a phenomenon constitutes the base on which to construct a model. How we proceed varies radically, depending on the problems and the context in which we are required to operate.

The most favorable context is certainly that of experimentation, in which we control experimental factors and observe the behavior of the variables of interest as those factors change.

In this context, we have a wide range of methods available. In particular, there is an enormous repertoire of statistical techniques for planning experiments, analyzing the results, and interpreting the outcomes.

It should be noted that "experimenting" does not signify that we imagine ourselves inside a scientific laboratory. To give a simple example: to analyze the effect of a publicity campaign in a local newspaper, a company selects two cities with similar socioeconomic structure, and applies the treatment (that is, it begins the publicity campaign) to only one of them. In all other aspects (existence of other promotional actions, etc.), the two cities may be considered equivalent. At a certain moment after the campaign, data on the sales of goods in the two cities become available. The results may be configured as an experiment on the effects of the publicity campaign, if all the factors required for determining sales levels have been carefully controlled, in the sense that they are maintained at an essentially equivalent level in both cities. One example in which factors are not controlled may arise from the unfortunate case of promotional actions by competitors that take place at the same time but are not the same in the two cities.

However, clearly an experiment is generally difficult in real-world environment, so it is much more common to conduct observational studies. These are characterized by the fact that because we cannot control all the factors relative to the phenomenon, we limit ourselves merely to observing them. This type of study also gives important and reliable information, again supported by a wide range of statistical techniques. However, there are considerable differences, the greatest of which is the difficulty of identifying causal links among the variables. In an experimental study in which the remaining experimental factors are controlled, we can say that any change in variable of interest Y as variable X (which we regulate) changes involves a causal relationship between X and Y. This is not true in an observational study, because both may vary due to the effect of an external (not controlled) factor Z, which influences both X and Y.

However, this is not the place to examine the organization and planning of experimental or observational studies. Rather, we are concerned with problems arising in the analysis and interpretation of this kind of data.

There are common cases in which the data do not fall within any of the preceding types. We often find ourselves dealing with situations in which the data were collected for different aims than those we intend to work on now. A common case occurs in business, when the data were gathered for contact or management purposes but are then used for marketing. Here, it is necessary to ask whether they can be recycled for an aim that is different from the original one and whether statistical analysis of data of this type can maintain its validity. A typical critical aspect is that the data may create a sample that is not representative of the new phenomenon of interest.

Therefore, before beginning data analysis, we must have a clear idea of the nature and validity of the data and how they represent the phenomenon of interest to avoid the risk of making disastrous choices in later analysis.

Bibliographic notes Two interesting works that clearly illustrate opposing styles of conducting real data analysis are those by Cox (1997) and Breiman (2001b). The latter is followed by a lively discussion in which, among others, David Cox participated, with a rejoinder by Breiman.

1.3 A Matter of Style

1.3.1 Press the Button?

The previous considerations, particularly those concluding the section, show how important it is to reflect carefully on the nature of the problem facing us: how to collect data, and above all how to exploit them. These issues certainly cannot be resolved by computer.

However, this need to understand the problem does not stop at the preliminary phase of planning but underlies every phase of the analysis itself, ending with interpretation of results. Although we tend to proceed according to a logic that is much more practical than in other environments, often resulting in black-box models, this does not mean we can handle every problem by using a large program (software, package, tool, system, etc.) in a large computer and pushing a button.

Although many methods and algorithms have been developed, becoming increasingly more refined and flexible and able to adapt ever more closely to the data even in a computerized way, we cannot completely discard the contribution of the analyst. We must bear in mind that "pressing the button" means starting an algorithm, based on a method and an objective function of which we may or may not be aware. Those who choose to 'press the button' without this knowledge simply do not know which method is used, or only know the name of the method they are using, but are not aware of its advantages and disadvantages.

More or less advanced knowledge of the nature and function of methods is essential for at least three reasons:

1. An understanding of tool characteristics is vital in order to choose the most suitable method.
2. The same type of control is required for correct interpretation of the results produced by the algorithms.
3. A certain competence in computational and algorithmical aspects is helpful to better evaluate the *output* of the computer, also in terms of its reliability.

The third point requires clarification, as computer output is often perceived as secure and indisputable information. Many of the techniques currently applied involve nontrival computational aspects and the use of iterative algorithms. The convergence of these algorithms on the solution defined by the method is seldom guaranteed by its theoretical basis. The most common version of this problem occurs when a specific method is defined as the optimal solution of a certain objective function that is minimized (or maximized), but the algorithm may converge on a optimal point which is local and not global, thus generating incorrect computer output without the user realizing it. However, these problems are not uniform among different methods; therefore, knowing the various characteristics of the methods, even from this aspect, has important applicative value.

The choice of style to be accomplished here, corroborated by practical experience, is that of combining up-to-date methods with an understanding of the problems inherent in the subject matter.

This point of view explains why, in the following chapters, various techniques are presented from the viewpoints not only of their operative aspects but also (albeit concisely) of their statistical and mathematical features.

Our presentation of the techniques is accompanied by examples of real-life problems, simplified for the sake of clarity. This involves the use of a software tool of reference. There are many such products, and in recent years software manufacturers have developed impressive and often valuable products.

1.3.2 Tools for Computation and Graphics

In this work, we adopt `R` (`R` Development Core Team, 2011) as the software of choice, because it constitutes a language and an environment for statistical calculations and graphical representation of data, available free at `http://www.r-project.org/` in *open-source* form. The reasons for this choice are numerous.

- In terms of quality, `R` is one of the best products currently available, inspired by the environment and language `S`, developed in the laboratories of AT&T.
- The fact that `R` is free is an obvious advantage, which becomes even more significant in the teaching context, in which—because it is easily accessible to all—it has an ideal property on which to construct a common working basis.
- However, the fact that it is free does not mean that it is of little value: `R` is developed and constantly updated by the `R` Development Core Team, composed of a group of experts at the highest scientific level.
- Because `R` is a language, it lends itself easily to programming of variants of existing methods, or the formulation of new ones.
- In addition to the wide range of methods in the basic installation of `R`, additional packages are available. The set of techniques thus covers the whole spectrum of the existing methods.

- R can interact in close synergy with other programs designed for different or collateral aims. In particular, cooperation between R and a relational database or tools of dynamic graphic representation may exist.
- This extendibility of R is facilitated by the fact that we are dealing with an *open-source* environment and the consequent transparency of the algorithms. This means that anyone can contribute to the project, both with additional packages for specific methods and for reporting and correcting errors.
- The syntax of R is such that users are easily made aware of the way the methods work.

The set of exploitable data mining methods by means of R are the same as those that underlie commercial products and constitute their engine. The choice of R as our working environment signifies that although we forgo the ease and simplicity of a graphic interface, we gain in knowledge and in control of what we are doing.

2

A–B–C

Everything should be made as simple as possible, but not simpler.

—Attributed to ALBERT EINSTEIN

2.1 OLD FRIENDS: LINEAR MODELS

2.1.1 Basic Concepts

Let us start with a simple practical problem: we have to identify a relationship that allows us to predict the consumption of fuel or, equivalently, the distance covered per unit of fuel as a function of certain characteristics of a car. We consider data for 203 models of cars in circulation in 1985 in the United States, but produced elsewhere. Twenty-seven of their characteristics are available, four of which are shown in figure 2.1: `city distance` (km/L), `engine size` (L), `number of cylinders`, and `curb weight` (kg). The data are marked in different ways according to `fuel type` (gasoline or diesel).

Some of the available characteristics are numerical: `city distance`, `engine size`, and `curb weight` are quantitative and continuous, and `number of cylinders` is quantitative and discrete. However, `fuel type` is qualitative; equivalent terms are *categorical variable* and *factor*.

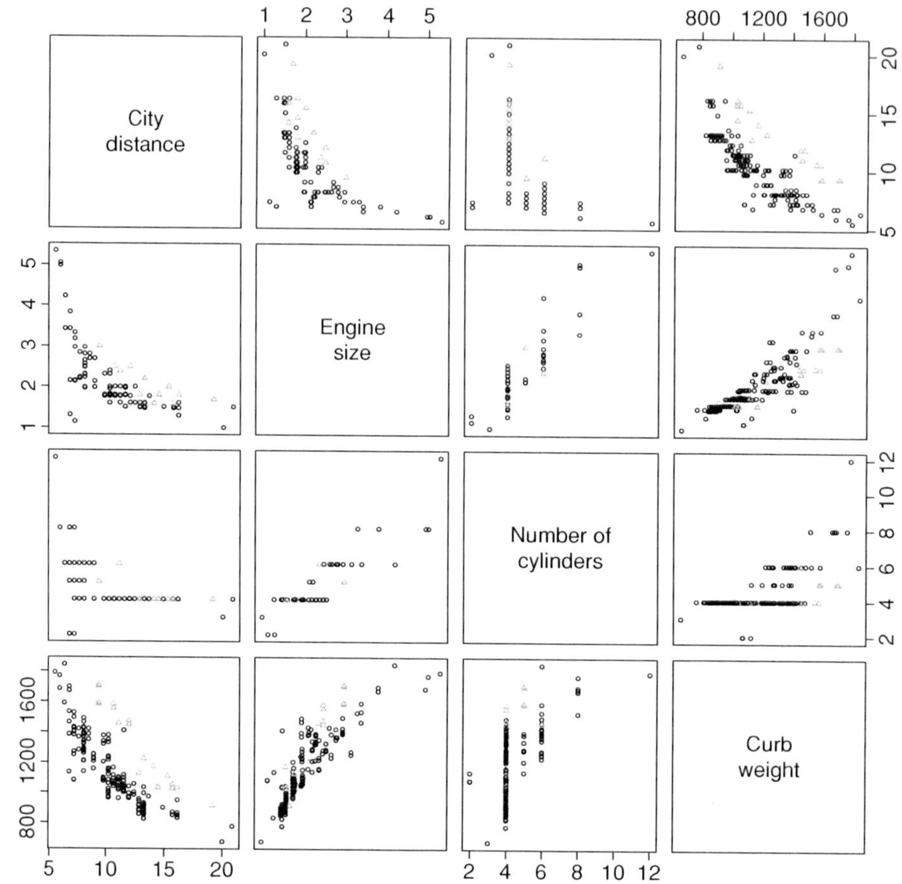

Figure 2.1 Matrix of scatterplots of some variables of car data, stratified by `fuel type`. Circles: gasoline; triangles: diesel.

In this case, when we are dealing with few data, we can represent them as a *scatterplot*, as in figure 2.1; in other cases, we would have to think of more elaborate representations.

In the first phase, for simplicity, we consider only two explanatory variables: `engine size` and `fuel type`, of which the former is quantitative and the latter qualitative. To study the relationship between quantitative variables, the first thing to make is a graphic representation, as in figure 2.2.

To study the relationship between two variables (for the moment leaving aside `fuel type`, which acts as a qualitative *stratification* variable), any statistics primer would first suggest a simple linear regression line, of the type

$$y = \beta_0 + \beta_1 x + \varepsilon \tag{2.1}$$

where y represents `city distance`, x `fuel type`, and ε is a nonobservable random 'error' term, which we assume to be of zero mean and constant but unknown variance σ^2. We also assume lack of correlation among error terms and

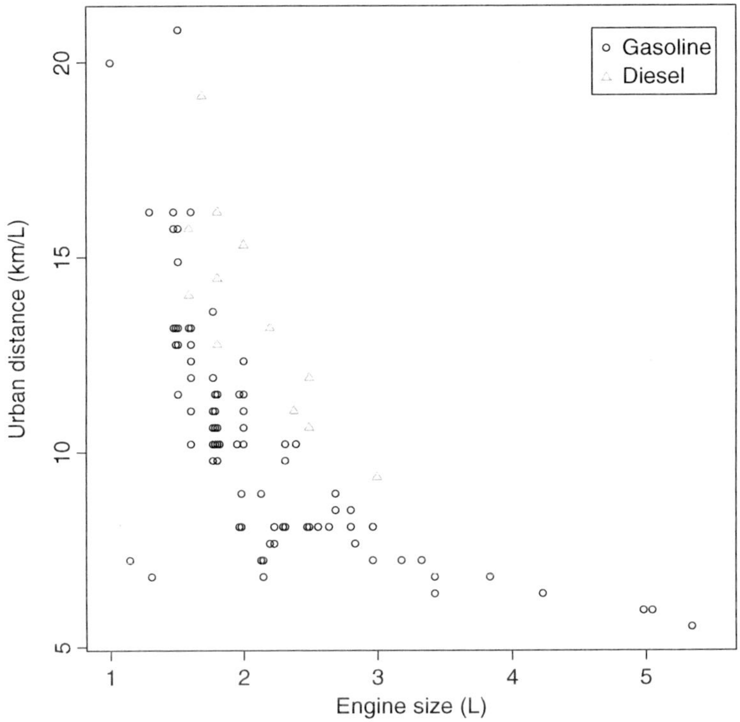

Figure 2.2 Car data, scatterplot of `engine size` and `city distance`, stratified by `fuel type`.

therefore also among observations y for differing units. This set of hypothesis is called 'of the second-order' because it involves mean, variance, and covariance, which are second-order moments.

We are looking for an estimate of unknown *regression parameters* β_0 and β_1 using n (in this case $n = 203$) pairs of observations, denoted by (x_i, y_i), for $i = 1, \ldots, n$.

Equation (2.1) is the simplest case for a more general formulation of the type

$$y = f(x; \beta) + \varepsilon, \tag{2.2}$$

which becomes (2.1) when f is the expression of the straight line and $\beta = (\beta_0, \beta_1)^\top$.

To estimate β, the *least squares criterion* leads us to identify the values for which we obtain the minimum, with respect to β, of the *objective function*

$$D(\beta) = \sum_{i=1}^{n} \{y_i - f(x_i; \beta)\}^2 = \|y - f(x; \beta)\|^2 \tag{2.3}$$

where the last expression uses matrix notation to represent vector $y = (y_1, \ldots, y_n)^\top$; $f(x; \beta) = (f(x_1; \beta), \ldots, f(x_n; \beta))^\top$; and $\|\cdot\|$ indicates the *Euclidean norm* of the vector, that is, the square root of the sum of squares of the elements.

The solution to this minimization problem is shown by $\hat{\beta}$, and we indicate the corresponding *fitted values*

$$\hat{y}_i = f(x_i; \hat{\beta}), \qquad i = 1, \ldots, n,$$

which, in the linear case (2.1), are of the type

$$\hat{y}_i = \hat{\beta}_0 + \hat{\beta}_1 x_i = \tilde{x}_i^\top \hat{\beta}$$

where $\tilde{x}_i^\top = (1, x_i)$.

From the same formula, we can also write the expression of the *predicted value*

$$\hat{y}_0 = \hat{\beta}_0 + \hat{\beta}_1 x_0$$

for a value x_0 of the explanatory variable, which does not necessarily correspond to any observation.

Clearly, however, the trend of the relationship in figure 2.2 does not lend itself to being expressed by a straight line. At this point, we can move in several alternative directions. The most immediate one is probably to consider a more elaborate form of function $f(x; \beta)$, for instance, a polynomial form

$$f(x; \beta) = \beta_0 + \beta_1 x + \cdots + \beta_{p-1} x^{p-1} \tag{2.4}$$

where β is now a vector with p components, $\beta = (\beta_0, \beta_1, \ldots, \beta_{p-1})^\top$. Using a polynomial function has the double advantage of (1) being conceptually and mathematically simple, and (2) offering simple treatment regarding the use of the least squares criterion.

Because (2.4) is *linear in the parameters*, it can be rewritten as

$$f(x; \beta) = X\beta \tag{2.5}$$

where X is an $n \times p$ matrix, called the *design matrix*, defined by

$$X = (1, x, \ldots, x^{p-1})$$

where x is the vector of the observations of the explanatory variable, and the various columns of X contain powers of order from 0 to $p-1$ of elements of x. The complete entry is therefore a particular case of a *linear model*

$$y = X\beta + \varepsilon \tag{2.6}$$

in which X refers to a polynomial regression, corresponding to (2.4).

In this formulation, the explicit solution to the minimization problem of (2.3) is

$$\hat{\beta} = (X^\top X)^{-1} X^\top y \tag{2.7}$$

with which the vector of fitted values is

$$\hat{y} = X\hat{\beta} = Py \tag{2.8}$$

where

$$P = X(X^\top X)^{-1} X^\top \tag{2.9}$$

is an $n \times n$ matrix, called the *projection matrix*. Properties $P^\top = P$, $PP = P$ hold, as does $\mathrm{tr}(P) = \mathrm{rk}(P) = p$.

The minimum value of (2.3) may be written in various equivalent forms

$$D(\hat{\beta}) = \|y - \hat{y}\|^2 = y^\top (I_n - P)y = \|y\|^2 - \|\hat{y}\|^2 \tag{2.10}$$

where I_n denotes the identity matrix of order n. Quantity $D = D(\hat{\beta})$ is called *deviance*, in that it is a quantification of the discrepancy between fitted and observed values.

From here, we also obtain the estimate of σ^2, usually given by

$$s^2 = \frac{D(\hat{\beta})}{n - p} \tag{2.11}$$

and this allows us to assess the variance of the estimates of β through

$$\widehat{\mathrm{var}}(\hat{\beta}) = s^2 \, (X^\top X)^{-1}. \tag{2.12}$$

The square root of the diagonal elements of (2.12) yields the *standard errors* of the components of $\hat{\beta}$—essential for inferential procedures, as we shall see shortly.

A somewhat more detailed explanation of linear model concepts and least squares is given in Appendix A.3.

In the case of the data in figure 2.2, it is plausible to use $p = 3$ or even $p = 4$. In any case, we still need one more element to treat the data effectively, and this is the qualitative variable `fuel type`. A nonnumerical variable must be conveniently encoded by *indicator variables*; if the possible *levels* assumed by the variable are k, then the number of required indicator variables is $k - 1$. In this case, we need a single indicator variable, because `fuel type` may have two levels, `diesel` and `gasoline`. There is an infinite number of choices, provided that each is associated with a single value of the indicator variable. One particularly simple choice is to assign value 1 to the level `diesel` and value 0 to the level `gasoline`; we indicate this new variable with I_A.

The simplest way to insert I_A into the model is additive, which is equivalent to presuming that the average difference of the distance covered by two

Table 2.1. CAR DATA: ESTIMATES AND ASSOCIATED QUANTITIES FOR MODEL (2.14)

	Estimate	**SE**	***t*-value**	***p*-value**
`(intercept)`	24.832	3.02	8.21	0.000
`(engine size)`	−10.980	3.53	−3.11	0.002
`(engine size)`^{2}	2.098	1.27	1.65	0.100
`(engine size)`^{3}	−0.131	0.14	−0.94	0.349
`fuel.diesel`	3.214	0.43	7.52	0.000

groups of diesel and gasoline cars is constant for any `engine size`. This simplified hypothesis is called the *additive hypothesis* of the effects. Also, if the additive hypothesis is not completely valid, this formulation constitutes a first approximation, which is often the most important part of the influence of the factor. This component, entered in an additive form, is therefore called the *main effect* of the factor.

This choice means that matrix X of (2.5) is now extended with the addition of a new column containing I_A. Function $f(x; \beta)$ and matrix X are therefore substituted by the new expressions

$$f(x;\beta) = \beta_0 + \beta_1 x + \cdots + \beta_{p-1} x^{p-1} + \beta_p I_A, \qquad X = (1, x, \ldots, x^{p-1}, I_A) \tag{2.13}$$

Correspondingly, we add a new component to vector β, which, given the specific form adopted by the dummy variable, represents the average deviation of the `distance covered` between diesel or gasoline cars.

Adopting this scheme for the data in figure 2.2, with $p = 4$, means that the linear model is specified in the form

$$y = \beta_0 + \beta_1 x + \beta_2 x^2 + \beta_3 x^3 + \beta_4 I_A \tag{2.14}$$

of which the estimates and *standard errors* are listed in table 2.1, together with the normalized value of estimate t = `estimate/(standard error)` and the corresponding p-value, or *observed significance level*, which we obtain if we can introduce the additional hypothesis of *normal or Gaussian distribution* for the error terms ε of (2.2). The estimated curves identified by these parameters are shown in figure 2.3.

To evaluate the goodness of fit, we need to calculate the *coefficient of determination*

$$R^2 = 1 - \frac{\texttt{(residual deviance)}}{\texttt{(total deviance)}} = 1 - \frac{\sum_i (y_i - \hat{y}_i)^2}{\sum_i (y_i - \bar{y})^2} \tag{2.15}$$

where $D(\hat{\beta})$ is calculated by (2.10) using X, the matrix corresponding to model (2.14); and $\bar{y} = \sum_i y_i/n$ indicates the arithmetic mean or *average* of y_i. In this specific case, we obtain $R^2 = 0.60$, which indicates a fair degree of correlation between observed and interpolated data.

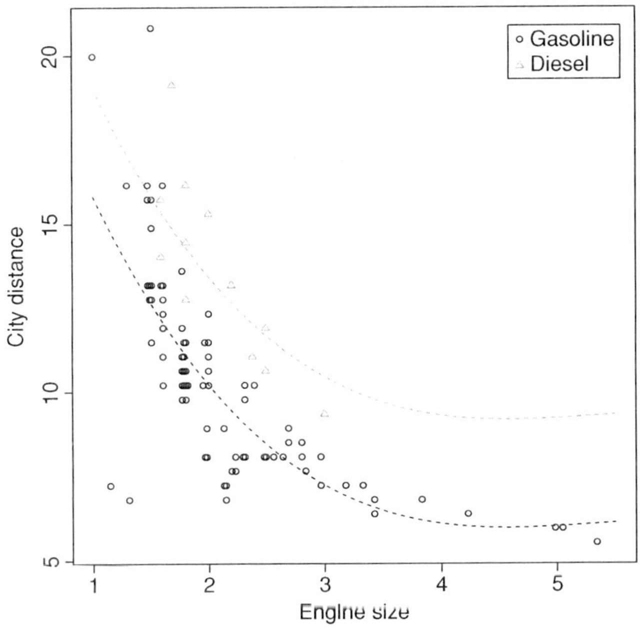

Figure 2.3 Car data: fitted curves relative to model (2.14).

However, we cannot reduce evaluation of the adequacy of a model to consideration of a single indicator. Other indications are provided by *graphical diagnostics*. There are several of these, and they all bring us back more or less explicitly to examination of the behavior of the *residuals*

$$\hat{\varepsilon}_i = y_i - \hat{y}_i, \qquad i = 1, \ldots, n, \tag{2.16}$$

which serve as surrogates of errors ε_i, which are not observable. The residuals have various aspects that we must evaluate according to various assumptions. Among the many diagnostic tools, two of the most frequently used are shown in figure 2.4.

Figure 2.4 (left) shows the *Anscombe plot* of the residuals with respect to the interpolated values, which would ideally have to present random scattering of all points if the selected model is to be deemed valid. In our case, it is evident that the variability of the residuals increases from left to right, signaling a probable violation of *homoscedasticity*—that is, var $\{\varepsilon_i\}$ must be a constant, say, σ^2, independent of index i—whereas here the graphic indicates something very different.

Figure 2.4 (right) shows the *quantile-quantile plot* for verification of the normality assumption for the distribution of ε_i. The y-axis gives the values of $\hat{\varepsilon}_i$, conveniently standardized and ordered in increasing terms, and the x-axis shows the corresponding expected values under the normality hypothesis, approximated (if necessary) for simplicity of calculation.

If the normal hypothesis is valid, we expect the observed points to lie along the bisector of the first and third quadrants. In this case, the data behave differently and do not conform to the normal hypothesis. In more detail, the central part of

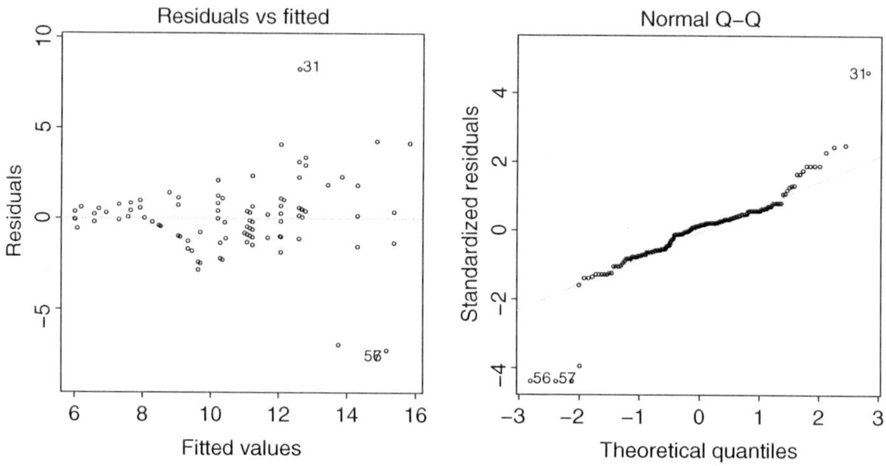

Figure 2.4 Car data: graphical diagnostics for model (2.14).

the diagram shows a trend that is quite satisfactory, although not ideal. The part of the graph that conforms least to expectations lies in the *tails* of the distribution, the portion outside interval $(-2, 2)$. Specifically, the observed residuals are of much larger absolute value than the expected ones, indicating *heavy tails* with respect to the normal curve.

Thus, using a simple linear model (2.14) suggests the following points, some of which, with necessary modifications, we find in other applications of linear models.

- The goodness of fit of the linear model of figure 2.3 is satisfactory on first analysis, especially if we want to use it to predict the `city distance` covered by a car of average `engine size` (i.e., between 1.5 and 3 L).
- The construction of the model is so simple, both conceptually and computationally, that in some cases, these methods can be applied automatically.
- Despite the superficially satisfactory trend of figure 2.3, the graphical diagnostics of figure 2.4 reveal aspects that are not satisfied.
- The model is not suitable for *extrapolation,* that is, for predicting the value of the variable outside the interval of observed values for the explanatory variables. This is seen in the example of the set of diesel cars with engines larger than 3 L, when the predicted values become completely unrealistic.
- The model has no grounding in physics or engineering, which leads to interpretive difficulties and adds paradoxical elements to the expected trend. For example, the curve of the set of gasoline cars shows a local minimum around 4.6 L, and then rises again!

This type of evaluation of a model's critical elements is not confined to linear models (see chapter 4).

2.1.2 Variable Transformations

We must explain what we mean by 'linear': these are models which are linear with respect to *parameters*, but we can use nonlinear variable transformations of both y and x_i, which may be different for different variables. In addition, we can use as many transformations as we need, for example, x_1 and x_2 can give place to $X = (1, x_1, x_2, x_1/x_2, e^{x_2^2+x_1})$. This flexibility of use, with respect to the basic formulation, is one of the successful features of linear models.

We already used this possibility in formulating polynomial model (2.14), which is a common variant, but we can also use many others, including transformations of the response variable. The theoretical structure remains unchanged, although in this case the objective function (2.3), and therefore the optimality criterion, work on the transformed scale.

In the foregoing examples, it is reasonable to consider fuel consumption per km as a response variable instead of distance covered. Hence, we can write

$$\texttt{consumption} = \beta_0 + \beta_1\,\texttt{(engine size)} + \beta_2 I_A + \varepsilon \tag{2.17}$$

where `consumption = 1/(distance covered)`. Obviously, here, error term ε and parameters β_j are not the same as those in (2.14), but the same hypotheses on the nature of the error component are retained. Figure 2.5 shows the scatterplot of the new variables, with two regression lines, the coefficients of which are listed in table 2.2.

Some simple observations may be made: (1) the trend of the points in figure 2.5 shows good alignment; (2) this is reinforced by the value of R^2, which is 0.64; (3) therefore, it is not necessary to draw on polynomials of higher order. However, it is useful to report the trend of the new estimated function on the original scale, which also allows comparisons with the previous estimate. The new estimated function is shown in figure 2.6, and is much more convincing, particularly in the edges of the explanatory variable `engine size`. To be comparable with model (2.14), R^2 is now recalculated to its original scale, giving a value of 0.56. The corresponding graphical diagnostics are shown in figure 2.7. Although the fit of figure 2.5 appears to be acceptable, the graphical diagnostics continue to be unsatisfactory.

Another type of transformation often used is the logarithm. In this case, it is also reasonable to transform both the explanatory variable and the response variable,

Table 2.2. Car Data: Estimates and Associated Quantities of Model (2.17)

	Estimate	SE	t-value	p-value
`(intercept)`	0.042	0.0035	11.94	0.000
`(engine size)`	0.029	0.0016	17.94	0.000
`fuel.diesel`	−0.025	0.0037	−6.78	0.000

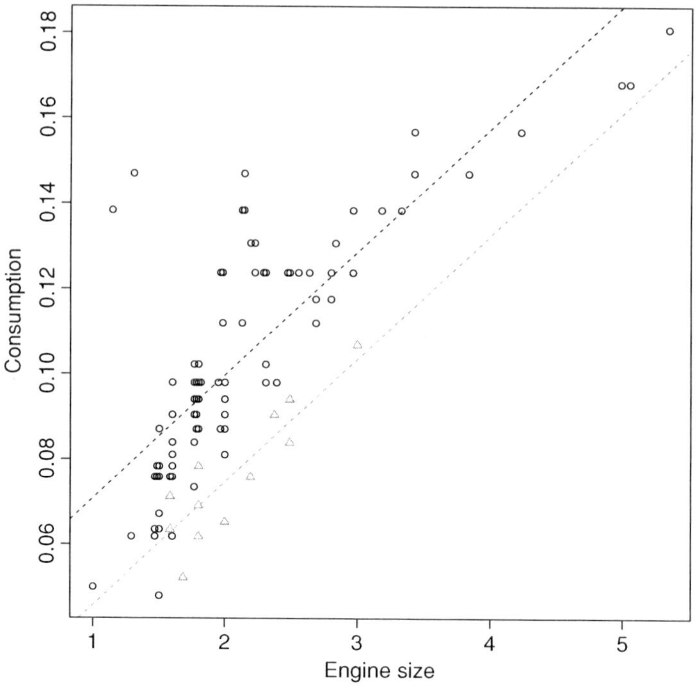

Figure 2.5 Car data: scatterplot of `engine size` and `consumption`, with regression lines of model (2.17).

aiming for the formulation

$$\log(\texttt{distance covered}) = \beta_0 + \beta_1 \log(\texttt{engine size}) + \beta_2 I_A + \varepsilon. \tag{2.18}$$

Logarithmic transformations are often used when intrinsically positive quantities are involved, such as `distance covered` and `engine size`. They have the advantage of allowing us to operate on variables that vary in $(-\infty, \infty)$, that is, the "right" support for linear models. In turn, this fact means that once the transformation is inverted, we are certain of obtaining positive quantities for the predicted values of the response variable. An additional advantage of logarithmic transformations is that they often correct the heteroscedasticity of the residuals.

Table 2.3 summarizes the fitted model, figure 2.8 shows the fitted curves on both transformed and original scales, and figure 2.9 shows the graphical diagnostics for the linear model. We can now deduce that model (2.18) is preferable to (2.14), but the graphical diagnostics remain substantially unsatisfactory.

Much of the inadequacy of model (2.18) is due to the persistence of heteroscedasticity in the residuals, as clearly shown in the left side of figure 2.9, as in figures 2.4 and 2.7. In turn, this heteroscedasticity is probably due to a *heterogeneity* in observed cases that is not adequately 'explained' by the explanatory variables.

To remedy this inconvenience, we have many other variables at our disposal. In particular, basic evaluations lead us to consider the `curb weight` of the car

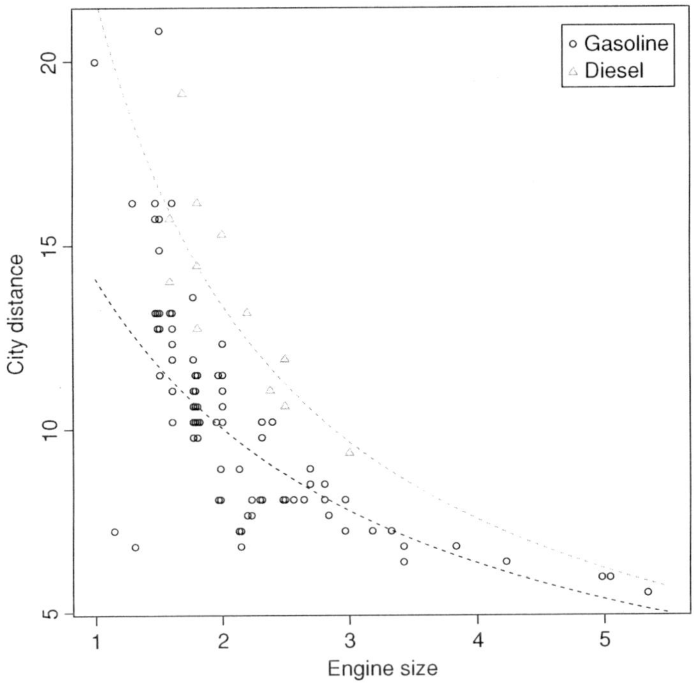

Figure 2.6 Car data: scatterplot of `engine size` and `distance covered` with curves fitted to model (2.17).

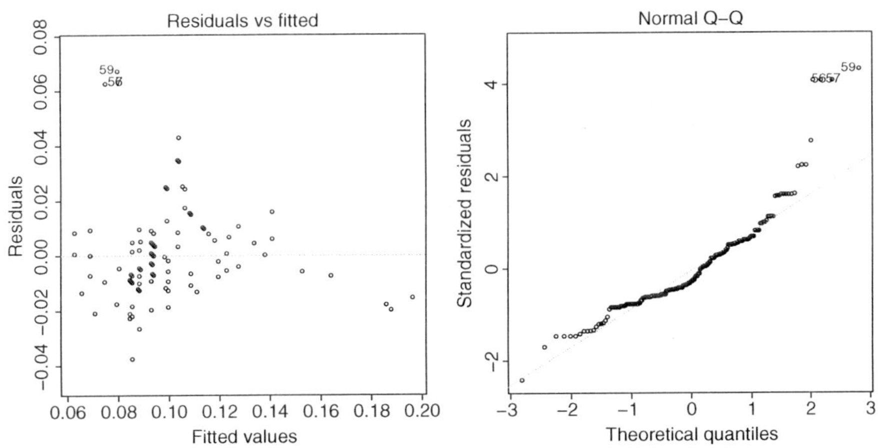

Figure 2.7 Car data: graphical diagnostics of model (2.17).

as an important variable. For reasons already mentioned with respect to the other two continuous variables, it makes sense to consider `curb weight` through its logarithmic transformation.

Another feature to take into account is the anomalous position of the two points in the bottom left corner of figure 2.2, which are never interpolated appropriately

Table 2.3. CAR DATA: ESTIMATES AND ASSOCIATED QUANTITIES OF MODEL (2.18)

	Estimate	SE	t-value	p-value
`(intercept)`	2.782	0.0295	94.30	0.000
log(`engine size`)	−0.682	0.0398	−17.13	0.000
`fuel.diesel`	0.278	0.0379	7.34	0.000

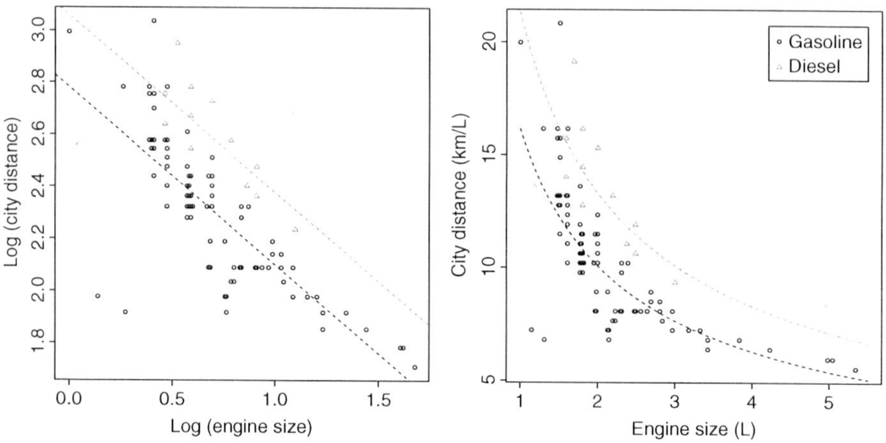

Figure 2.8 Car data: scatterplots and fitted curves of model (2.18) on transformed (left) and natural scales (right).

by any of the regression curves. They turn out to correspond to four cars, all with two-cylinder engines, and they are the only ones to have this characteristic. We must therefore add a new indicator variable, I_D, to the model, with a value of 1 if the engine has two cylinders and 0 otherwise.

Combining the considerations of the last two paragraphs, we can formulate the new model

$$\log(\texttt{distance covered}) = \beta_0 + \beta_1 \log(\texttt{engine size}) + \beta_2 \log(\texttt{curb weight}) + \beta_3 I_A + \beta_4 I_D + \varepsilon \tag{2.19}$$

for which table 2.4 lists the summary outcome of the estimation process. The value of R^2 is 0.88, and the corresponding value on the original scale is 0.87. These values are evidently much more convincing than the previous ones, even though the number of parameters has not been increased to any great extent. In addition, the graphical diagnostics of the residuals of figure 2.10 give a much better picture, although the residual distribution is slightly *skewed*, highlighted by the mild convexity of the trend of the the quantile-quantile plot in the top right panel.

In this case, we have added two extra graphic panels, containing the scatterplots of the residuals (transformed into the square roots of their absolute values) with

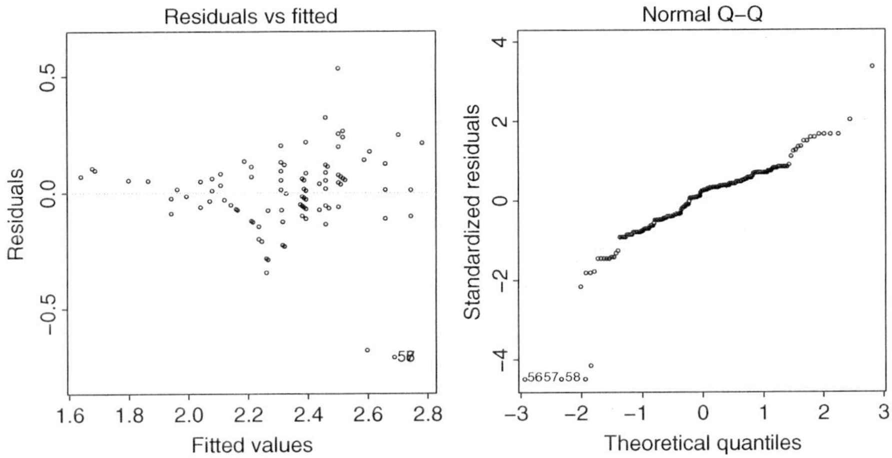

Figure 2.9 Car data: graphical diagnostics for model (2.18).

Table 2.4. Car Data: Estimates and Quantities for Model (2.19)

	Estimate	**SE**	***t*-value**	***p*-value**
`(intercept)`	9.07	0.475	19.08	0.000
log(`engine size`)	−0.18	0.051	−3.50	0.001
`fuel.diesel`	0.35	0.022	15.93	0.000
`cylinders.2`	−0.48	0.052	−9.30	0.000
log(`curb weight`)	−0.94	0.072	−13.07	0.000

respect to the estimated values, and the *Cook distance* for every observation. The *Cook distance* allows us to evaluate the effect on $\hat{\beta}$ produced by removing (x_i, y_i) from the set of observations, and this perturbation of $\hat{\beta}$ is linked to a corresponding perturbation of $\hat{y}$. Therefore, the Cook distance provides an indicator of the *influence* of this observation on the fitted model. Both diagrams are entirely satisfactory in that they show neither heteroscedasticity of residuals nor *influential observations*.

The meaning and interpretation of the numerical values in table 2.4 are largely according to expectations, in the sense that `curb weight`, `engine size`, and `fuel type` all correspond to common knowledge of the distance covered by a car, or rather, its logarithmic transformation, as examined here.

However, a specific comment must be made regarding factor I_D, the coefficient of which has a negative sign and is of considerable statistical significance—in outstanding contrast with intuitive expectations, as a car with two cylinders should in fact consume less than the others, that is, it should have a positive β_4 coefficient in the prediction of log(`distance covered`).

The explanation of this apparently paradoxical behavior is due to the structure of the relationships between *all* the variables involved, not only between the response and explanatory variables. In particular, figure 2.1 shows that the

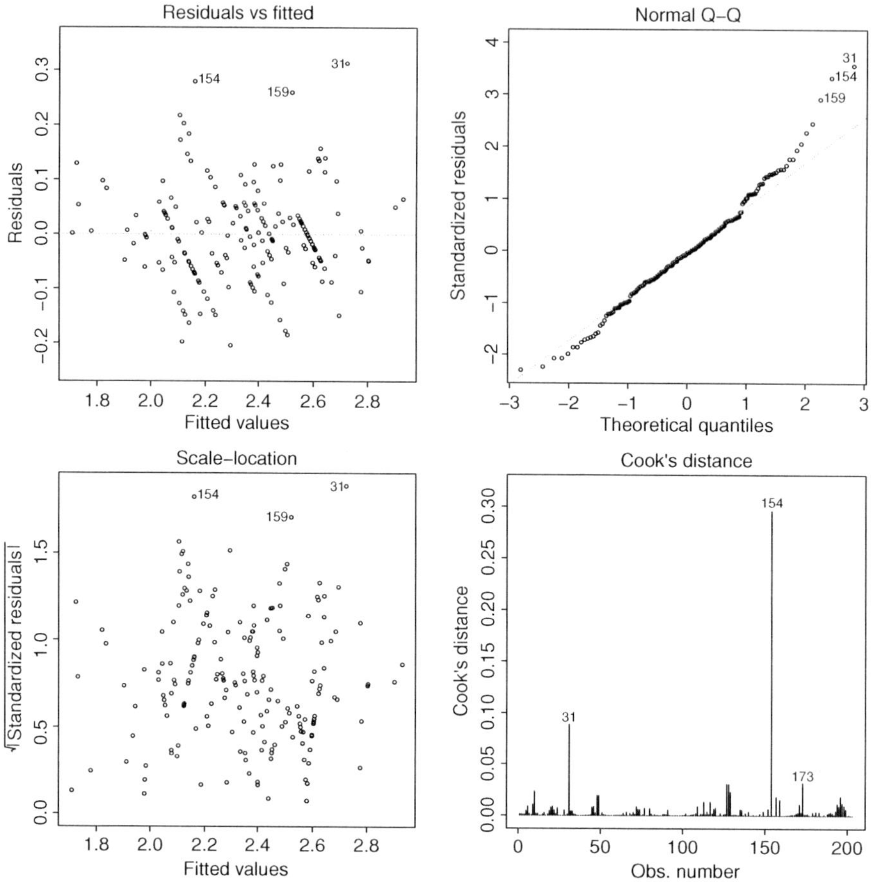

Figure 2.10 Car data: graphical diagnostics for model (2.19).

`curb weight` of the two-cylinder cars is similar to that of four-cylinder ones and much higher than those of three-cylinder cars, and this group of cars also behaves anomalously with respect to the general trend in the scatterplots for other variables.

There are many ways of dealing with this type of situation. The simplest is adopted here: the indicator variable I_D of the anomalous group is inserted among the explanatory variables. Thus, the value of the estimate -0.48 for β_4 is not interpreted in the sense that two-cylinder cars generally have a log(`distance covered`) that is 0.48 lower than that of the others: this is due to the particular way the fact of having two cylinders links up with the other explanatory variables, mainly `curb weight`.

2.1.3 Multivariate Responses

In some cases, there are several response variables of interest, for the same sets of units and explanatory variables. An immediate example comes from the car data themselves, and here it is interesting to consider not only `city distance` but also `highway distance`, so we examine the same set of explanatory variables in both responses.

If there are q response variables, we can construct a matrix Y, the columns of which contain these q variables. In our car example, $q = 2$ and

$$Y = (\texttt{(city distance), (highway distance)}).$$

If we create q models of linear regression, each of type (2.6), using the same regression matrix X for each, we obtain

$$Y = XB + E \tag{2.20}$$

where B is the matrix formed of q columns of dimension p, each representing the regression parameters for the corresponding column of Y, and matrix E is made up of error terms. Here, too, each of its columns refers to the corresponding column of Y, with the condition that

$$\mathrm{var}\left\{\tilde{E}_i\right\} = \Sigma$$

where $\tilde{E}_i^\top$ represents the ith row E, for $i = 1, \ldots, n$, and Σ is a variance matrix of dimensions $q \times q$ independent of i, which expresses the correlation structure between the error components and therefore also between the response variables. Equation (2.20) constitutes a model of *multivariate multiple linear regression*, where the term 'multivariate' refers to q response variables and 'multiple' to p explanatory variables.

The natural extension of the least squares criterion to the case of q response variables is given by the sum of q terms of type (2.3). Because this sum is minimal when each additive term is minimal, the solution to the multivariate least squares problem is

$$\hat{B} = (X^\top X)^{-1} X^\top Y \tag{2.21}$$

which is simply the juxtaposition of q vectors estimated for each response variable. The corresponding estimate of Σ is

$$\hat{\Sigma} = \frac{1}{n-p} Y^\top P Y$$

of which the diagonal gives the terms equivalent to s^2 of (2.11), yielding standard errors, as in the scalar case, from (2.12).

Bibliographical notes

The treatment of linear models appears in a variety of styles and levels; we only mention a few references. For an introduction focusing on applicative use, see Weisberg (2005) and Cook & Weisberg (1999), who deal with extended aspects of graphical representation and the use of graphical diagnostics. A more formal treatment of linear models is in chapter 4 of Rao (1973). For the operational aspects, we refer to Venables & Ripley (2002, ch. 6). Classical methods for analysing multivariate response variables are provided by Mardia et al. (1979).

2.2 Computational Aspects

Computational aspects take on a very important role in data mining. Let us start by referring to the linear models that represent their most simple algebraic formulation.

The main element to be calculated is the estimate of β in (2.7), and then the other quantities associated with it—in particular, estimate s^2 of σ^2 given by (2.11) and the relative standard errors of the components of $\hat{\beta}$.

2.2.1 Least Squares Estimation by Successive Orthogonalization

As we saw in section 2.1.1, the solutions to least squares problems (2.7) and related quantities are all based on inversion of the $(X^\top X)$ matrix, and the most frequently used method of inverting symmetric matrices is based on Cholesky factorization. The solutions to least squares problems by this method has a computational cost of $p^3 + np^2/2$ elementary operations (see, e.g., Trefethen & Bau, 1997, Lecture 11).

However, a matrix can be inverted only if all its rows and columns are linearly independent—that is, in this case, if there is no linear dependence between the columns of X. Clearly, if some columns of X are almost linearly dependent, the solution of $(X^\top X)^{-1}$ will probably be computationally unstable. The best situation is when all the columns of X are orthogonal to each other, so that the inverse is obtained very efficiently. The Gram-Schmidt procedure, shown in algorithm 2.1, transforms the original variables sequentially, by successive orthogonalization yielding a new formulation of X with orthogonal columns, so that the inverse of $X^\top X$ is easily obtained.

The algorithm may be written in matrix form by considering the QR decomposition of X as the product of an $n \times p$ orthogonal matrix Q, usually normalized so that $Q^\top Q = I$, and a $p \times p$, upper triangular matrix R. The least squares solution is therefore

$$\hat{\beta} = R^{-1}Q^\top y \qquad \text{and} \qquad \hat{y} = QQ^\top y$$

where the inversion of R is easy because it is a upper triangular matrix.

The computational cost of least squares fitting by QR decomposition requires approximately $2np^2$ operations, about twice that of direct inversion by Cholesky decomposition when $n \gg p$ and about the same when $p = n$. Depending on the number of variables and records available, we choose the most appropriate algorithm.

2.2.2 When n is Large

However, when n is large, the solutions presented in the last section become difficult to ascertain, because they involve handling matrices of dimensions $n \times p$, which is time consuming. When n is really very large, even simply loading X into memory may be problematic.

Algorithm 2.1 Gram-Schmidt algorithm for least squares estimates

1. Start: Initialize $z_0 = x_0 = 1_n$.
2. Cycle for $j = 1, 2, \ldots, p-1$: regress x_j on $z_0, z_1, \ldots, z_{j-1}$, to produce coefficients:

$$\hat{\gamma}_{kj} = \frac{z_k^\top x_j}{z_k^\top z_k}$$

$k = 0, \ldots, j-1$ and residual vector $z_j = x_j - \sum_{k=1}^{j-1} \hat{\gamma}_{kj} z_k$.

3. Regress y on residual vector z_{p-1} to give estimate $\hat{\beta}_{p-1}$.

A simple method of overcoming this problem is as follows. The elements necessary for calculating (2.7) are only

$$W = X^\top X, \qquad u = X^\top y$$

of dimensions $p \times p$ and $p \times 1$, respectively, where W is the symmetric matrix, so we can write

$$\hat{\beta} = W^{-1} u. \tag{2.22}$$

Also, putting

$$X = \begin{pmatrix} \tilde{x}_1^\top \\ \tilde{x}_2^\top \\ \vdots \\ \tilde{x}_n^\top \end{pmatrix}$$

where $\tilde{x}_i^\top$ is the ith row of X, we obtain

$$W = \sum_{i=1}^{n} \tilde{x}_i \tilde{x}_i^\top, \qquad u = \sum_{i=1}^{n} \tilde{x}_i y_i.$$

We can also write

$$W_{(j)} = W_{(j-1)} + \tilde{x}_j \tilde{x}_j^\top, \qquad u_{(j)} = u_{(j-1)} + \tilde{x}_j y_j, \qquad \text{for } j = 2, \ldots, n.$$

where $W_{(j)}$ is the matrix formed by the first j summands of W, and $u_{(j)}$ is defined analogously, starting from

$$W_{(1)} = \tilde{x}_1 \tilde{x}_1^\top, \qquad u_{(1)} = \tilde{x}_1 y_1.$$

It is now clear that W and u can be calculated by reading the data of a single *record* at a time and increasing the sums gradually as the data are read, with a construction involving memory use independent of n. At this point, $\hat{\beta}$ can be calculated by exploiting an algorithm for the inversion of symmetric matrices. The most frequently used method is based on Cholesky decomposition. If some of the columns of X are made up at least partially of variables obtained by transforming the original data, such transformations can be performed progressively as the data are read.

The previous procedure may also be extended to calculate s^2 and the standard errors of β with a memory use independent of n.

2.2.3 Recursive Estimation

When the data flow continuously (i.e., are a *data stream*) and we must update the estimates in real time, we need an algorithm that updates them recursively.

The previous setting allows us to solve this problem, as there are no restrictions on n. However, it does behave in a way that for every data read cycle, we must reinvert matrix W, of dimensions $p \times p$, and this may be problematic if p is not small and the data flow fast. We can also improve our procedure by suitably manipulating the formulas.

Let us presume that we have calculated the least squares estimates for the set of the first n observations and that we have

$$\hat{\beta}_{(n)}, \qquad V_{(n)} = W_{(n)}^{-1} = (X_{(n)}^\top X_{(n)})^{-1}$$

where n as a subscript reminds us that the quantities refer to the first n observations.

On reading the $(n+1)$th observation, formed by y_{n+1} and $\tilde{x}_{n+1}$, we must update the estimates and other connected quantities. We write

$$X_{(n+1)} = \begin{pmatrix} X_{(n)} \\ \tilde{x}_{n+1}^\top \end{pmatrix}, \qquad W_{(n+1)} = X_{(n+1)}^\top X_{(n+1)} = (X_{(n)}^\top X_{(n)} + \tilde{x}_{n+1}\tilde{x}_{n+1}^\top)$$

and use the Sherman-Morrison formula (A.2) to invert $W_{(n+1)}$, obtaining

$$V_{(n+1)} = V_{(n)} - h\, V_{(n)}\tilde{x}_{n+1}\tilde{x}_{n+1}^\top V_{(n)}$$

where $h = 1/(1 + \tilde{x}_{n+1}^\top V_{(n)}\tilde{x}_{n+1})$. After due substitutions in (2.22), we obtain the recursive expression

$$\begin{aligned} \hat{\beta}_{(n+1)} &= V_{(n+1)}\,(X_{(n)}^\top y + \tilde{x}_{n+1}y_{n+1}) \\ &= \hat{\beta}_{(n)} + \underbrace{h\, V_{(n)}\tilde{x}_{n+1}}_{k_n}\ \underbrace{(y_{n+1} - \tilde{x}_{n+1}^\top\hat{\beta}_{(n)})}_{e_{n+1}} \\ &= \hat{\beta}_{(n)} + k_n\, e_{n+1} \end{aligned} \tag{2.23}$$

where e_{n+1} represents the *prediction error* of y_{n+1} based on the estimate of β obtained from the first n observations. We thus have the new quantities $\hat{\beta}_{(n+1)}$ and

$V_{(n+1)} = (X_{(n+1)}^{\top} X_{(n+1)})^{-1}$, with which we can resume the updating cycle from the beginning.

Making use of (A.2) in a similar fashion, we can also obtain a corresponding recursive form to calculate the sum of the squares of the residuals (2.10), that is,

$$Q_{n+1}(\hat{\beta}_{(n+1)}) = Q_n(\hat{\beta}_{(n)}) + h\, e_{n+1}^2 \tag{2.24}$$

where $Q_{n+1}(\cdot)$ is calculated with matrix $X_{(n+1)}$ and response vector $y_{(n+1)}$, and, analogously, $Q_n(\cdot)$ refers to the first n observations. Equations (2.24) and (2.11) give estimate s_{n+1}^2, which, multiplied by $V_{(n+1)}$, yields the standard errors of $\hat{\beta}_{(n+1)}$.

The updating rule (2.23) takes the form of a *linear filter*, in which new estimate $\hat{\beta}_{(n+1)}$ is obtained by modifying old estimate $\hat{\beta}_{(n)}$ according to prediction error e_{n+1}, weighted with the *gain* k_n of the filter. Using the terminology typical of the field of *machine learning*, we say that the estimator "learns from its errors" by adjusting the current estimate each time, according to error e_{n+1}.

This scheme therefore calculates only a single inversion of the $p \times p$ matrix at first, and then we simply have to update the estimates and related quantities. When n is very large, as when we work with a continuous data stream, we can further simplify the procedure, introducing an approximation that becomes negligible as n increases. As in this case, the first p observations have little influence on the total, and we can begin in whatever way we like—for example, with $\hat{\beta}_{(p)}$ equal to the zero vector and $V_{(p)}$ to the identity matrix of order p, which essentially corresponds to following only step 6 of algorithm 2.2. In this way, the values of $\hat{\beta}$ are not the correct ones, but they tend to became so gradually as n increases.

This sequence of previous operations is shown schematically in algorithm 2.2. The Diag$(\cdot)$ notation is used to indicate the diagonal elements of a general square matrix.

Bibliographical notes

An authoritative coverage of the computational aspects of least squares estimation is given by Golub & Van Loan (1983). The algorithm of recursive least squares was presented by Plackett (1950), who also refers to the original work of Gauss of 1821.

2.3 Likelihood

2.3.1 General Concepts

Up to now we have reviewed cases in which the variable of interest (y) was continuous and the problem of studying the relationship between y and explanatory variables $(x_1, \ldots, x_{p-1})$ could be managed through the least squares criterion. The latter finds its field of application more appropriate when the range of y is $(-\infty, \infty)$. The most correct usage of associated inferential techniques is possible if the distribution of error terms ε, and thus also of y, is normal or Gaussian, at least approximately.

Algorithm 2.2 Recursive linear least squares

1. Let $W_{(p\times p)} \leftarrow 0, u_{(p\times 1)} \leftarrow 0, Q \leftarrow 0$.
2. Cycle for $n = 1, \ldots, p$:

   a. read nth record: $x \leftarrow \tilde{x}_n, y \leftarrow y_n$,
   b. $W \leftarrow W + x\,x^\top$,
   c. $u \leftarrow u + x\,y$.

3. $V \leftarrow W^{-1}$.
4. $\hat{\beta} \leftarrow V\,u$.
5. Cycle for $n = p+1, p+2, \ldots$:

   a. read nth record: $x \leftarrow \tilde{x}_n, y \leftarrow y_n$,
   b. $h \leftarrow 1/(1 + x^\top V\,x)$,
   c. $e \leftarrow y - x^\top \hat{\beta}$,
   d. $\hat{\beta} \leftarrow \hat{\beta} + h\,V\,x\,e$,
   e. $V \leftarrow V - h\,V\,x\,x^\top V$,
   f. $Q \leftarrow Q + h\,e^2$,
   g. $s^2 \leftarrow Q/(n-p)$,
   h. $\text{std.err}(\hat{\beta}) \leftarrow s\,\text{Diag}(V)^{1/2}$.

For many other cases, to fit a model to data, we need a more general criterion than that of least squares. From both theoretical and practical points of view, the preferred criterion for statistical estimation of model parameters is that of *maximum likelihood,* which substantially comprises least squares as a special case.

This criterion requires specification of a parametric family of probability distributions, dependent on a parameter θ (possibly p-dimensional) that must be estimated from available data. This probability distribution represents the law governing random variable Y from which empirical value y was observed. The distribution is identified by its probability density function in the case of continuous variables, or by the probability function for discrete variables. We usually use the notation $p(t; \theta)$ to indicate this probability or density function, where t varies in the set of possible values of Y.

With these hypotheses, we define the *likelihood function* as

$$L(\theta) = c\,p(y; \theta) \tag{2.25}$$

where c is an arbitrary positive constant, but fixed once and for all. Because $p(t; \theta)$ is evaluated in observed value y, the term on the left-hand side is a function only of θ; however, in some cases we use the notation $L(\theta; y)$ to show that it depends on observations.

Equation (2.25) therefore constitutes a family of functions, indexed by c. As c plays a significant role only for the development of theoretical results but has no effect either on the use of $L(\theta)$ or on the properties of the associated inferential techniques, in the following we keep $c = 1$.

Because $p(y;\theta)$ is essentially positive, it makes sense to define the *log-likelihood function* as

$$\log L(\theta) = \log p(y;\theta) \tag{2.26}$$

setting $\log L(\theta) = -\infty$ if $p(y;\theta) = 0$.

We obtain the estimate of θ according to the *maximum likelihood criterion* by maximizing (2.25) or, equivalently, (2.26). We can also write

$$\hat{\theta} = \arg\max_{\theta} L(\theta) = \arg\max_{\theta} \log L(\theta) \tag{2.27}$$

although this notation is not completely rigorous, because the existence and uniqueness of the maximum of L are not guaranteed. However, in the regular cases used in practice, this ambiguity does not occur because a unique global maximum exists.

The actual maximization of L can be explicitly obtained only in simple cases. In many others, we have to return to *numerical analysis* methods to identify it. In regular cases, we have to resolve the system of *likelihood equations*

$$\frac{\partial}{\partial\theta} \log L(\theta) = 0 \tag{2.28}$$

and then verify that the resulting solution corresponds to a maximum point. It is, in fact, quite simple to check whether we have a local maximum, but its definition (2.27) requires selection of the *global* maximum point. This can sometimes (but not always) be resolved by exploiting the mathematical properties of $p(y;\theta)$. We therefore see that this method can cause computational problems, at least in the case of complex models.

Every estimate must be accompanied by quantification of its precision, and this requires evaluation of its variance. One of the advantages of the maximum likelihood method is that we have a general scheme available for it, starting from *Fisher's observed information matrix*

$$\mathcal{J}(\hat{\theta}) = -\frac{\partial^2}{\partial\theta\,\partial\theta^\top} \log L(\theta)\bigg|_{\theta=\hat{\theta}} \tag{2.29}$$

of which the inverse gives an approximation to var $\{\hat{\theta}\}$, in conditions that can be verified in most practical cases. We can therefore obtain standard errors for $\hat{\theta}$ through

$$\text{std.err}(\hat{\theta}) = \text{Diag}(\mathcal{J}(\hat{\theta})^{-1})^{1/2}$$

where the Diag$(\cdot)$ notation indicates the diagonal elements of a square matrix.

Combining these facts with the additional property of estimates of maximum likelihood, that is, they have an approximately normal distribution when sample size is sufficiently high, we obtain

$$\hat{\theta}_r \pm z_{\alpha/2}\,\text{std.err}(\hat{\theta}_r) \tag{2.30}$$

to construct *confidence intervals* of at least approximate level $1-\alpha$ for the rth component θ_r of θ; here, $z_{\alpha/2}$ indicates the quantile of level $1-\alpha/2$ of distribution N(0,1).

This construction of a confidence interval for θ_r is associated with the construction of a procedure for testing the *hypothesis*

$$H_0 : \theta_r = a$$

for a specified value a. For fixed *statistical significance level* α, *Wald's test* criterion leads to rejection of hypothesis H_0 when $|t| > z_{\alpha/2}$, because we put

$$t = \frac{\hat{\theta}_r - a}{\text{std.err}(\hat{\theta}_r)}, \tag{2.31}$$

and (2.30) is consequently called a Wald-type confidence interval.

Equivalently, we can calculate the *p-value*, or *observed significance level*, approximated by $2\Phi(-|t|)$, which is compared with α.

When we are interested in testing a hypothesis on the components of θ expressed by q constraints of the type

$$H_0' : g_j(\theta) = 0, \quad (j = 1, \ldots, q), \tag{2.32}$$

where g_j are differentiable functions, against the alternative that at least one equality is false, the foregoing method cannot be used. Instead, we use the criterion of the *likelihood ratio*, defined by the test function

$$w = 2\,\{\log L(\hat{\theta}) - \log L(\hat{\theta}_0)\} \tag{2.33}$$

where $\hat{\theta}_0$ indicates the maximum likelihood estimate subject to q constraints (2.32).

For a fixed significance level α, the criterion leads to rejection of hypothesis H_0' when observed value w is greater than the $1-\alpha$ quantile of distribution χ_q^2. Here again, we can calculate the p-value, now expressed by

$$p = \mathbb{P}\{X^2 > w\}$$

where $X^2 \sim \chi_q^2$, at least approximately and compare p with α. The distributive properties associated with the procedure are exact in the case of normal distribution of observations and hypothesis H_0' expressed by linear constraints; in other cases, these properties are approximate.

Note that the two testing procedures based on (2.31) and (2.33) are connected. When they are both applicable, they give identical (or at least approximately equal) results. This is because hypothesis H_0 corresponding to a single linear constraint may be expressed as $H_0' = \theta_r - a = 0$, and $2\Phi(-|t|)$ is at least approximately equal to $\mathbb{P}\{X^2 > w\}$, where $w = t^2$ and $X^2 \sim \chi_1^2$.

2.3.2 Linear Models with Gaussian Error Terms

Discussion of the regression models of section 2.1 was based on specifying for error term ε only hypotheses up to second-order moments (i.e., mean, variance, and covariance), but without formulating a complete hypothesis on the nature of the distribution of ε, and therefore of the response variable.

As already mentioned, the distributive hypothesis that assumes normal or Gaussian distribution for ε, with independence between components for separate observations, is by far the most common and historically consolidated. Combining this fact with the contents of section 2.1.1 gives us $\varepsilon \sim N(0, \sigma^2)$. Therefore, regarding random variable Y_i, which generates the ith observation of model (2.2), we write

$$Y_i \sim N(f(x_i; \beta), \sigma^2), \qquad \text{for } i = 1, \ldots, n$$

and the corresponding log-likelihood function is

$$\log L(\beta, \sigma^2) = -\frac{n}{2}\log\sigma^2 - \frac{1}{2\sigma^2}D(\beta)$$

where $D(\beta) = \|y - f(x; \beta)\|^2$ is defined as in (2.3). This means that the maximization of likelihood with respect to β corresponds to the minimization of $D(\beta)$, and therefore the estimates of maximum likelihood coincide with those of least squares. To estimate σ^2, the maximum likelihood estimate,

$$\hat{\sigma}^2 = D(\hat{\beta})/n$$

is similar to s^2 of (2.11); the difference in the denominator tends to be relatively negligible as n gradually increases. It also follows that

$$-2\log L(\hat{\beta}, \hat{\sigma}^2) = n\log\{D(\hat{\beta})/n\} + n.$$

The new estimates are thus effectively the same as the least square ones, but the new formulation means that we have access to all the inferential apparatus mentioned in section 2.3.1.

The principal type of regression model is linear, which may be expressed as (2.6). In this framework, one common practical problem is testing the significance of regression parameters β; in particular, hypotheses of the type $H_0 : \beta_r = 0$ are commonly involved. In this case, the distribution of test function (2.31) can be calculated exactly, by Student's t distribution, like the p-value in the tables of section 2.1. The approximation error caused by avoiding the exact calculation of the p-value is not important for sample sizes larger than a few dozen.

If the q constraints (2.32) are expressed by linear relations on parameters, quantity (2.33) takes the form:

$$w = \frac{\|y - \hat{y}_0\|^2 - \|y - \hat{y}\|^2}{\hat{\sigma}^2} = \frac{\|\hat{y} - \hat{y}_0\|^2}{\hat{\sigma}^2} \tag{2.34}$$

where $\hat{y}_0$ is the vector of interpolated values under the q constraints. Each of the terms

$$D = D(\hat{\beta}) = \|y - \hat{y}\|^2 \quad \text{and} \quad D_0 = D(\hat{\beta}_0) = \|y - \hat{y}_0\|^2$$

in (2.34) represent a deviance, respectively, of the unconstrained model and that with q constraints.

The approximated distribution of reference for w is χ^2_q, according to the general results of section 2.3.1. In the specific case of Gaussian error terms, we can also obtain the exact distribution, which is usually expressed in terms of the transformation

$$F = w\,\frac{n-p}{n\,q} = \frac{\|\hat{y} - \hat{y}_0\|^2/q}{s^2} \tag{2.35}$$

which, as null distribution, is Snedecor's F with $(q, n-p)$ degrees of freedom, if p is the number of parameters in the nonconstrained model. Also in this case, the approximation error due to the asymptotic distribution for calculating the p-value is not important for sample sizes exceeding a few dozen.

2.3.3 Binary Variables with Binomial Distribution

In the case of binary response variables, let us denote one possible outcome as "success" and the other as "failure." When π denotes the probability of success in a single observation, the probability distribution of the total number of successes Y out of n independent observations in constant conditions is given by the binomial distribution of index n and probability parameter π. If y denotes the observed value of Y, the corresponding log-likelihood function is

$$\log L(\pi) = \text{constant} + y\,\log(\pi) + (n-y)\log(1-\pi), \qquad (0 \le \pi \le 1).$$

The estimate of maximum likelihood and its standard error are, respectively,

$$\hat{\pi} = y/n, \qquad \text{std.err}(\hat{\pi}) = \sqrt{\hat{\pi}\,(1-\hat{\pi})/n}$$

and the corresponding maximum of the log-likelihood function is

$$\log L(\hat{\pi}) = \text{constant} + y\,\log\hat{\pi} + (n-y)\log(1-\hat{\pi}).$$

where we mean that $0\log 0 = 0$, for continuity.

One frequent practical problem arises when we wish to examine a population stratified into two groups, say, 1 and 2, and denote the probability of success in a single observation from each group by π_1 and π_2, respectively. In this case, the log-likelihood function depends on two parameters and is

$$\begin{aligned}\log L(\pi_1, \pi_2) &= \text{constant} + y_1 \log(\pi_1) + (n_1 - y_1)\log(1 - \pi_1)\\ &\quad + y_2 \log(\pi_2) + (n_2 - y_2)\log(1 - \pi_2)\end{aligned}$$

where y_1 and y_2 denote the number of successes and n_1 and n_2 the sizes of two samples from the subpopulations.

In the previous notation, the test hypothesis was $H_0 : \pi_1 - \pi_2 = 0$, and the null hypothesis thus imposes $q = 1$ constraints of type (2.32) on the parameters. The likelihood ratio test statistic is therefore

$$w = 2\{\log L(\hat{\pi}_1, \hat{\pi}_2) - \log L(\hat{\pi}, \hat{\pi})\}$$

where $\hat{\pi}_j = y_j/n_j$, for $j = 1, 2$, and $\hat{\pi} = (y_1 + y_2)/(n_1 + n_2)$ is the estimate of common values of π. Observed value w is compared with approximate reference distribution χ_1^2.

By analogy with the framework of section 2.3.2, quantity w is also described as *deviance*, because here too it expresses the discrepancy between the formulated hypothesis and the general case and is usually indicated by the same symbol, D, of (2.10). We bear in mind here the fact that there is no parameter of scale σ^2. The same applies to the likelihood test: the concept of deviance has a much more general value than in the given example, because it may also refer to cases with J groups and the formulated hypothesis may not be that of equality of π for all groups, but corresponds to q constraints. Under this assumption π is estimated by $\hat{\pi} = \sum y_j / \sum n_j$. Some simple manipulations yield

$$D = 2\left\{\log L(\hat{\pi}_1, \ldots, \hat{\pi}_J) - \log L(\hat{\pi}, \ldots, \hat{\pi})\right\} \tag{2.36}$$

$$= D_0 - D_1 \tag{2.37}$$

where

$$D_1 = -2\log L(\hat{\pi}_1, \ldots, \hat{\pi}_J) = -2\sum_{j=1}^{J}\left\{y_j \log \hat{\pi}_j + (n_j - y_j)\log(1 - \hat{\pi}_j)\right\}, \tag{2.38}$$

and D_0 is similarly obtained when all $\hat{\pi}_j = \hat{\pi}$. Obviously, if the number of subgroups of which we want to test the equality of probability of success is $q + 1$, and correspondingly the number of constraints imposed on π is q, the number of degrees of freedom changes, and the approximate reference distribution is χ_q^2.

For a numerical illustration, we consider the data of the Brazilian bank (described in Appendix B.3) and split the degree of satisfaction into two levels, `high` and `low`, stratified into two subpopulations of `old people` and `young` people, according to age (over and under 45). The observed frequencies are shown below.

		young	old	total
satisfaction	**low**	84	34	118
	high	225	157	382
	total	309	191	500

This gives us the estimates of the probability of high satisfaction, $\hat{\pi}_1 = 225/309 = 0.728$ (SE $= 0.025$) for the young group, and $\hat{\pi}_2 = 157/191 = 0.822$ (SE $= 0.028$) for the old group, and the estimate without age stratification is $\hat{\pi} = 382/500 = 0.764$ (SE $= 0.019$). The corresponding calculation of the likelihood ratio test, that is, of deviance, gives $D = 2\,(273.21 - 270.25) = 5.96$, p-value 0.015, indicating the influence of age class on the degree of satisfaction.

Bibliographic notes

For a general treatment of statistical inference, at various levels, see Cox & Hinkley (1979), Casella & Berger (2002), or Wasserman (2004). For treatment of likelihood-based inference, see Azzalini (1996).

2.4 Logistic Regression and GLM

In the previous numerical example, we concluded that the young customers of the bank are significantly less satisfied than the older ones. Because the variable age can be used in numerical form, it seems preferable to use it in a nondichotomized way. To do this, we need a tool that allows us to study the relation between a quantitative variable and a dichotomous one, like satisfaction.

This situation is still a study of the relation between variables, but in this case the dichotomous nature of the response variable advises against the use of linear regression. A simple extension of the idea of linear regression to the new problem is logistic regression, which connects probability π of the event of interest to a set $x = (x_1, x_2, \ldots, x_{p-1})$ of explanatory variables in the following form. Response variable Y for any given subject is now a Bernoulli random variable, whose probability of success $\pi(x)$ depends on the covariates. If we indicate by $\eta(x)$ a combination of covariates, linear on the parameters, of the type

$$\eta(x) = \beta_0 + \beta_1 x_1 + \cdots + \beta_{p-1} x_{p-1} \tag{2.39}$$

similar to those used in § 2.1, and define the *logistic function*:

$$\ell(\eta) = \frac{e^{\eta}}{1 + e^{\eta}} \tag{2.40}$$

the model of logistic regression is given by

$$\pi(x) = \ell(\eta(x)) = \frac{e^{\eta(x)}}{1 + e^{\eta(x)}} \tag{2.41}$$

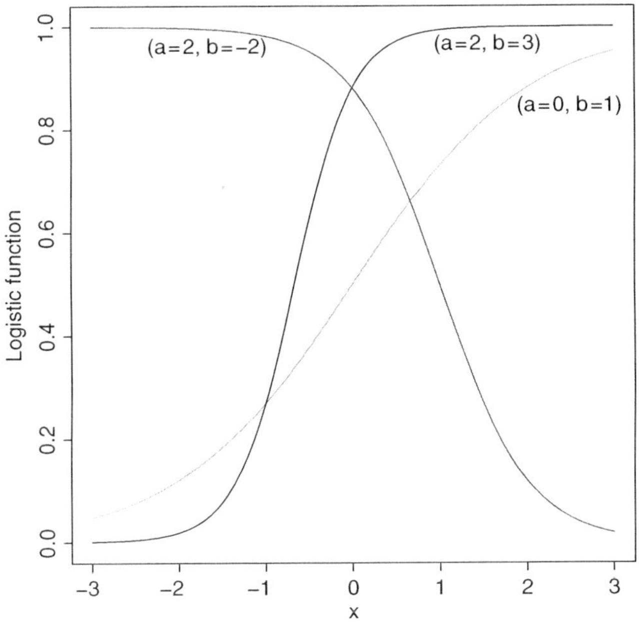

Figure 2.11 Logistic function for some choices of pair (β_0, β_1) when $\eta(x) = \beta_0 + \beta_1 x$.

where we note that the probability of the event of interest depends on x, through *linear predictor* $\eta(x) = x^\top \beta$.

Figure 2.11 shows some examples of the behavior obtainable in this way when we have only one explanatory variable and $\eta(x) = \beta_0 + \beta_1 x$, for some choices of pair (β_0, β_1). The specific pair $(0, 1)$ corresponds to $\ell(x)$ defined by (2.40).

The scheme of logistic regression is one of the family of *generalized linear models* (GLM) in which the relationship between the explanatory variables and the response variable may be expressed as

$$g\Big(\mathbb{E}\{Y|x_1, \ldots, x_{p-1}\}\Big) = x^\top \beta = \eta(x) \tag{2.42}$$

for an appropriate choice of *link function* $g(\cdot)$. The notation $\mathbb{E}\{Y|x_1, \ldots, x_{p-1}\}$ used here indicates that the values of variables x_j are predetermined or that we operate conditionally on the assumed values of the variables.

For this family of models, the probability distribution of Y conditional on covariates $x_1, \ldots, x_{p-1}$ must belong to a specific set of distributions. Although, mathematically speaking, this set is quite narrow, in practical terms it covers all the commonly employed families of distributions—Gaussian, gamma, binomial, Poisson, inverse Gaussian, and negative binomial. For this form, there is a clearly structured inference theory based on likelihood and deviance that, in this framework, plays an important role.

In general, we cannot express the maximum likelihood estimate of a GLM explicitly as a function of the observed data, and we must therefore use an iterative

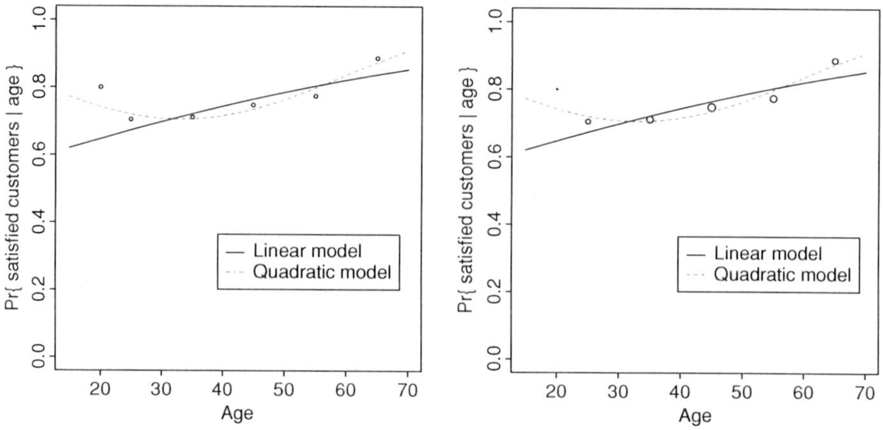

Figure 2.12 Bank data: frequencies of satisfied customers according to age and estimated curves of logistic regression. Left: circles have same diameter; right: circles have areas proportional to size of group.

numerical procedure. However, there is a very efficient and reliable iterative algorithm to obtain the estimates, through an appropriate sequence of estimates called *iterated weighted least squares*.

The case of logistic regression obtains when $g(\cdot)$ in (2.42) is

$$g(\pi) = \text{logit}\,\pi = \log\frac{\pi}{1-\pi} \tag{2.43}$$

that is, the inverse function of (2.40), and Y has Bernoulli distribution of parameter π, which is a function of the explanatory variables, that is, $\pi(x) = \ell(\eta(x))$. In the previous example, in which the response variable was dichotomous, the index of the binomial distribution was 1, but extension to the case of m observations made at value x is immediate, and therefore Y is a binomial with index m and parameter $\pi(x)$. It is common to use the quantity

$$\text{odds} = \frac{\pi}{1-\pi}$$

with inverse function

$$\pi = \frac{\text{odds}}{1+\text{odds}}.$$

If we examine the Brazilian bank data in greater detail, without aggregating the age values, the picture emerging from section 2.3.3 changes considerably. Figure 2.12 shows the fitted curve of the relative frequencies of satisfied customers according to age; the two panels are equal, except for the different way of representing the observed values. Figure 2.12 shows that customers' behavior does vary appreciably with age, in the sense that younger customers behave more like older ones than customers in the intermediate age classes.

Table 2.5. BANK DATA: SUMMARY OF LOGISTIC REGRESSION MODEL, QUADRATIC (UPPER) AND LINEAR (LOWER)

MODEL WITH QUADRATIC COMPONENT

	Estimate	SE	*t*-value	*p*-value
`(intercept)`	2.0356	1.2734	1.60	0.110
`age`	−0.0700	0.0602	−1.16	0.245
`age`^{2}	0.0011	0.0007	1.56	0.120

$D = 0.795$ with 3 d.f.

MODEL WITHOUT QUADRATIC COMPONENT

	Estimate	SE	*t*-value	*p*-value
`(intercept)`	0.1490	0.3829	0.39	0.697
`age`	0.0230	0.0084	2.73	0.006

$D = 3.302$ with 4 d.f.

We can now apply this method to study of the relationship between the probability of high satisfaction and the age of the bank's customers. The latter variable is available in the form of a central value of the respective age class, which we now indicate by x, of which possible values are (20, 25, 35, 45, 55, 65). The points in figure 2.12 represent the observed relative frequencies at the values of x, and the dotted curve is obtained by adapting model (2.40) as follows:

$$\eta(x) = \beta_0 + \beta_1 x + \beta_2 x^2$$

where selection is based on preliminary inspection of the data. Note that for the class of the youngest customers, the trend is opposite that of intermediate classes. Table 2.5 lists the estimate operations, which also show that the quadratic component has a Wald test *p*-value of 0.12, which is not significant.

We can also evaluate the importance of component β_2 by comparing the two deviances of the model with and without a quadratic component. The difference between them is $D = D_1 - D_2 = 3.302 - 0.795 = 2.507$, a value that is exceeded with a probability of 0.11 by variable χ^2_1, where the degrees of freedom are calculated by the difference $4 - 3$ between the degrees of freedom of the two ingredients. This value is not perfectly identical, but is basically equivalent, to that obtained by the Wald test.

Removing the quadratic component yields a model the relevant values of which are shown in the lower part of table 2.5, and the estimated curve is that which is continuous in figure 2.12.

It is initially surprising that the quadratic component is not necessary for a proper description of the relationship, in view of the very high frequency of the younger group. In fact, this deceptive impression comes from the type of graphical representation used, which does not consider group size. The right panel of figure 2.12 uses a more appropriate representation, in that the areas of the points are proportional to the size of the various groups, providing a visual impression that includes information. The choice of the model without a quadratic component no longer seems surprising, as the first group is of negligible size.

Bibliographic notes

A classical reference for these models is given in McCullagh & Nelder (1989). The work of Azzalini (1996) includes a shorter treatment of generalized linear models. Specific coverage of logistic regression, with particular attention to applicative aspects, is given by Hosmer & Lemeshow (1989).

EXERCISES

2.1 In model (2.14), applied to car data, remove the cubic term and estimate the new model. Observe that the quadratic term becomes significant. Explain this result.

2.2 Use the estimate of linear model (2.14) and extrapolate predicted values for gasoline cars with `engine size` in the interval $(1, 7)$. Comment on the results.

2.3 For model (2.17), value R^2 ranges from 0.64 to 0.56 when calculated from the original data instead of the transformed data, falling below value 0.60 of the model (2.14). Explain and comment on these differences.

2.4 Extend model (2.17), inserting variables `curb weight` and I_D, and compare the result of the fit of the new model with that of (2.19).

2.5 For model (2.18), reproduce the two graphs of figure 2.8.

2.6 For model (2.18), give a critical analysis of the elements in table 2.3 and associated graphs along the lines of the discussion at the end of section 2.1.1.

2.7 Fit an appropriate linear model to predict `highway distance` for car data, in two ways: (a) using the variables described in this chapter; (b) using any variables listed in Appendix B.2.

2.8 Complete the details of the statements at the end of section 2.2.2 by calculating s^2 and standard errors, using (2.10) or any other method.

2.9 Check the correctness of the Sherman-Morrison formula (A.2).

2.10 Check the correctness of the formulas provided by recursive updating of the least squares estimates.

2.11 Prove (2.24).

2.12 What is the difference between the confidence interval of the value of the function and the prediction interval, both relative to the next observation?

2.13 The curves of figure 2.11 are all monotone, whereas one of those in figure 2.12 is not. Explain this discrepancy.

3

Optimism, Conflicts, and Trade-offs

Pluralitas non est ponenda sine necessitate.

—William of Ockham

3.1 Matching the Conceptual Frame and Real Life

A solidly based and rich theory of statistical inference, of which we have only mentioned a few key components, underlies the methods described in chapter 2. This theory is characterized by a number of properties that hold only if the model is chosen according to a conceptual foundation that must preexist the availability of the data, and the model itself is appropriate, at least for the purposes of the analysis in question. The related inferential paradigm was developed within a specific research context, with important connections with the foregoing experimental and scientific settings.

However, certain applied problems, which are often encountered, do not fit this scheme very well. A particularly common critical point is the absence of an adequate background theory, which prevents us from formulating a reliable model before inspecting the data. Preliminary exploration of the data is therefore often required to identify the most suitable model; this approach is even adopted as a general course of action. Chapter 2 describes some examples.

In our areas of application, the inferential paradigm must be adapted to some extent, because no proper sampling design exists.

Similar to the procedure described in chapter 2—that is, exploratory inspection of the data to choose the most suitable models—we consider diagnostic methods to at least partially verify the appropriateness of the choice of model. These diagnostics cover one aspect of the problem, but the issue of assessing the validity of a model is much broader. We explore this topic in the next sections.

3.2 A Simple Prototype Problem

We consider here a very simple example serving as the prototype for much more complex and realistic circumstances. Let us presume that yesterday we observed $n = 30$ pairs of data (x_i, y_i), for $i = 1, \ldots, n$, shown in the scatterplot of figure 3.1. The data were generated artificially by an equation such as

$$y = f(x) + \varepsilon \tag{3.1}$$

where ε is an error component with distribution $N(0, \sigma^2)$ and $\sigma = 10^{-2}$; $f(x)$ is a function which we leave unspecified—the only requirement is that this function should follow an essentially regular trend. Clearly, to generate the data, we had to choose a specific function (not a polynomial), but we do not disclose our choice.

Say we wish to obtain an estimate of $f(x)$ today that allows us to predict y as new observations of x become available. A reasonable choice consists of using the techniques mentioned in chapter 2, particularly the polynomial regression in (2.4).

If we have no information to guide us in choosing the degree of the polynomial, we first consider all possible degrees from 0 to $n - 1$, thereby introducing p

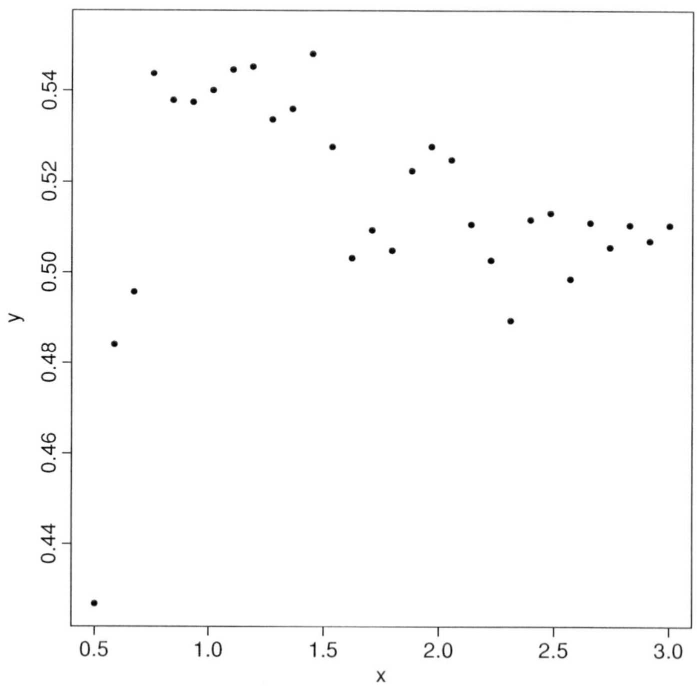

Figure 3.1 Yesterday's data: scatterplot.

parameters ranging from 1 to n, in addition to σ. For brevity, figure 3.2 only shows the fitted curves for only some values of p. Obviously, the polynomial fit improves as p increases, as shown in figure 3.3, in which the residual deviance (2.10) and coefficient of determination R^2 (2.15) are plotted as functions of p.

A special case exists when $p = n$, corresponding to a polynomial that exactly interpolates the observed data, with residual deviance 0 and $R^2 = 1$. Such a case is apparently ideal, but it corresponds to the unacceptable situation shown in the last plot of figure 3.2. The nearly vertical lines are simply the visualized portions of the very large fluctuations the 29-degree polynomial must follow to interpolate all the observed points exactly.

As already mentioned, we need to use an estimate of $f(x)$ to predict values of y for new data $\{y_i, i = 1, \ldots, n\}$ produced by the same generating mechanism, but these will become available tomorrow. To simplify the process, we assume that these y_i are associated with the same x_i used for yesterday's data. We now evaluate the quality of the prediction using yesterday's fit of the polynomials for the new y_i, as if we could obtain tomorrow's data today. Figure 3.4 shows tomorrow's data with the predictions from the previously fitted polynomials. It is noteworthy that the higher-degree polynomials fluctuate and no longer fit the new points, whereas for smaller values of p, an increase in the degree of the polynomial improves the fit of the general trend. This improvement gradually ceases as the increase in degree causes the polynomial to follow random fluctuations in yesterday's data, not observed in the new sample. Figure 3.5 summarizes and quantifies this information by showing that the residual deviance decreases to a certain point and then increases, whereas index R^2 peaks and then falls.

The concepts of deviance and R^2 are used here in a way that extends beyond their common definitions, since the sum of the squares of the quantities involved are computed by using data other than those used for the fit.

3.3 If We Knew $f(x)$...

In a general sense, when we formalize the observations of section 3.2, we can say that we want to estimate $f(x)$ using a generic estimator $\hat{y} = \hat{f}(x)$, which, in our example, can be provided by one of the 30 fitted polynomials.

We start by considering a specific value x' for x. If we knew the mechanism used to generate the data precisely, that is, $f(x')$, we could calculate a few quantities of interest for the quality of estimator $\hat{y}$. An important goodness-of-fit index is given by the *mean squared error*

$$\mathbb{E}\{[\hat{Y} - f(x')]^2\} = \left[\mathbb{E}\{\hat{Y}\} - f(x')\right]^2 + \text{var}\{\hat{Y}\}, \tag{3.2}$$

where $\hat{Y}$ denotes the parent random variable of $\hat{y}$. When $f(\cdot)$ is a polynomial with fixed degree p, (3.2) can be explicitly obtained; see exercise 3.2.

Because we are interested in more than one single point x', we consider the sum of the mean squared errors for all the n values of x. Representing the resulting value as a function of p, which is an indicator of *model complexity*, we obtain the

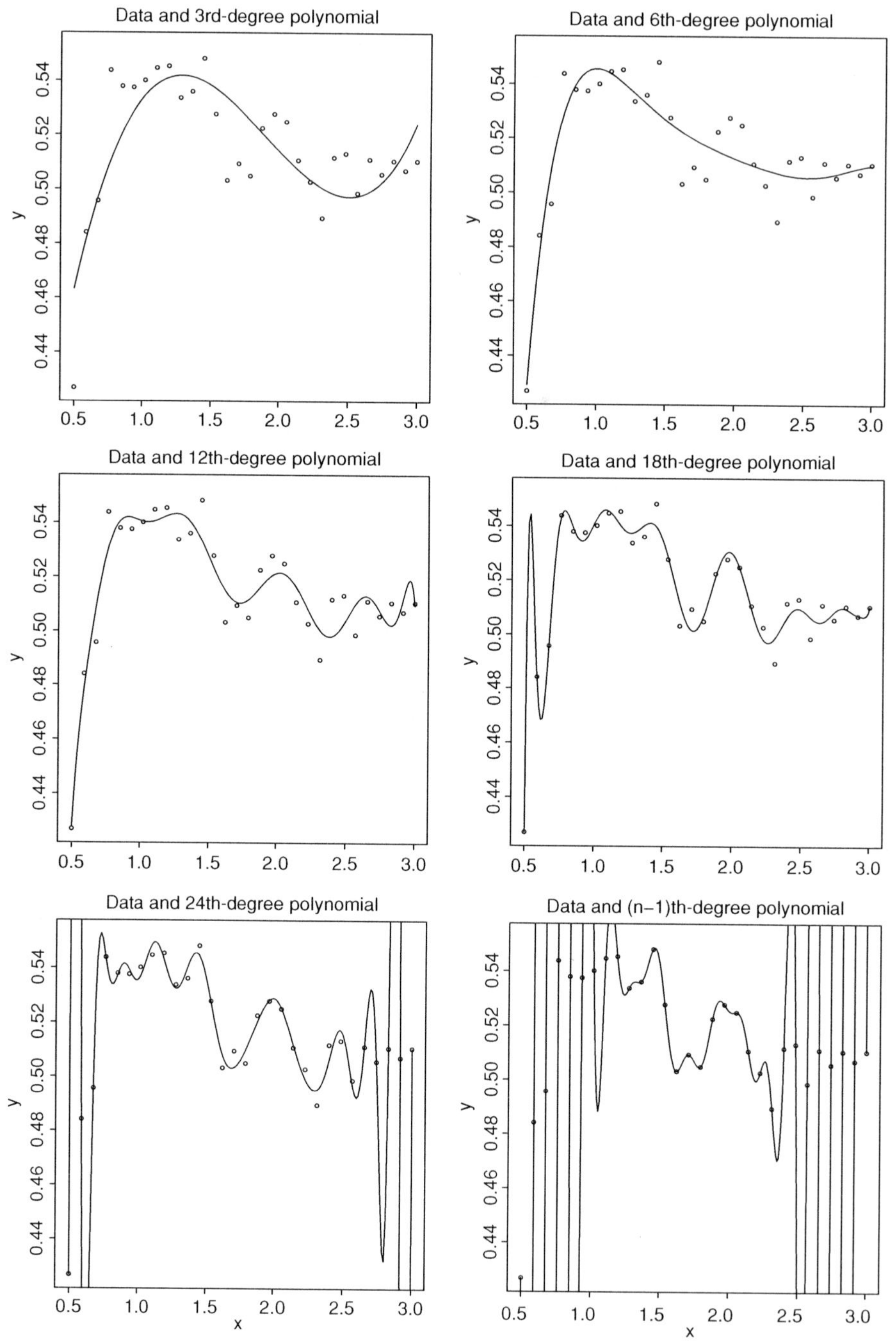

Figure 3.2 Yesterday's data: interpolations with polynomials of various degrees.

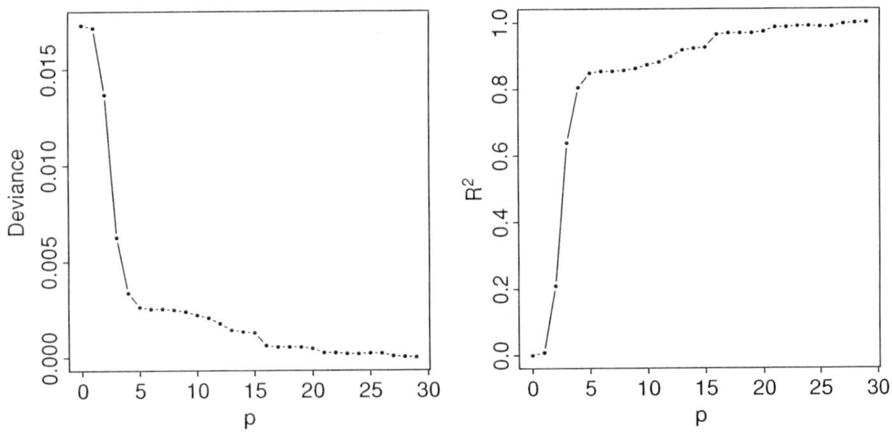

Figure 3.3 Yesterday's data: deviance and R^2 coefficient when p varies.

plot shown in figure 3.6. Note that when p increases, the mean squared error first decreases and then increases, thus providing the level of 'complexity' corresponding to a minimum mean squared error—in this case, for $p = 5$.

In the foregoing treatment, we used the family of polynomials as a set of models in which complexity was controlled by a certain parameter p, which was precisely the polynomial degree. Polynomials are not the only possible choice; the Fourier series is another that comes to mind. In any case, the final message remains unaltered, even when the family of models chosen is changed. When complexity increases, there is usually an initial gain followed by a loss.

When we further consider (3.2), the components of which are such that

$$\mathbb{E}\{[\hat{Y} - f(x')]^2\} = \text{bias}^2 + \text{variance}, \tag{3.3}$$

we see that this decomposition applies not only in the case of polynomial regression, but also in general. Figure 3.7 shows how these two components contribute to the mean squared error for this example. When model complexity, quantified by p, is low, bias is high and variance is moderate; when p increases, bias decreases but variance increases. As mentioned in section 3.2, when p increases, the polynomials fit the data better, but when p becomes too large, they follow random fluctuations in the data. In this case, variance increases without any important gain in bias. In these situations, the model *overfits* the data and involves an excess of *optimism* in evaluating the prediction error.

This behavior is found in much more general situations involving models with increasing complexity. Bias and variance are conflicting entities, and we cannot minimize both simultaneously. We must therefore choose a *trade-off between bias and variance*. This situation guides the developments that follow.

A bias component is essentially due to lack of knowledge of the data-generating mechanism. If this mechanism were known, we could set up an appropriate parametric model, such as a polynomial of specified degree, and the bias would be null or at most negligible. This is typical of parametric models when they are

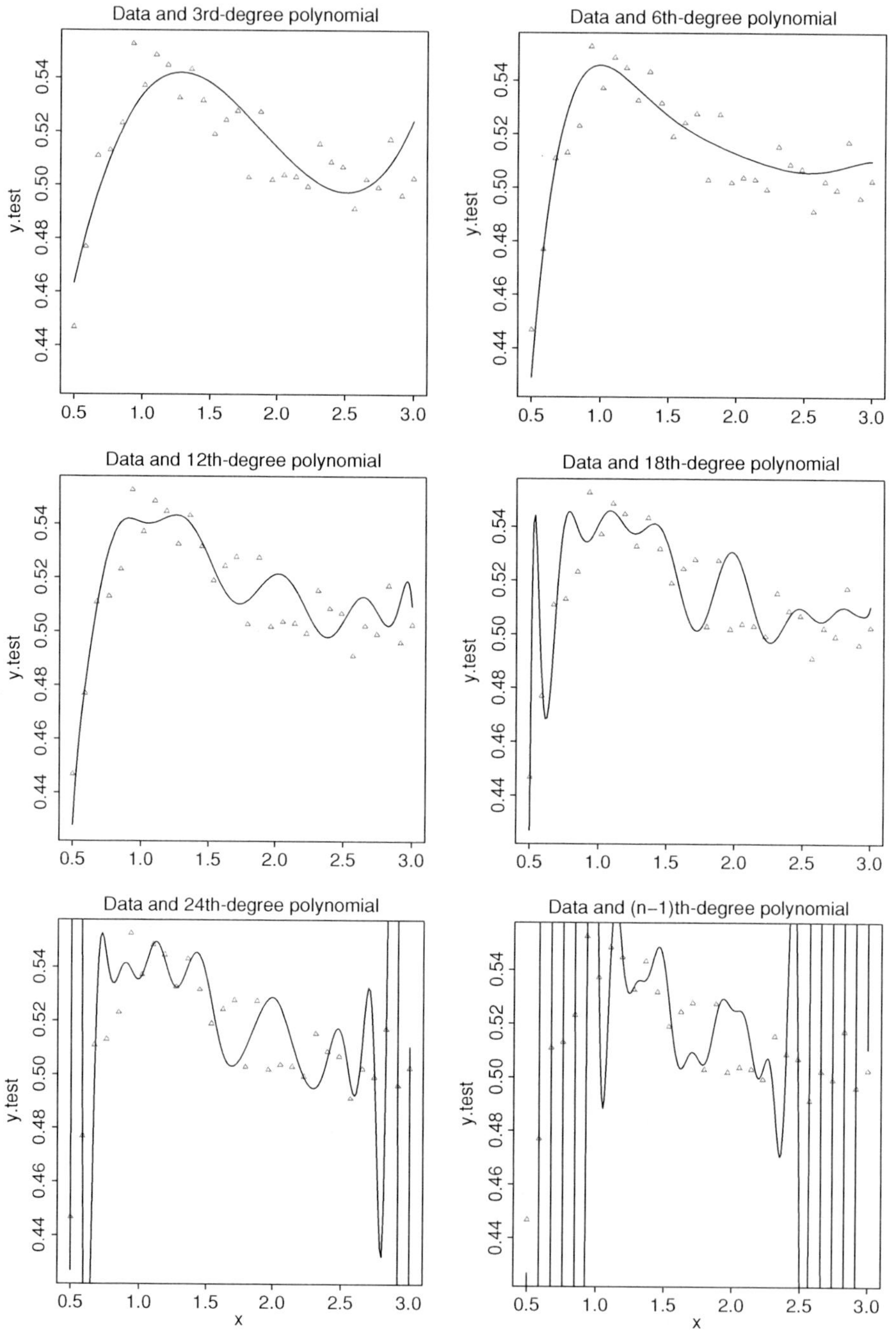

Figure 3.4 Tomorrow's data: interpolation with polynomials obtained by fitting yesterday's data.

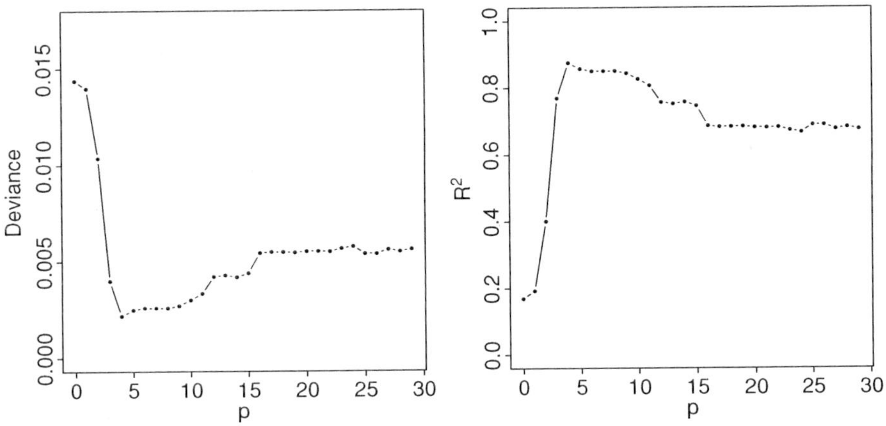

Figure 3.5 Tomorrow's data: deviance and R^2 coefficient as a function of p.

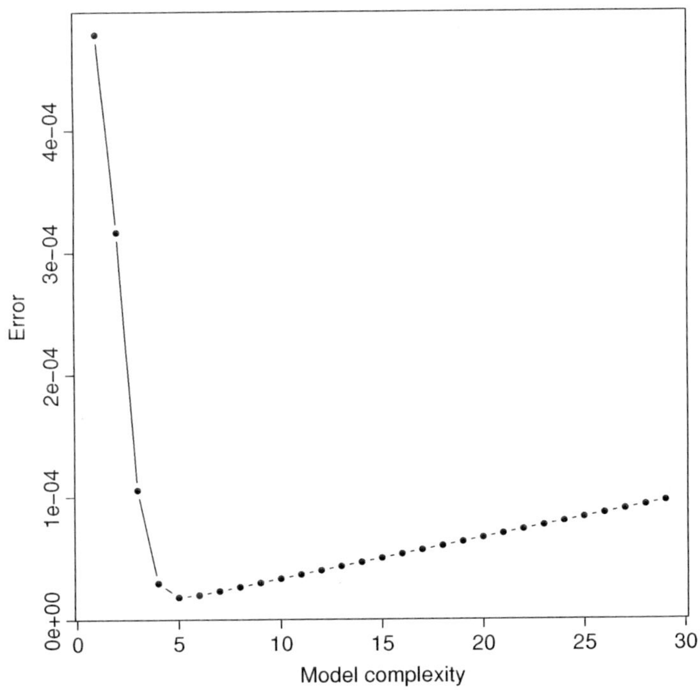

Figure 3.6 Yesterday's data: mean squared error as a function of p.

correctly specified. Instead, the context in which we are working obliges us to use an essentially *nonparametric approach*, although we used parametric tools (polynomials) as building materials for the sake of simplicity.

3.4 But as We Do not Know $f(x)$...

We just concluded that we must expect a trade-off between error and variance components. In practice, however, we cannot do this because, of course, $f(x)$ is unknown.

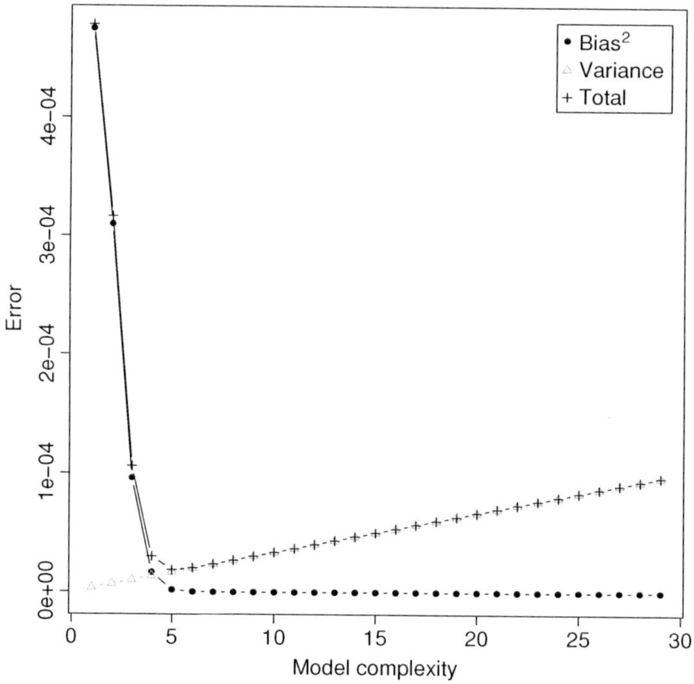

Figure 3.7 Yesterday's data: mean squared error as a function of p, decomposed into bias and variance.

We have seen that overfitting is a trap that must be avoided. Overfitting occurs when a model closely fits some nonessential features of the observed sample. If these characteristics are not structural to the phenomenon under study, they will not recur in a new sample. As this problem originates because we calculate deviance with the same data with which we fitted the model, one way of avoiding this trap is to evaluate the model with other data.

In our example, the models fitted to "yesterday's" data can be compared with those of "tomorrow," yielding the plot of figure 3.8, which shows the residual deviance for various polynomials fitted to yesterday's data. Clearly, the deviance calculated for tomorrow's data provides a reasonable indication of the complexity of the model, essentially analogous to that given in figure 3.6; equally clearly, the two figures do not have the same nature: one is an approximation of the other, and the curve obtained with tomorrow's data is also affected by the variability of the new data. Nevertheless, the indication provided by the deviance of tomorrow's data does not suffer from the drawback we wished to avoid, and its message is essentially valid, with a minimum point at $p = 4$.

3.5 Methods for Model Selection

We must confess here that we cheated. We do not in fact have two sets of data, one for yesterday and one for tomorrow. We have 60 observations, randomly divided into two groups of 30 observations each, but we acted as we did to illustrate

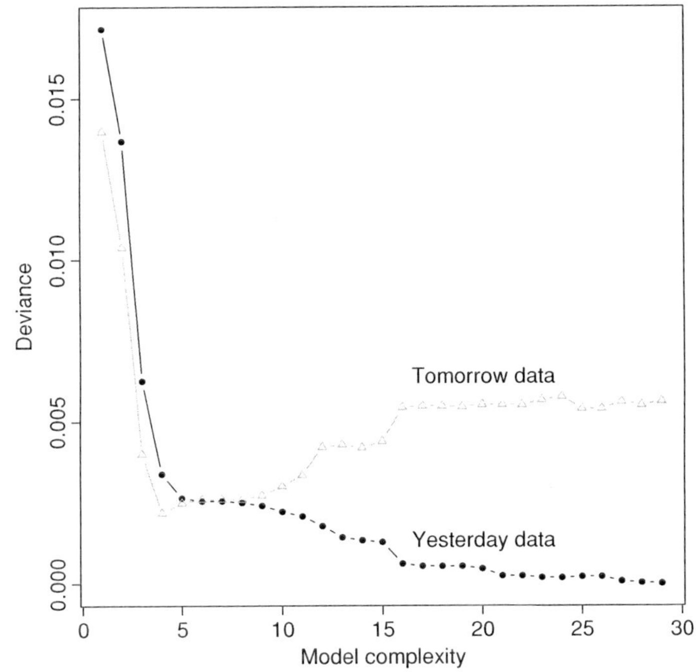

Figure 3.8 Yesterday's and tomorrow's data: residual deviance as a function of degree p of polynomials fitted to yesterday's data.

the problem. We now consider the principal tools used in model selection by identifying the trade-off between bias and variance.

3.5.1 Training Sets and Test Sets

Dividing the data into two groups circumvents the overfitting problem and allows us to reach a plausible solution for choosing p.

This approach is not our invention but is a common procedure in this kind of context. A randomly selected portion of data, called *training set*, is used to fit the various candidate models. The remaining portion, the *test set*, is used to evaluate the performance of the available models and choose the most accurate one.

Clearly, this scheme reduces the sample size used for fitting the model, which may be inadvisable when sample sizes are already small. Having too few data is not a concern in the context of data mining; having too many might be the problem. Instead, in the current context, it is more important to neutralize or at least diminish any estimation bias, as already noted.

Because the same test set can be used to evaluate many different models, there is a risk that the final assessment, obtained at the end of the entire process, is still somewhat biased and too optimistic, because of the same mechanism that acts when we use the training set. For this reason, and because the data are abundant, a third set, called the *validation set*, is often created for use at the end of analysis for final evaluation of the prediction error.

There is no precise rule on how to select the size of these sets, but the table that follows gives some commonly used reference values for proportions with two or three subsets.

Portion of data for:	**training**	**test**	**validation**
	50%	25%	25%
	75%	25%	0
	67%	33%	0

3.5.2 Cross-validation

Recall the procedure described in section 3.5.1 and presume that we use 75% of the data for training and 25% for testing the models. However, for greater accuracy, we do not want to assign only *that* specific 25% of the data to the role of test set. In addition, if n is not very large and we only use 75% of the data to fit the model, the estimate will be further impoverished, whereas we would like to take better advantage of available information.

One way of partially overcoming this arbitrariness is to split the data into four equal parts and use three portions *in rotation* for training the model and the remaining portion for testing it. We then *cross* the role of the data sets: one of the portions used as the training set is now used as a test set, and the test set is incorporated into the training set with the other two portions. Obviously, this scheme requires four iterations of the training and testing procedures.

Because this scheme results in four different estimates, which probably do not differ by much, an average or some other combination of them can be used. Analogously, we have four different figures similar to figure 3.8, and use these to obtain an "average curve," from which we can determine the minimum point.

It is intuitive that the procedure becomes progressively more accurate if, instead of 4 parts sized $n/4$, we use k portions of size n/k and repeat the operations k times. This is more effective when large values of k are used.

The maximum possible value for k is n. To fit the model, $n - 1$ observations are used, and the remaining observation is used for testing. This procedure is known as *leave-one-out* cross-validation and is described in detail in algorithm 3.1. Once we have rotated the only datum serving as test set, we must perform a total of n fitting operations. Clearly, the computational burden of this procedure increases considerably as n increases.

Fortunately, in many cases, it is possible to obtain estimates of a model using data deprived of a single observation, by means of simple operations based on estimates obtained from the complete data set. In particular, in the case of a linear model such as (2.6), in which the fitted values are given by (2.8), the following relationship holds

$$y_i - \hat{y}_{-i} = (y_i - \hat{y}_i)/(1 - P_{ii}), \tag{3.4}$$

so that we can obtain interpolated value $\hat{y}_{-i}$ for the ith observation without using the observation itself, but only using value (or values) x_i. Here, P_{ii} is the ith diagonal element of projection matrix (2.9). In this way, we obtain all the interpolated

Algorithm 3.1 Cross-validation (*leave-one-out*)

1. Read n records of x and y.
2. Cycle for $p = 0, 1, \ldots, \max_p$:

   a. cycle for $i = 1, \ldots, n$:

      i. fit the model of degree p by eliminating the ith observation,
      ii. obtain prediction $\hat{y}_{-i}$ for y_i corresponding to point x_i,
      iii. obtain residual $e_i \leftarrow (y_i - \hat{y}_{-i})$,

   b. calculate $D^*(p) \leftarrow \sum_{i=1}^{n} e_i^2$.

3. Choose p so that $D^*(p)$ is minimum.

values for the n possible subsets of training data by a simple modification of interpolated value $\hat{y}_i$ and using matrix P, which needs to be calculated in any case.

Algorithm 3.1 for the 60 observations considered so far and the simplified formula (3.4) produces figure 3.9, which indicates $p = 4$ is the preferred value.

We introduced the cross-validation criterion on a purely intuitive basis. There are theoretical results guaranteeing that when n diverges, this procedure certainly leads us to select the most appropriate model. However, we should add that for small sample sizes, this method often gives a very variable choice for p.

3.5.3 Criteria Based on Information

The main statistical method applied for estimating the unknown parameters of a model is to maximize the log-likelihood. However, when the model itself is not fixed in advance and is chosen from a sometimes large set of alternative models, we cannot simply proceed by maximizing the likelihood function for each alternative model; we must also take into account the different number of parameters, introducing a suitable penalty. Criteria that follow this logic can be traced back to objective functions such as

$$\text{IC} = -2 \log L(\hat{\theta}) + \text{penalty}(p), \tag{3.5}$$

where penalty(p) quantifies the penalty assigned to a model incorporating p parameters.

The choice of the specific penalty function identifies a particular criterion. Clearly, this function must be positive and must increase with p. A more specific indication is supplied by the following considerations. When we compare two nested models by test function (2.33) when the restricted model has one parameter

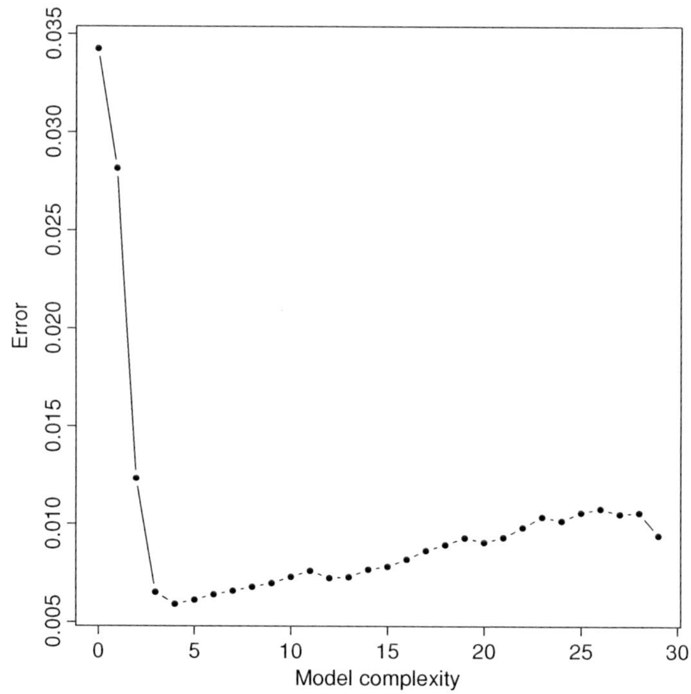

Figure 3.9 Yesterday's and tomorrow's data: cross-validation model selection.

less than does the other—that is, the nested model specifies one variance on the $p + 1$ parameters of the larger model—we know that asymptotically

$$2\left(\log L_{p+1} - \log L_p\right) \sim \chi_1^2,$$

when the $(p + 1)$th parameter is actually redundant. Here, we denote the maximum achieved by the likelihood of the two compared models as L_{p+1} and L_p. Thus, the insertion of an unnecessary parameter leads the mean of $-2 \log L$ to decrease by one unit, so that the penalty for p parameters must be strictly greater than p.

This approach to model selection was introduced by Akaike (1973), who proposed the now famous Akaike information criterion (AIC). Akaike suggested minimizing the *Kullback-Leibler divergence*:

$$KL(p_*(\cdot), p(\cdot; \theta)) = \mathbb{E}_{p_*}\left\{\log \frac{p_*(Y)}{p(Y;\theta)}\right\} = \mathbb{E}_{p_*}\left\{\log p_*(Y)\right\} - \mathbb{E}_{p_*}\left\{\log p(Y;\theta)\right\}$$

between true distribution $p_*(y)$ and fitted model $p(y; \theta)$. This quantity may be interpreted as a measure of the divergence between the distribution of future data generated by random variable Y and that predicted by the model. It is clear that to minimize KL, we can only act on the second term of the last expression and must therefore consider a value that maximizes $\log p(Y; \theta)$, that is, the maximum likelihood estimate. As this estimate $\hat{\theta}_y$ is a function of past observations, say, y, and as we want to use $p(\cdot; \hat{\theta}_y)$ to predict the behavior of the model on future data generated from random variable Y, we also need to take into account the variability connected with the estimating procedure. This leads us to consider the quantity

$$\mathbb{E}_y\left\{\mathbb{E}_Y\left\{\log p(Y; \hat{\theta}_y)\right\}\right\}.$$

Calculation of this expression requires some assumptions, as well as analytic approximations. According to Akaike's initial formulation, after appropriate analytical developments, we obtain

$$-2 \log p(y; \hat{\theta}) + 2p$$

as an estimate of the quantity of interest $\mathbb{E}_{p*}\{\log p(Y; \theta)\}$, multiplied by the conventional factor -2, which is inserted by alignment with the consolidated notations related to likelihood, in particular (2.33).

Akaike's original work was followed by several other proposals, differing in their assumptions and the way they approximate certain quantities. Some of them are shown in the table that follows, which provides some alternative penalties to be included in (3.5).

Criterion	Author	Penalty(p)
AIC	Akaike	$2p$
AIC$_c$	Sugiura, Hurvich-Tsay	$2p + \dfrac{2p\,(p+1)}{n-(p+1)}$
BIC/SIC	Akaike, Schwarz	$p \log n$
HQ	Hannan-Quinn	$c\,p \log\log n, \quad (c > 2)$

Note that the difference between AIC and AIC$_c$ tends to be negligible when n is large, because AIC$_c$ is a corrected AIC for small sample sizes. The last two criteria use a penalty that increases with increasing n and were generated by theoretical considerations quite different from the first two criteria. Although the logical framework of these information-based criteria and the procedures for hypothesis testing is not really the same, in practice they are often employed as if they were competing on the same ground.

An important advantage of information-based criteria with respect to the likelihood ratio test is that they can also be applied to families of unnested models, provided that the arbitrary constant in likelihood function (2.25) is set at 1. The disadvantage is that any evaluation of the probability of error of the procedure is not available.

Figure 3.10 illustrates the results obtained with these criteria for the data used so far. In this case, all four criteria suggest the same choice: $p = 4$. Obviously, there is no need to split the data into two sets, hence we use all 60 observations.

To conclude this short review of methods for model selection, figure 3.11 plots the fitted curve, which in this case is the same for all methods, with $p = 4$. This represents the basic trend of the phenomenon in a plausible fashion.

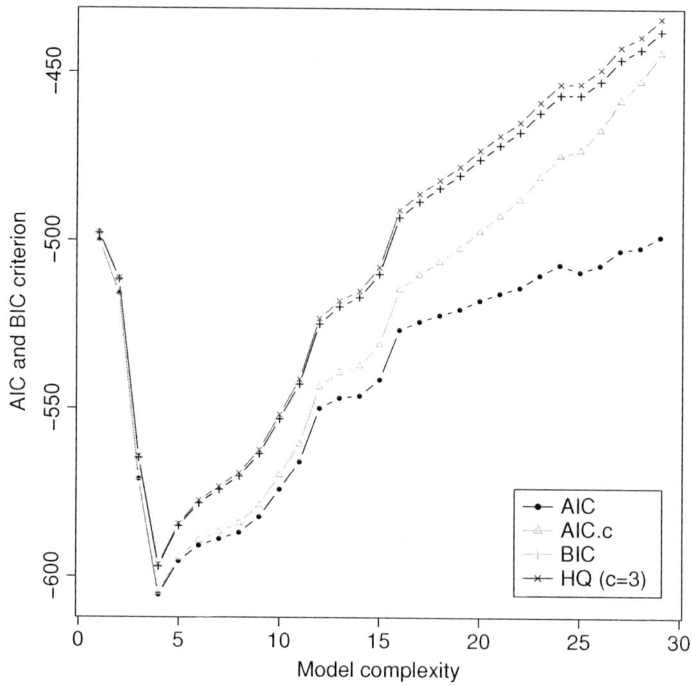

Figure 3.10 Yesterday's and tomorrow's data: various information criteria as a function of model complexity.

Bibliographical notes

Although the first examples of the use of cross-validation methods are quite old, the introduction and systematic study of this criterion are attributed to Stone (1974). The AIC appeared for the first time in Akaike (1973). An extended discussion of AIC-related criteria is given in Burnham & Anderson (2002). Recent specific coverage of model selection is to be found in Claeskens & Hjort (2008).

3.6 Reduction of Dimensions and Selection of Most Appropriate Model

We can now devise automatic procedures for model selection by examining a set of alternative models fitted to a certain data set. Implementation of these procedures is made easier, and for this reason is particularly widespread, when various models are all of the same type, differing only in their set of explanatory variables. Therefore, in practice, these are *variable selection* procedures.

3.6.1 Automatic Selection of Variables

For the sake of simplicity, we refer to the problem discussed in section 3.2 and to model (3.1). The set of models competing for $f(x)$ consists of a family of polynomial functions. The explanatory variables are the powers of x, so that the generic covariate, say, x_j, is such that $x_j = x^j$, with a degree ranging from 0 to a fixed maximum q (e.g., $q = n - 1$). In previous sections, we argued on the assumption that when we use a polynomial of a certain degree, all the terms of lower degree are

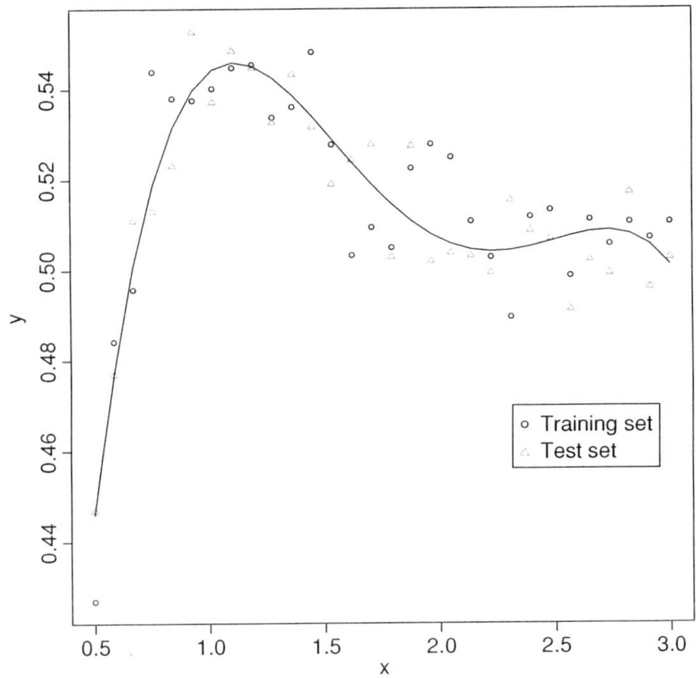

Figure 3.11 Yesterday's and tomorrow's data: fitted curve with $p = 4$.

inserted in the regression curve. However, this requirement is not necessary in the following.

In a generic regression context, we consider a set of explanatory variables such as

$$S = \{1,\ x_1,\ x_2,\ \ldots,\ x_q\}, \tag{3.6}$$

where the inclusion of constant 1 is not a formal need but is in fact almost universally applied. For each choice of a subset of S, it is straightforward to obtain estimates for regression coefficients $\hat{\beta}_j$, other connected quantities such as deviance, and if we assume Gaussian errors, log-likelihood, and AIC. However, the regression model is not the only one available: GLM linear or other parametric models can also be used, despite the possible greater computational burden.

An automatic procedure for variable selection aims at identifying the subset of S that minimizes the AIC or a similar criterion. Obviously, this operation requires fitting many models and must be done by computer. Even so, the related computational burden is huge if q is not small and we have to go through all the possible subsets and look for the *optimal subset.*

Thus, if q is not small, it is more common to use a simplified procedure known as *stepwise selection,* or some variant of this name. We begin with a certain model, identified by a certain subset S_0 of S, and then add the member of S not included in S_0. Alternatively, we eliminate the member of S_0, which gives the lower value of

AIC of all the operations of this type. We thus obtain a subset, S_1, which contains one element more or one element less than S_0. This operation is repeated, this time starting from S_1, looking for the optimal variation. The result is subset S_2, and the procedure is repeated until we reach a set, S_*, which cannot be improved by either reduction or enlargement. This is the selected subset.

When starting, subset S_0 is the minimum size that we want to consider with respect to S, for example, $S_0 = \{1\}$, the final outcome will obviously be a subset $S_* \supseteq S_0$, and the procedure is called *forward selection*. However, when $S_0 = S$, the final selection will be $S_* \subseteq S_0$ and the procedure is called *backward selection*.

The use of these automatic selection techniques is particularly justified when a large number of the explanatory variables is available and detailed analysis of all of them is not feasible. Another reason is the lack of suggestions and guidelines provided by the original applied problem. Both of these conditions often occur in the context of data mining.

However, it should be noted that these procedures, although they use inferential tools with well-known probabilistic characteristics as functioning ingredients, are out of that context in practice. For example, it is very difficult to establish which properties (in terms of actual precision of standard errors) are associated with the various estimates. This is because they do not refer to a predetermined model with respect to data, which is the basic condition for evaluating standard errors. Obviously, these observations also apply to other situations where the model is selected using the same data, but they are more relevant in cases when multiple individual models are evaluated.

Bibliographical notes

Stepwise regression is described and discussed, for example, in Weisberg (2005, section 10.3), Miller (2002, chapter 3), and Izenman (2008, section 5.7). Chapter 8 of Afifi & Clark (1990) gives a detailed presentation of automatic selection procedures.

3.6.2 Principal Component Analysis

Another strategy for selecting a model is based on reducing the dimension of the explanatory variables, transforming them in some way into a set of new variables of smaller number, but at the same time trying to lose only information that is not important in predicting the response variable.

The simplest possibility is to consider linear transformations of explanatory variables that have some sort of optimality property. *Principal component analysis* (PCA) is probably the most frequently used technique for deriving a reduced set of new variables by linear combination of the original variables that explains most of the variability of those variables.

We consider matrix X, obtained from the set of explanatory variables in (3.6), as the sampling determination of a multivariate random variable and, for ease of explanation, assume that it has mean 0 and covariance matrix Σ. If the variables are not centered around 0, it is always possible to calculate deviations from the mean and obtain zero mean variables. The variance of linear combination $Z = X\alpha$ is

$\text{var}\{Z\} = \alpha^\top \Sigma \alpha$. We must find a vector of weights α, so that $\text{var}\{Z\}$ is the largest among all normalized linear combinations of the columns of X, by imposing a scale restriction on α. This leads to the principal component criterion

$$\max_\alpha \text{var}\{Z\} = \max_\alpha \alpha^\top \Sigma \alpha \qquad (3.7)$$

$$\text{subject to} \quad \|\alpha\| = 1.$$

Once the first component has been selected, with coefficients α_1, we look for another linear combination, orthogonal to the first one, maximizing the variance

$$\max_\alpha \text{var}\{Z\} = \max_\alpha \alpha^\top \Sigma \alpha \qquad (3.8)$$

$$\text{subject to} \quad \|\alpha\| = 1 \text{ and } \alpha^\top \alpha_1 = 0.$$

The other components are defined in a similar fashion by requiring orthogonality with all the previous components.

The mathematical solution of this problem is given by the spectral decomposition of Σ: $\text{var}\{Z_1\} = \text{var}\{X\alpha_1\} = \lambda_1$ is the largest eigenvalue of Σ and α_1 is the corresponding eigenvector; $\text{var}\{Z_2\} = \lambda_2$ and α_2 correspond to the second-largest eigenvalue and the related eigenvector, and so on. Solution α_1 is called the *first vector of principal loadings*, combination $Z_1 = X\alpha_1$ is the *first principal component*, and so on for α_2, Z_2, and so on.

Because Σ is not usually known, in practice spectral decomposition is obtained on estimate $\hat{\Sigma}$. We denote by z_j the observed value of Z_j. Principal components have a simple geometrical interpretation, because Z_1 is the projection of the data on the longest observed direction—that is, the direction having the largest variance among all such normalized projections—Z_2 is the projection on the second longest direction orthogonal to the first one, and so on. This is illustrated in figure 3.12 for the two-dimensional case.

The sum of the eigenvalues is equal to the trace of the covariance matrix, so that the sum of the variances of the components is the same as that of the original variables, and $\sum_{i=1}^{k} \lambda_i / \sum_{i=1}^{p+1} \lambda_i$ is the fraction of total variance explained by the first k components, and fraction $\lambda_i / \sum_{i=1}^{p+1} \lambda_i$ measures the importance of the ith component in explaining total variability. If the percentage of total variance explained by the first k components is large enough, we can eliminate the remaining components and take only the first k to describe variability among explanatory variables, using them as new independent variables of the model.

PCA for dimension reduction in prediction problems is often used to solve the problem of multicollinearity among explanatory variables. This technique is also used when there are more independent variables than observations, a typical problem found in some data mining applications such as gene expression problems, where a large number of genes (variables) is typically observed for a small number of samples (observations).

Despite the considerable merit of this technique (reducing the number of variables used), it suffers from the fact that the new variables are often not as

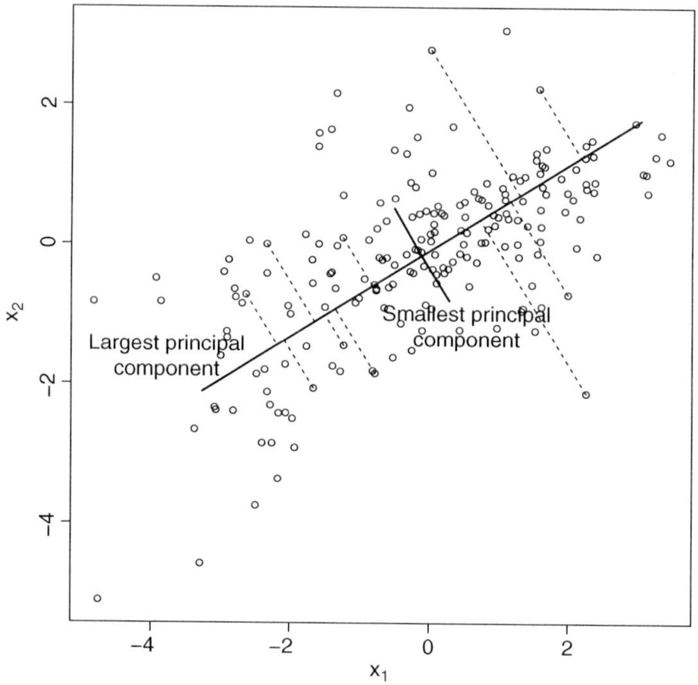

Figure 3.12 Principal components for a set of simulated data in two dimensions. Length of each solid segment is proportional to variance λ_i explained by each component. Dashed segments: perpendicular distances from first component for some observations.

easy to interpret as the original ones. However, suitable modifications of PCA are often used as tools to identify latent but interpretable data structures. The methods used to find unobservable but interpretable variables, based on principal components or not, are usually grouped under the name of *factor analysis*.

A substantial body of literature proposes many other types of combinations of variables. Some of these maintain their linear structure and change the optimization criteria (3.7), (3.8), and so on; of these, *canonical correlation analysis* maximizes the correlation between two groups of variables, and *independent component analysis* requires the components to be statistically independent instead of orthogonal. However, the linearity required is a limitation to the procedure, because it does not allow for different combinations and reductions of data. Other methods have been proposed to allow for nonlinear transformations. For example, *principal curves and surfaces* provide smooth one- and two-dimensional curved approximations to a set of data points.

Bibliographical notes

PCA was introduced by Pearson (1901) and developed by Hotelling (1933). It is now one of the most frequently used techniques in exploratory multivariate analysis. Depending on the field of application, it is also called the discrete Karhunen-Loève transform, the Hotelling transform, or proper orthogonal decomposition. Detailed presentations are discussed in all works on multivariate analysis, for example, Mardia et al. (1979) or Johnson & Wichern (1998). A standard account of PCA is the work of Jolliffe (2002). Generalizations of PCA, such as independent component analysis (ICA) and principal curves and surfaces, are discussed in many data mining and multivariate analysis works, such as Hastie et al. (2009, sections 14.5 and 14.7) and Izenman (2008, sections 15.3 and 16.3).

3.6.3 Methods of Regularization

When a large number of covariates is available, least squares estimates of a linear model often have low bias but high variance when compared with models with a smaller number of variables. As we have seen, methods of variable selection and dimension reduction may help improve prediction accuracy by allowing for larger bias but smaller variance.

These methods may be unattractive for reasons of computational burden (variable selection) or interpretation (dimension reduction), as discussed in the previous sections. A different approach is to modify the estimation method by abandoning the requirement of an unbiased estimator of the parameters, and instead considering the possibility of using a biased estimator, which may have smaller variance. There are several such estimators, most based on regularization: all the variables are left in the model, but when the model is fitted, their coefficients shrink.

The idea is to obtain a shrinkage toward the mean, so that usually the intercept is not penalized. We can therefore operate in two steps: first, we obtain the average of y as estimate for the intercept; then we replace each y_i with $y_i - \bar{y}$, and the x_{ij} with centered variables $x_{ij} - \bar{x}_j$ (for $j = 1, \ldots, p - 1$). For the rest of this section, without loss of generality, X is the new matrix with $p - 1$ columns, the first constant column 1_n having been eliminated, and there is no longer any intercept to be estimated.

Ridge regression is probably the most common shrinkage method. Consider linear model (2.6), $y = X\beta + \varepsilon$, for which ridge regression coefficients minimize a constrained form of (2.3)

$$\sum_{i=1}^{n} \{y_i - x_i^\top \beta\}^2 \qquad \text{subject to} \qquad \sum_{j=1}^{p-1} \beta_j^2 \leq s \tag{3.9}$$

An equivalent formulation of this problem can be obtained with the Lagrange form, so that the ridge regression coefficients minimize the penalized residual sum of squares

$$D_{\texttt{ridge}}(\beta, \lambda) = \sum_{i=1}^{n} (y_i - x_i^\top \beta)^2 + \lambda \sum_{j=1}^{p-1} \beta_j^2 = \|y - X\beta\|^2 + \lambda \beta^\top \beta \tag{3.10}$$

where λ is uniquely determined by s. The solution is $\hat{\beta}_\lambda = (X^\top X + \lambda I)^{-1} X^\top y$, where I is the identity matrix. Estimator $\hat{\beta}_\lambda$ is biased but for some values of $\lambda > 0$ may have a smaller mean squared error than the least squares estimator.

Note that $\lambda = 0$ gives the least squares estimator and, if $\lambda \to \infty$, then $\hat{\beta} \to 0$. Ridge regression is particularly useful when explanatory variables are collinear, as even a small $\lambda > 0$ makes solution $\hat{\beta}_\lambda$ numerically and statistically stable. Parameter λ should be adaptively chosen, for example, by cross-validation or the other methods discussed in section 3.5.

Ridge regression has a simple geometrical interpretation according to PCA, because it projects response variable y on the principal components and then shrinks the coefficients of low-variance components by more than those of high-variance components. It is, in fact, often (although not always) reasonable to expect that the response variable will vary more in the direction of high variance of explanatory variables. Therefore, when compared with principal component transformation of explanatory variables, ridge regression shrinks the coefficients of the principal components, relatively more shrinkage being applied to the smaller components than the larger ones, whereas principal component regression discards the components with smaller eigenvalues (see, for example, Hastie et al., 2009, section 3.4.1).

The choice of an alternative penalty to be added to the sum of squares (2.3) may provide a shrinkage method that, in addition to parameter restriction, requires some coefficients to be zero. When the quadratic constraint in (3.9) is replaced by absolute value constraint $\sum_{j=1}^{p-1} |\beta_j| \leq s$ and a sufficiently small s is chosen, constrained minimization of the sum of squares sets some coefficients exactly at 0, by performing a kind of continuous model selection. This shrinkage method is called *lasso* and minimizes

$$\sum_{i=1}^{n}(y_i - x_i^\top \beta)^2 \qquad \text{subject to} \qquad \sum_{j=1}^{p-1} |\beta_j| \leq s \tag{3.11}$$

or, in Lagrange form:

$$D_{\texttt{lasso}}(\beta) = \sum_{i=1}^{n}(y_i - x_i^\top \beta)^2 + \lambda \sum_{j=1}^{p-1} |\beta_j| = \|y - X\beta\|^2 + \lambda \sum_{j=1}^{p-1} |\beta_j|.$$

The solutions are nonlinear in y and, because of the nature of the constraint, they may be solved by quadratic programming. As for ridge regression, regularization parameter s (or λ) should also be adaptively chosen, according to the methods discussed in section 3.5.

When we compare the coefficient estimates obtained by ridge regression and lasso, we observe that if inputs are orthogonal, ridge regression coefficients are obtained from multiplication of least squares coefficients by a constant between 0 and 1, whereas lasso translates them toward 0 by a constant, as shown in figure 3.13 for the simple case when the columns of the X matrix are orthonormal; note that stepwise regression truncates small estimated coefficients at 0.

The appealing characteristics of lasso are offset by the complicated quadratic programming algorithm required to estimate the coefficients. In recent years,

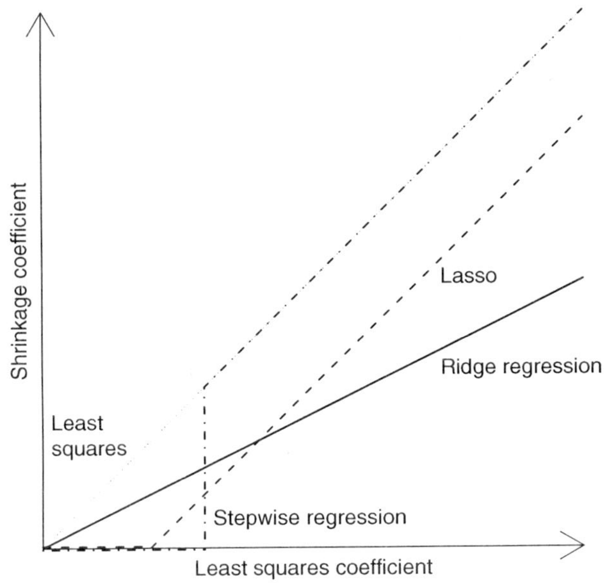

Figure 3.13 Transformed coefficient with respect with least squares coefficient for ridge regression, lasso, and stepwise regression for orthonormal case.

several faster algorithms have been proposed, one based on pathwise coordinate descent. The most elegant and efficient algorithm is based on least-angle regression (LAR), a modification of the Gram-Schmidt algorithm to estimate least squares coefficients of model (2.6) by successive orthogonalization; see algorithm 2.1.

As we saw in section 3.6.1, forward stepwise regression adds one variable at a time to the model by identifying the variable to be included in that model at each step. LAR uses a similar strategy, but adds to the model only that portion of information included in a variable which is needed, as we show in algorithm 3.2. LAR starts by adding to the model the variable most correlated with the response and, rather than fit this variable by least squares, chooses the coefficient by moving its value continuously between 0 and the least squares value. As the estimated coefficient moves between them, the correlation between the variable and the residuals decreases in absolute value. At some point in this evolution of the first coefficient, the correlation between the variable and the residuals becomes equal to the correlation between another variable and the same residuals. This second variable is then included in the model, and its coefficient is chosen together with the first one, by moving them in the direction of their least squares coefficient, until some other variable has as much correlation with the current residuals. The process continues until all the variables are included in the model and we obtain the least squares coefficients.

The LAR algorithm is of comparable computational complexity to the least squares fit, which can be computed in $p^3 + np^2$ operations.

The interesting aspect of LAR is its simple relationship with lasso: a modification of the algorithm can generate its entire sequence path. In fact, it is enough to add

Algorithm 3.2 Least-angle regression with lasso modification

Let A be the set of active covariates indices, X_A the matrix with the active covariates, and β_A the coefficients vector for these variables.

1. Start: $r \leftarrow y$, $\hat{\beta}_j \leftarrow 0, j = 1 \ldots, p$. Assume x_j standardized. $A \leftarrow \emptyset$.
2. Find predictor, say, x_{j_1} most correlated with r. Update $A \leftarrow A \cup \{j_1\}$.
3. Increase β_{j_1} in the direction of $\text{sign}(\text{corr}\{r, x_{j_1}\})$ until some other competitor x_{j_2} has as much correlation with current residual as x_{j_1} does. Update $A \leftarrow A \cup \{j_2\}$.
4. Cycle for $k = 3, \ldots, p$;

   a. Update residuals $r \leftarrow y - X_A\hat{\beta}_A$.
   b. Move $\hat{\beta}_A$ in the joint least squares direction for the regression of r on X_A (i.e., equiangular between the variables already in X_A) until some other competitor x_{j_k} has as much correlation with the current residual. Update $A \leftarrow A \cup \{j_k\}$.
   c. [lasso modification:] If a nonzero coefficient reaches 0 (e.g., changes its sign), remove that variable from set A and recompute the current equiangular (joint least squares) direction.

5. Stop when $\text{corr}\{r, x_j\} = 0$, for all j, that is, least squares solution.

a new step to the algorithm by indicating that if a nonzero coefficient becomes 0, the corresponding variable must be removed from the model. The best joint least squares direction is then recomputed, requiring the algorithm to start again from this new best direction. Clearly, the number of steps in the lasso-modified LAR algorithm (which is called LARS) may be larger than that of the LAR algorithm itself, but the order of magnitude of computations remains the same.

Bibliographical notes

Hoerl & Kennard (1970) proposed ridge regression to solve the problem of the instability of the least squares estimator in linear models, and since then the method has been presented and discussed in many works. Lasso was proposed by Tibshirani (1996) and the LAR procedure by Efron et al. (2004). They are also presented in detail in Hastie et al. (2009, section 3.4), Miller (2002, sections 3.10–3.11) and Izenman (2008, sections 5.5–5.9).

Exercises

3.1 Prove (3.2).

3.2 Write (3.2) in explicit form when $\hat{y}$ is a polynomial of degree p (for $p = 0, 1, \ldots, n-1$), and $x' = 2$.

3.3 Consider a linear regression model with p parameters fitted to a training set $(x_1, y_1), \ldots, (x_n, y_n)$ randomly selected from the available data, in which β are linear regression coefficients. The mean squared error on this set is $R_{train}(\beta) = \frac{1}{n}\sum_{i=1}^{n}(y_i - \beta^\top x_i)^2$. A test set $(x_{m+1}, y_{m+1}), \ldots, (x_{m+n}, y_{m+n})$ is also available with mean squared error $R_{test}(\beta) = \frac{1}{m}\sum_{i=1}^{m}(y_{n+i} - \beta^\top x_{n+i})^2$. Show that $\mathbb{E}\left\{R_{train}(\hat{\beta})\right\} \leq \mathbb{E}\left\{R_{test}(\hat{\beta})\right\}$, in which expectations are taken with respect to all the random elements in each expression.

3.4 Obtain (3.4) by applying the results of section 2.2.3.

3.5 If q in (3.6) is 9, what is the size of the set of all possible models that can be fitted?

4

Prediction of Quantitative Variables

4.1 Nonparametric Estimation: Why?

Let us go back to the car data used in chapter 2 and examine the problem of predicting city distance by making use of the other available variables, in particular, engine size and weight. The method used in section 2.1 was parametric, in that we have assumed that function f of (2.2), which expresses the relationship between the response and the covariates, is a member of a parametric class of functions and that the parameter estimate $\hat{\beta}$ denotes the chosen member of the class.

The simplest example of this approach is by use of the regression line specified in (2.1) in the case of a single covariate. However, we saw that this formulation is not sufficient—for instance, for the data shown in figure 2.2—and this requires more elaborate formulations, that is, polynomials, transformations of response variables, nonlinear transformations of covariates, and so on.

An alternative route is to make no reference to either the framework of linear models or any other parametric formulation for f, but to estimate f in a nonparametric way—that is, without assuming that f belongs to a specific parametric class of functions and assuming only some mathematical regularity conditions. Consequently, there is no longer any need to transform the variables in a nonlinear way.

The nonparametric approach to regression turns out to be particularly effective, mainly (but certainly not only) when there is a considerable amount of data, as is often the case in our type of applications. In fact, with a large amount of data, we always have enough empirical evidence to "falsify" any parametric model, except when dealing with the "true" model and, as mentioned in section 1.2.1, this very rarely occurs. The reason for this failure lies in the attempt to summarize all the data in a limited number of parameters, but this difficulty can be overcome with tools that offer great flexibility.

The main aim of this chapter is to explore these tools. Because the approach lends itself to several very different formulations, we only select the main ones here. We also note that the existence of diverse formulations signifies that the "free" expression of the data just mentioned is not in fact completely free: there are various methods available, and using one rather than another may produce different results, at least partially or in certain circumstances. Again, it is up to us to choose the tool best adapted to the specific problem.

4.2 Local Regression

4.2.1 Basic Formulation

We are interested in examining the relationship that links two quantities, represented by variables x and y, using a formula of the type

$$y = f(x) + \varepsilon \tag{4.1}$$

where ε is a random, nonobserved error term. Without loss of generality, we can assume that $\mathbb{E}\{\varepsilon\} = 0$ because a possible nonzero value can be included in $f(x)$. This formulation is similar to that of (2.2), but we do not presume that f is a member of a specific parametric class. We limit ourselves to looking for an estimate of $f(x)$, presuming only some regularity conditions.

Consider a general but fixed point x_0 of real numbers. We want to estimate $f(x)$ of (4.1) at point x_0.

If $f(x)$ is a derivable function with a continuous derivative at x_0, then, based on development of the Taylor series, $f(x)$ is locally approximated by a line passing through $(x_0, f(x_0))$, that is,

$$f(x) = \underbrace{f(x_0)}_{\beta_0} + \underbrace{f'(x_0)}_{\beta_1}(x - x_0) + \text{remainder}$$

where the remainder is a quantity with an order of magnitude less than $|x - x_0|$.

Transferring this idea to the context of statistical estimation, we estimate $f(x)$ in a neighborhood x_0 by means of a criterion that takes advantage of this fact, according to n observation pairs (x_i, y_i) for $i = 1, \dots, n$. The remainder term is incorporated in ε.

Let us therefore introduce a criterion analogous to (2.3), but we now weigh observations based on their distance from x_0, which is

$$\min_{\beta_0, \beta_1} \sum_{i=1}^{n} \left\{ y_i - \beta_0 - \beta_1(x_i - x_0) \right\}^2 w_i \tag{4.2}$$

where weights w_i are chosen so that they are largest when $|x_i - x_0|$ is smallest. Formula (4.2) is a particular form of the *weighted least squares criterion*,

Table 4.1. Some Common Choices for Kernels

Nucleus	$w(z)$	Support
Normal	$\frac{1}{\sqrt{2\pi}} \exp\left(-\frac{1}{2}z^2\right)$	$\mathbb{R}$
Rectangular	$\frac{1}{2}$	$(-1, 1)$
Epanechnikov	$\frac{3}{4}(1-z^2)$	$(-1, 1)$
Biquadratic	$\frac{15}{16}(1-z^2)^2$	$(-1, 1)$
Tricubic	$\frac{70}{81}(1-\lvert z\rvert^3)^3$	$(-1, 1)$

a generalization of least squares when a set of weights is available. Following this criterion, the estimates of the parameters $\beta = (\beta_0, \beta_1)^\top$ are

$$\hat{\beta} = (X^\top W X)^{-1} X^\top W y.$$

where X is a $n \times 2$ matrix whose ith row is $(1, (x_i - x_0))$, and W is the $n \times n$ diagonal matrix with w_i as diagonal elements. Because weights w_i are constructed with a "local" perspective around x_0, the resulting estimation method is called *local regression*. Minimization problem (4.2) is resolved by $\hat{\beta}$ and the estimate of $f(x_0)$ is $\hat{f}(x_0) = \hat{\beta}_0$.

One way to select the weights is to set

$$w_i = \frac{1}{h} w\left(\frac{x_i - x_0}{h}\right)$$

where $w(\cdot)$ is a symmetric density function around the origin, which in this context, is called a *kernel*, and h (with $h > 0$) represents a scale factor, which is called *bandwidth* or *smoothing parameter*. Some of the more common choices for kernel $w(\cdot)$ are listed in table 4.1. It is convenient to think of the normal kernel, corresponding to density $N(0, 1)$, which we use from now on.

Figure 4.1 exemplifies the result of nonparametric estimation in the case of data for distance covered in relation to car engine size. The top-left panel presents the data, already seen in chapter 2. The top-right panel illustrates how the estimate works, highlighting the system of weights relative to specific point $x_0 = 3$, for the particular choice $h = 0.5$ with normal kernel, as indicated by the dashed curve. The shaded area distinguishes the *smoothing window* on the x-axis, whose points have an overall relative weight of 95% in (4.2). The other points on the continuous curve were obtained by shifting the weights indicated by the dashed curve along the x-axis and reapplying (4.2).

Expression (4.2) depends on weights w_i, which in turn depend on elements h, $w(\cdot)$, and x_0. Even with h and kernel $w(\cdot)$ fixed, the minimization problem depends on x_0, and estimating $f(x)$ for different choices of x requires many minimization operations. Repeating the minimization operation is not a problem, as we can

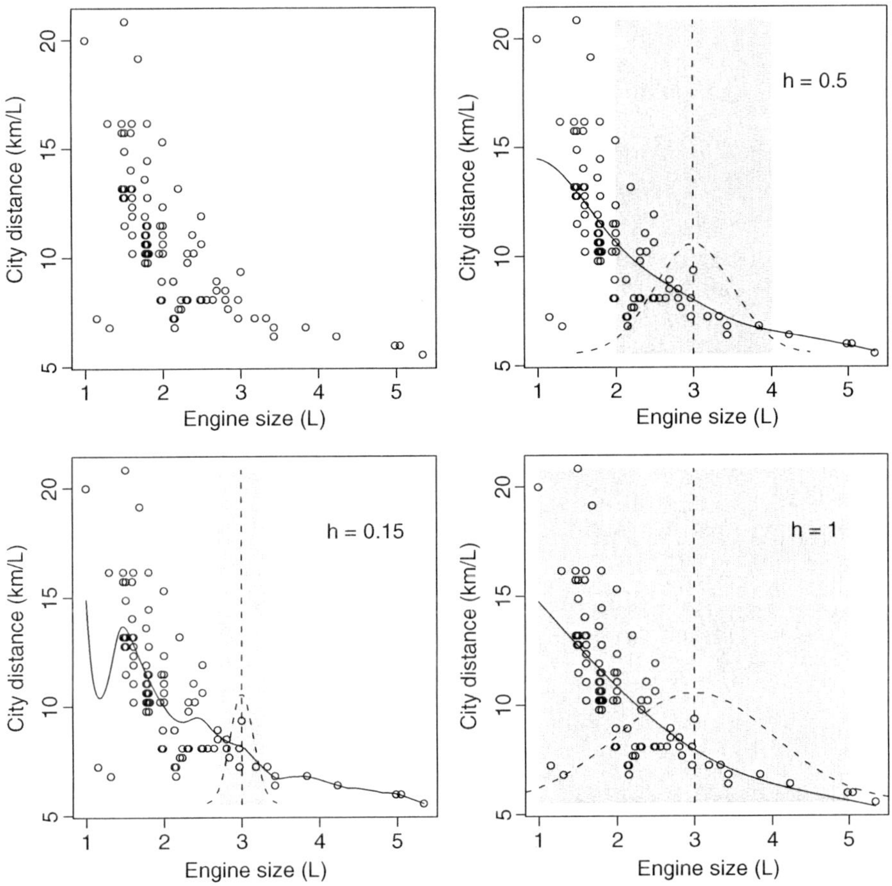

Figure 4.1 Car data: estimates with local regression of relationship between `engine size` and `city distance` for some choices of h.

show that the estimate relative to a general point x can be obtained from the explicit formula:

$$\hat{f}(x) = \frac{1}{n}\sum_{i=1}^{n} \frac{\{a_2(x;h) - a_1(x;h)(x_i - x)\}\, w_i\, y_i}{a_2(x;h)\, a_0(x;h) - a_1(x;h)^2}, \tag{4.3}$$

where $a_r(x;h) = \{\sum(x_i - x)^r w_i\}/n$, for $r = 0, 1, 2$. We are therefore dealing with an estimate that is noniterative and linear in the y_i, and can therefore write

$$\hat{f}(x) = s_h^\top y$$

for a suitable vector $s_h \in \mathbb{R}^n$ depending on h, x and $x_1, \ldots, x_n$.

We do not usually estimate $f(x)$ at a single point, but on a whole set of m values (generally equally spaced) that span the interval of interest for variable x. We can calculate each of the m estimates by a single matrix operation of the type

$$\hat{f}(x) = S_h y \tag{4.4}$$

where S_h is an $m \times n$ matrix, called *smoothing matrix*; x is now the vector (in $\mathbb{R}^m$) of the x-axis where we estimate function f; and $\hat{f}(x)$ is the corresponding estimation vector.

If n is very large, we can reduce the size of matrix S_h by regrouping variable x into classes, and therefore use an $m \times n'$ matrix, with $n' \ll n$.

The choice to approximate a function $f(x)$ locally by a straight line may be relaxed by fitting a polynomial locally. Degree 0 and degree 2 are the alternatives in actual use. When a polynomial of degree 0 is used, the estimate of each point is a weighted mean of the data in a neighborhood of that point. However, a modification of this procedure with degree 0, called k-nearest-neighbor and described later (section 4.2.4), is typically preferred. A polynomial of degree 2 is an appropriate choice when the data show sharp peaks and troughs, because this variant is more suitable for producing steep curves.

4.2.2 Choice of Smoothing Parameters

The problem of the choice of h and $w(\cdot)$ remains. The latter is not critical, as many studies on the subject have shown, and we can use any kernel listed in table 4.1. At most, there is a slight benefit in using continuous functions and some computational advantages in the choice of kernels with limited support.

The truly important aspect is the choice of smoothing parameter h. One direct indication of the effect of the choice of h is provided by the last two panels in figure 4.1. Lowering value h clearly produces curve $\hat{f}$, which is closer to the local behavior of the data and is therefore rougher, because the allocated weights system works on a smaller window and is more affected by local data variability. In the other direction, the increase in h produces the opposite effect: the window on which the weights operate widens and the curve becomes smoother.

To understand which ingredients regulate the behavior of $\hat{f}$, particularly in relation to h, we must study the formal properties of $\hat{f}$. Limiting ourselves to quite simple working hypotheses, let us assume that var $\{\varepsilon_i\} = \sigma^2$ is a positive constant common to all observations and that the observations are not correlated. Under suitable regularity conditions for f, we can prove that for h sufficiently close to 0 and n sufficiently large, the approximations

$$\mathbb{E}\left\{\hat{f}(x)\right\} \approx f(x) + \frac{h^2}{2}\sigma_w^2 f''(x), \qquad \text{var}\left\{\hat{f}(x)\right\} \approx \frac{\sigma^2}{n\,h}\frac{\alpha(w)}{g(x)} \tag{4.5}$$

hold, where $\sigma_w^2 = \int z^2 w(z)\,\mathrm{d}z$, $\alpha(w) = \int w(z)^2\,\mathrm{d}z$, and $g(x)$ indicates the density from which the x_i were sampled.

These expressions show that bias is a multiple of h^2 and the variable is a multiple of $1/(n\,h)$. Therefore, although we would like to choose $h \to 0$ to bring down the bias, this makes the variance of the estimate diverge. For $h \to \infty$, the opposite occurs: the variance is reduced, but the bias diverges. Relations (4.5) are valid in the somewhat restrictive hypotheses previously mentioned, but the same type of indication is essentially valid with weaker hypotheses: the resulting formulas are more complex, but the qualitative indication is similar.

At this point, we can also verify the same contrast between the bias and variance of the estimate already seen in chapter 3, in another context. As in that case, we must adopt a trade-off solution, balancing bias and variance in some way.

In a certain sense, the optimal solution is implicit in relations (4.5). That is, minimizing the sum of the variance and the square of the bias, as indicated in (3.3), the asymptotically best choice for h is

$$h_{\text{opt}} = \left(\frac{\alpha(w)}{\sigma_w^4 f''(x)^2 \, g(x) \, n} \right)^{1/5}. \tag{4.6}$$

However, this expression is not directly useful because it involves unknown terms $f''(x)$ and $g(x)$, although it does supply at least two important elements:

- it tells us that h must tend to 0 as $n^{-1/5}$, and therefore that it decreases very slowly;
- if we substitute this h_{opt} into the mean and variance expressions (4.5), it tells us that the mean squared error tends to 0 at a rate of $n^{-4/5}$; therefore this method of nonparametric estimation is intrinsically less efficient than a parametric one with a rate of decrease of n^{-1}, when the parametric model is satisfactory.

This last remark has much broader validity than is apparent here, in the sense that the basic indication is also valid for other methods of nonparametric estimation (see later).

Operatively, to choose h, we therefore take different routes to those in (4.6), or at least we do not use it directly. A somewhat rudimentary but effective method is to try some values and select by eye which seem most appropriate, as we did for figure 4.1. There are, however, more formal processes, which follow lines similar to those of section 3.5.

In particular, the methods of cross-validation and AIC_c (section 3.5) are in current use, having been suitably adapted to the problem. Specifically, the AIC_c variant

$$\text{AIC}_c = \log \hat{\sigma}^2 + 1 + \frac{2\,\{\text{tr}(S_h) + 1\}}{n - \text{tr}(S_h) - 2}$$

is proposed, inspired by section 3.5.3; see Hurvich et al. (1998). Here

$$\hat{\sigma}^2 = \frac{1}{n} \sum_i \left(y_i - \hat{f}(x_i) \right)^2 = \frac{1}{n} y^\top (I_n - S_h)^\top (I_n - S_h)\, y$$

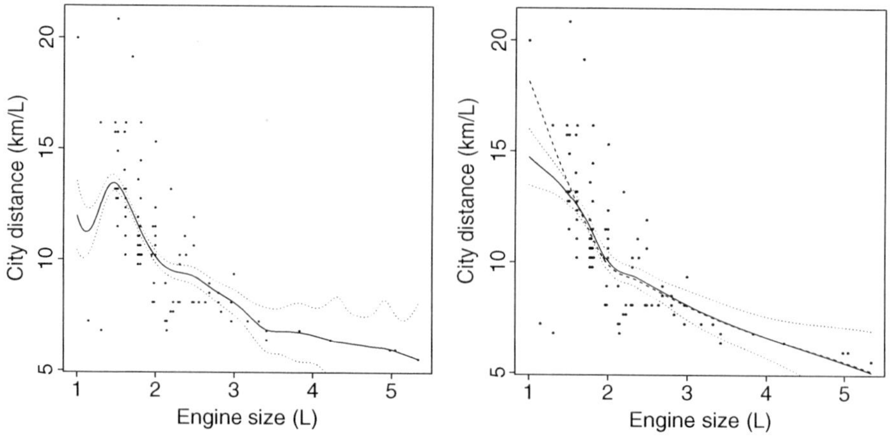

Figure 4.2 Car data: estimation by local regression with h chosen by AIC$_c$ (left) and by `loess` method (right).

is the estimate of residual variance σ^2, and $\text{tr}(S_h)$ indicates the trace of matrix S_h in (4.4). This trace is a substitutive measure of the number of parameters involved, for reasons that will be clarified in section 4.7.1.

The first panel in figure 4.2 presents the result of local regression with $h = 0.21$, chosen by the AIC$_c$ criterion represented by the continuous curve, but removing the values corresponding to the four anomalous points (shown as two single points bottom-left). The meaning of the dotted curves will be explained shortly.

To conclude, we note that the linearity of the estimation process with respect to y_i, established at the end of section 4.2.1, is valid when h is fixed independently of the data. However, if h is chosen on the basis of the same data, as commonly occurs, then the method is no longer linear.

4.2.3 Variability Bands

To make inferences, it is useful to develop a tool that is similar to the confidence interval, to give the estimate of $f(x)$ an indicator of its reliability. To construct such an interval, we must refer to a pivotal quantity, at least approximately, of the type

$$\frac{\hat{f}(x) - f(x) - b(x)}{\sqrt{\text{var}\left\{\hat{f}(x)\right\}}} \sim N(0, 1) \tag{4.7}$$

where $b(x)$ indicates the bias of the estimate, of which the main term is approximated by the final term of the first expression of (4.5); analogously, the variance in the denominator is approximated by the second expression of (4.5). Note that for the asymptotically optimal bandwidth (4.6), the bias has the same order of magnitude as the denominator of (4.7). Therefore, the bias term cannot be neglected in this framework, in contrast with what happens in a parametric context.

Of the various quantities in play, all, in some way, can be computed at least approximately, except term $f''(x)$, which is included in bias $b(x)$. This means that constructing a confidence interval is not feasible, even in an approximate form.

Instead of looking for extremely complicated corrections to remedy the problem, a current solution is to construct *variability bands* of the type

$$\left(\hat{f}(x) - z_{\alpha/2}\,\mathrm{std.err}(\hat{f}(x)),\ \hat{f}(x) + z_{\alpha/2}\,\mathrm{std.err}(\hat{f}(x))\right)$$

where $z_{\alpha/2}$ is the $1-\alpha/2$ quantile of distribution $N(0,1)$ and $\mathrm{std.err}(\hat{f}(x))$ the denominator of (4.7). Strictly speaking, the previous expression is clearly that of an interval, but once the expression is applied to every point on the x-axis, it gives rise to two bands. The result is shown by the dotted curves in the left panel of figure 4.2.

Two observations are necessary: (1) for every fixed x, the previous interval does not constitute a confidence interval, for the reasons already mentioned, but only provides an indication of the local variability of the estimate; (2) even if bias $b(x)$ were not present, the interval thus constructed would have a confidence level of approximately $1-\alpha$ for $f(x)$ to *each* fixed value of x, but not globally for the entire curve.

4.2.4 Variable Bandwidths and `loess`

There are several variations to the basic method of local regression as described up to now. The most common variation regards the use of a nonconstant bandwidth along the x-axis, but according to the level of sparseness of observed points. If again we look at figure 4.1, it is reasonable to use larger values of h when x_i are more scattered (mainly for $x > 3$).

These intuitive considerations are confirmed by expression (4.6), in which the presence of $g(x)$ in the denominator shows that when density $g(x)$ is low, that is, when observations x_i are sparse, we must use a larger value of h to keep $\mathrm{var}\left\{\hat{f}(x)\right\}$ the same.

One technique, which arose from these considerations, is `loess`, which is very similar to the local regression in section 4.2.1. A distinctive feature of `loess` is that it expresses the smoothing parameter by means of the fraction of effective observations for estimating $f(x)$ at a certain point on the x-axis; this fraction is kept constant. To understand how this works, let us look at the top-right panel in figure 4.1. When we estimate $f(x)$ at another point on the x-axis, with local regression the weights system and associated colored area are shifted horizontally, and we do not take into account the level of local sparseness of points on the x-axis. Instead, `loess` widens or narrows the window, so that the fraction of observations involved remains constant.

We can now see that the degree of smoothing is regulated by the fraction of points used, just like the bandwidth. Therefore, this fraction constitutes the smoothing parameter in `loess`.

Another typical aspect of `loess` is that it combines the ideas of local regression and *robust estimation*, which means that we substitute the quadratic function

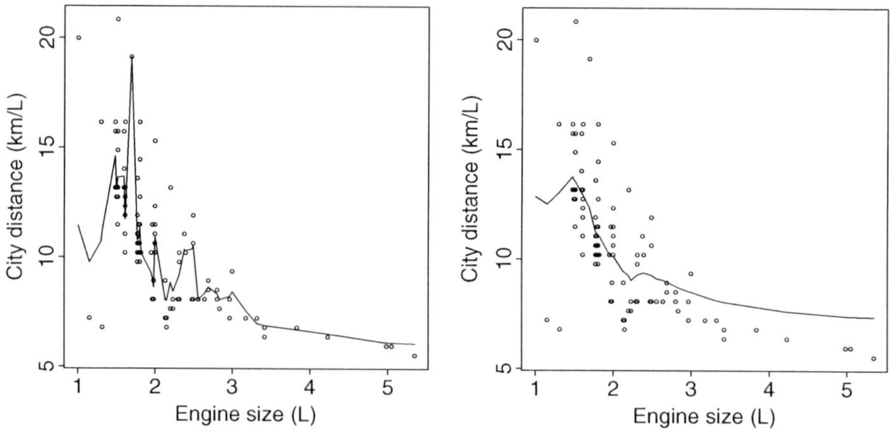

Figure 4.3 Car data: estimation by k-nearest-neighbor with $k = 10$ (left) and $k = 60$ (right).

of (4.2) with another objective function that limits the effect of *anomalous observations,* commonly called *outliers.*

Again according to the *robustness* considerations of the procedure, `loess` uses a limited support kernel, generally tricubic (see table 4.1), which also has the advantage of more clearly distinguishing between used and unused points in the estimate.

The second panel of figure 4.2 shows the result of the estimate and relative variability bands obtained through `loess` for the car data, with a fraction of 75% of the observations and a quadratic objective function such as (4.2) as smoothing parameter. The same panel also shows the estimated curve when the robust variant is used: this is the dashed curve that goes beyond the variability bands of the nonrobust method.

The local regression of degree 0, when a nonconstant bandwidth along the x-axis is chosen, is very simple and quite commonly used. The estimate of the function at each point is obtained as the average of a fixed number of closest observations around that point. This method is called *k–nearest–neighbor,* where k denotes the number of observations averaged by the estimate. We use k to indicate the decreasing complexity of the procedure because, when $k = n$, the estimate is simply the average of all available observations, giving a constant fit over the entire x-axis. Instead, when $k = 1$, the value of y of the closest observation is used at every single point as an estimate of the function, producing a very rough curve.

As an example, figure 4.3 displays the k-nearest-neighbor predicting function of `city distance` with `engine size` at $k = 10$ and $k = 60$.

4.2.5 Extension to Several Dimensions

The formulation of section 4.2.1 may also be applied when two or more covariates, say, p, are used. Let us begin with the simplest case of two variables, x_1 and x_2, and presume that a relationship of the type

$$y = f(x_1, x_2) + \varepsilon$$

holds, where $f(x_1, x_2)$ is now a function from $\mathbb{R}^2$ to $\mathbb{R}$.

The available data are now made up of the same y_i as previously and of points $x_i = (x_{i1}, x_{i2}) \in \mathbb{R}^2$, for $i = 1, \ldots, n$. To estimate f corresponding to a specific point, $x_0 = (x_{01}, x_{02})$, a natural extension of the criterion (4.2) takes the form

$$\min_{\beta_0, \beta_1, \beta_2} \sum_{i=1}^{n} \left\{ y_i - \beta_0 - \beta_1(x_{i1} - x_{01}) - \beta_2(x_{i2} - x_{02}) \right\}^2 w_i \tag{4.8}$$

where weights w_i are now to be determined as a function of a suitable distance between x_i and x_0. A common way of choosing w_i is to set

$$w_i = \frac{1}{h_1 h_2}\, w\left(\frac{x_{i1} - x_{01}}{h_1}\right) w\left(\frac{x_{i2} - x_{02}}{h_2}\right)$$

which is a simple extension of what we saw in section 4.2.1. Clearly, this expression involves two smoothing parameters, h_1 and h_2, to take into account the different variability of x_1 and x_2.

From a computational point of view, we can also tackle this problem as a variation of weighted least squares. If we indicate by X the $n \times 3$ matrix of which the ith row is

$$(1,\ x_{i1} - x_{01},\ x_{i2} - x_{02}),$$

$y = (y_1, \ldots, y_n)^\top$ and $W = \mathrm{diag}(w_1, \ldots, w_n)$, then the solution of the previous minimum problem is the first element, which corresponds to β_0, of $(X^\top WX)^{-1}X^\top Wy$. Obviously, this calculation is repeated for every choice of point x_0, and tendentially the number of these points is now higher than in the scalar case of section 4.2.1.

Figure 4.4 shows the results obtained for the car data with x_1 = `engine size`, x_2 = `curb weight`, and y = `city distance`, in two forms of representation: perspective and level curves. To avoid extrapolating the estimate where we have no observations, it is limited to the convex hull of the observed points of (x_1, x_2).

Formally, most of the results can be easily extended to the multivariate case, where the formulation is of the type

$$y = f(x) + \varepsilon = f(x_1, \ldots, x_p) + \varepsilon. \tag{4.9}$$

Definition of the estimation method seen for $p = 1$ and $p = 2$ extends naturally to the case of general p, meaning that there is no need to repeat the discussion of various connected aspects, such as the choice of h, and so on.

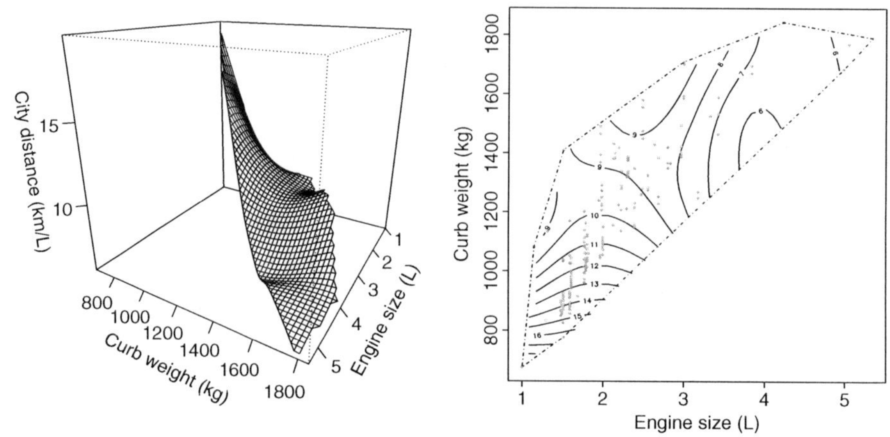

Figure 4.4 Car data: estimation of `city distance` by local regression with two covariates, `engine size` and `curb weight`.

Bibliographical notes

A more detailed but still fairly informal presentation of the nonparametric approach through local regression is given in Bowman & Azzalini (1997). The book is associated with the `R` package, an evolution from an earlier version of software written in `S-plus` and with a set of additional scripts. These tools have been used extensively here; specifically figure 4.1 is based on one of the scripts associated with the book. For more advanced mathematical coverage of the subject, see Fan & Gijbels (1996) and Wand & Jones (1995). Loader (1999) extends the local regression approach by combining it with the likelihood concept, particularly within generalized linear models, and supplies other software tools for the `S-plus` and `R` environments. `Loess` was originally proposed by Cleveland (1979) and further developed by Cleveland & Devlin (1988) and is described in Cleveland et al. (1992).

4.3 The Curse of Dimensionality

In practice, we rarely go much beyond two dimensions in nonparametric regression. The first reason is the poor conceptual manageability of the resulting object: although the idea of a function with 6 or 26 variables is not conceptually different from one with 2 variables, it is actually difficult to visualize mentally and graphically. Interpreting the results is also difficult.

A second and perhaps more important aspect is that with increasing dimension p of the space in which the covariates are placed, the observed points scatter very quickly. To understand the essence of the problem intuitively, think of $n = 500$ points on the x-axis, randomly set over an interval that, without loss of generality we may presume to be unit interval $(0, 1)$. If we use these n points to estimate function $f(x)$, we obtain a reliable estimate, thanks to the small average distance that separates them. If the same number n of points is then distributed in square $(0, 1)^2$ of plane (x_1, x_2), they are much less close to each other. If we then move to higher dimensions, say, p, the dispersion of n points in space $\mathbb{R}^p$

increases very rapidly, and the quality of the obtainable estimate correspondingly worsens.

To compensate for the increased space between the points, we need a number of points of the order of magnitude n^p. However, although it is common to use a sample of size $n = 500$, it is much more uncommon to have 500^5 units available, and practically impossible to have 500^{10}, even in a data mining context. These are the sizes that are in some way equivalent to estimating function f nonparametrically when the number of covariates is 5 or, respectively, 10.

This situation of substantial impossibility in estimating function f accurately when p is large is called the *curse of dimensionality*. For a more detailed explanation of how the scatter of the points increases with p, and for other similar issues, see Hastie et al. (2009; section 2.5).

A further critical aspect with increasing p is the increased computational cost, at least when a substantial increase in n also occurs.

These problems are not confined to the specific technique of local regression, but they are substantially valid for all nonparametric estimation techniques, as they are due to the dimension and dispersed nature of the data with respect to the number of points from which we wish to estimate the function and not so much to the method chosen for data processing.

To overcome the problem of the curse of dimensionality, one strategy is to carry out a preliminary operation to reduce the number p of the covariates, transforming them into a reduced set of new variables but at the same time losing as little of their informative content as possible.

The simplest and probably most frequent way of achieving this is to extract some of the *principal components* of the original covariates. Therefore, once the complete set of principal components has been constructed, a suitable number of them are chosen, keeping a sufficient proportion of the original variability and the number of new variables low. For a discussion of the advantages and disadvantages of PCA, see section 3.6.2.

Therefore, in the following section, what we indicate as covariates may not represent the original variables but those constructed through principal components or other methods of dimensionality reduction (see section 3.6.2).

Bibliographical notes

The concept of the curse of dimensionality was introduced by Bellman (1961). Hastie et al. (2009; section 2.5) give a very detailed description of it in the context of data mining and also discuss a number of additional issues.

4.4 Splines

The term *spline* originally meant the flexible strips of wood used to shape ships' hulls. Some points on the cross-section of the hull were chosen, and the rest of the curve of the hull was derived by forcing the wooden strips to pass through such points, leaving them free to fit into the rest of desired curve according to their natural tendency. This gave rise to a regular curve with preassigned behavior in certain positions.

4.4.1 Spline Functions

The term *spline* is used in mathematics to construct piecewise polynomial functions, according to a logic that partly replicates the mechanism just described, to approximate functions of which we know the value only at certain points.

We choose K points $\xi_1 < \xi_2 < \cdots < \xi_K$, called *knots*, along the x-axis. A function $f(x)$ is constructed so that it passes exactly through the knots and is free at the other points, with the constraint that it presents regular overall behavior. In this sense, the function behaves like splines used in shipyards.

The following strategy is followed: between two successive knots, say, in the interval (ξ_i, ξ_{i+1}), curve $f(x)$ coincides with a suitable polynomial, of prefixed degree d, and these sections of polynomials meet at point ξ_i $(i = 2, \ldots, K-1)$, in the sense that the resulting function $f(x)$ has a continuous derivative from degree 0 to degree $d-1$ in each of the ξ_i.

The degree that is almost universally used is $d = 3$, and we therefore speak of cubic splines. The reason for this is that the human eye cannot perceive discontinuity in the third derivative. The foregoing conditions are therefore written as

$$f(\xi_i) = y_i, \quad \text{for } i = 1, \ldots, K$$

$$f(\xi_i^-) = f(\xi_i^+), \quad f'(\xi_i^-) = f'(\xi_i^+), \quad f''(\xi_i^-) = f''(\xi_i^+), \quad \text{for } i = 2, \ldots, K-1$$

where $g(x^-)$ and $g(x^+)$ indicate the left and right limits of a function $g(\cdot)$ at point x.

The framework of the problem requires the following set of conditions: each of the $K-1$ cubic components requires four parameters; there are K constraints of the type $f(\xi_i) = y_i$, and $3\,(K-2)$ continuity constraints of the function and the first two derivatives.

As the difference between coefficients and constraints is 2 units, the system of conditions does not univocally identify a function. We must therefore introduce two additional constraints.

Many proposals have been made to define these additional constraints, most of which concern the outmost interval or the extreme points of the function. A particularly simple choice consists of constraining the second derivatives of the polynomials in the two extreme intervals to 0, $f''(\xi_1) = f''(\xi_K) = 0$, which means that the two extreme polynomials are straight lines. The resulting function $f(x)$ is called the *natural cubic spline*.

4.4.2 Regression Splines

The previous tool is also useful in statistics, in various forms, in the study of relations between a covariate x and a response y, for which we use n pairs of observations (x_i, y_i) for $i = 1, \ldots, n$.

Let us begin by applying these ideas to parametric regression. We return to model (2.2), where $f(x; \beta)$ is hypothesized to be a spline function. Then we divide the x-axis into $K+1$ intervals separated by K knots, $\xi_1, \ldots, \xi_K$, and interpolate the n points with criterion (2.3), where the β coefficients are now the nonconstrained parameters of the $K + 1$ constituent polynomials.

With respect to section 4.4.1, there is a certain difference in that the spline function coefficients can no longer be chosen according to constraints of the type $f(\xi_j) = y_j$, because K and n are no longer linked and $K \ll n$. This means that we have to use a fitting criterion between the data and the interpolated function, for example, the least squares criterion or a similar one.

If we use cubic splines, the total number of cubic parameters is $4(K + 1)$ subject to $3K$ continuity constraints, and therefore β has $K + 4$ components. The solution to the minimum problem (2.3) may be rewritten in the equivalent form

$$f(x; \beta) = \sum_{j=1}^{K+4} \hat{\beta}_j h_j(x) \tag{4.10}$$

where

$$h_j(x) = x^{j-1} \qquad \text{for } j = 1, \ldots, 4,$$
$$h_{j+4}(x) = (x - \xi_j)^3_+, \qquad \text{for } j = 1, \ldots, K$$

and $a_+ = \max(a, 0)$. The solution is thus represented by a suitable linear combination of *basis functions* $\{h_j(x), j = 1, \ldots, K + 4\}$, composed partly of low-order powers of x and partly of functions of the type $\max(0, (x - \xi)^3)$.

The number K of knots and their position along the x-axis need to be chosen. Because K is viewed as a tuning parameter, regulating the complexity of the model, the strategies proposed in section 3.5 apply. Once K has been set, when no information is available about the shape of the function to be estimated, a reasonable choice for knot positions is uniformly along the x_i range. Alternatively, the quantiles of the empirical distribution of the x_i are chosen as knots.

Figure 4.5, which concerns our yesterday's data, illustrate regression splines. We used $K = 2$ knots, marked by vertical dotted lines. As well as the standard solution for the degree $d = 3$, we also constructed those for $d = 1$ and $d = 2$ for purposes of illustration only. Obviously, in the last two cases, the basis function changes: in particular, for $d = 1$, the basis is represented by

$$h_1(x) = 1, \quad h_2 = x, \quad h_{j+2}(x) = (x - \xi_j)_+, \qquad \text{for } j = 1, \ldots, K.$$

The right panel of figure 4.5 shows function $(x - \xi_1)_+$ as an example of the characteristic component of this approach.

4.4.3 Smoothing Splines

Another way of using spline functions in studying the relationship between variables is to introduce an approach to nonparametric estimation as an alternative to local regression.

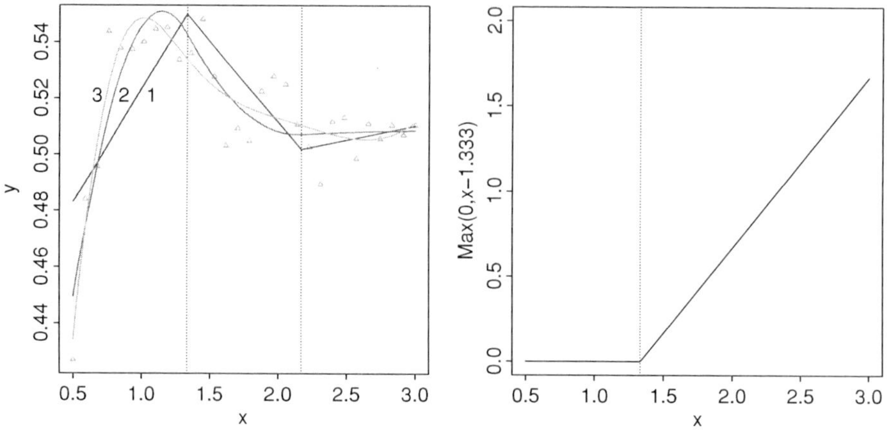

Figure 4.5 Yesterday's day: interpolated functions for $d = 1, 2, 3$ (left) and the component function $(x - \xi_1)_+$ (right).

Let us consider the penalized least squares criterion

$$D(f, \lambda) = \sum_{i=1}^{n} [y_i - f(x_i)]^2 + \lambda \int_{-\infty}^{\infty} \{f''(t)\}^2 \, dt \tag{4.11}$$

where λ is a positive penalization parameter of the roughness degree of curve f, quantified by the integral of $f''(x)^2$, and therefore acts as a smoothing parameter.

If $\lambda \to 0$, there is no penalization for the roughness of $f(x)$, so the previous criterion is not influenced by $f(x)$ outside points $x_1, \ldots, x_n$, and the optimal solution $\hat{f}(x_i)$ is the arithmetic mean of the y_i corresponding to each fixed x for each of the observed x_i but is not determined for other values of x. If $\lambda \to \infty$, the penalty is maximal and means adapting a line imposing $f''(x) \equiv 0$. The overall result is the least squares line. Therefore, the role of λ is qualitatively similar to that of h in the case of local regression.

A noteworthy mathematical result (Green & Silverman 1994) shows that the solution to the minimization problem (4.11) is represented by a *natural cubic spline*, whose knots are distinct points x_i. The solution may be written as

$$\hat{f}(x) = \sum_{j=1}^{n_0} \theta_j N_j(x)$$

where n_0 is the number of distinct x_i and the $N_j(x)$ are natural cubic splines basis functions.

We can rewrite

$$D(f, \lambda) = (y - N\theta)^\top (y - N\theta) + \lambda \theta^\top \Omega \theta$$

where N is the matrix in which the jth column contains the values of N_j corresponding to the n_0 distinct values of x, and Ω is the matrix of which the

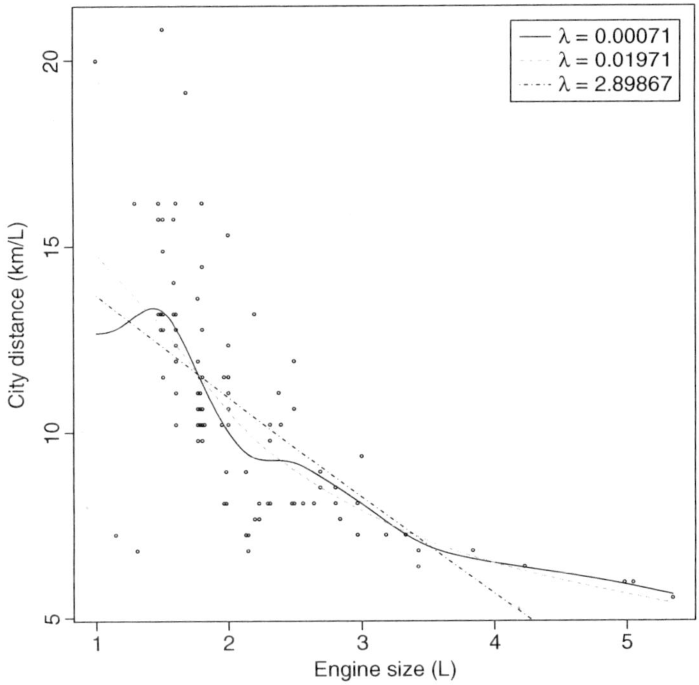

Figure 4.6 Car data: Estimate of `city distance` according to `engine size` by a smoothing spline for three choices of λ.

generic element is $\int N_j''(t)\, N_k''(t)\ dt$. The solution of the optimization problem is given by

$$\hat{\theta} = (N^\top N + \lambda\Omega)^{-1} N^\top y \tag{4.12}$$

which clearly depends on the choice of smoothing parameter λ.

If this expression of $\hat{\theta}$ is substituted into that of $f(x)$, we have $\hat{y} = \tilde{S}_\lambda\, y$ for a certain matrix $\tilde{S}_\lambda$ of dimension $n_0 \times n_0$, that is, we are dealing with another *linear smoother*. In this case, we speak of *smoothing splines*.

However, from a computational point of view, we do not proceed with (4.12), which involves a matrix of order n_0. There are much more efficient algorithms, for which we refer readers to the specialized literature (see the bibliographical notes). In addition, when the quantity of data is very large, we can reduce the number of knots used, without loss of accuracy, as we did for local regression at the end of section 4.2.1.

Again, figure 4.6 shows what is obtained when this procedure is applied to the car data for three choices of parameter λ. We can also use the criteria discussed earlier for the choice of smoothing parameter λ (in sections 3.5 and 4.2.2), but here we choose three values that highlight the effect of variations in parameter λ.

4.4.4 Multidimensional Splines

Extending splines to cases with two or more covariates is not as automatic as for the other smoothing techniques presented in this chapter. Extension of cubic smoothing splines, for example, are *thin-plate splines,* obtained by a generalization of (4.11) in which, in the penalty function, the second derivative of function f is substituted by the Laplacian. Due to the elevated computational complexity involved, thin-plate splines are hard to use with more than two covariates. In the simple case when we have a pair of covariates $x = (x_1, x_2)^\top \in \mathbb{R}^2$, the roughness penalty function in (4.11) becomes

$$\int\int_{\mathbb{R}^2} \left\{ \left(\frac{\partial^2 f(x)}{\partial x_1^2} \right)^2 + 2 \left(\frac{\partial^2 f(x)}{\partial x_1 \partial x_2} \right)^2 + \left(\frac{\partial^2 f(x)}{\partial x_2^2} \right)^2 \right\} \, dx_1 \, dx_2.$$

The solution to optimization problem (4.11) with this penalty function can be proved to have the form

$$f(x) = \hat{\beta}_0 + \hat{\beta}^\top x + \sum_{j=1}^{n} \alpha_j h_j(x)$$

where $h_j(x) = \eta(\|x - x_j\|)$, $\eta(z) = z^2 \log z^2$, and estimates $\hat{\alpha}_j$, $\hat{\beta}_0$, and $\hat{\beta}$ are determined by substituting $f(x)$ in (4.11) and minimizing with respect to the parameters.

Figure 4.7 presents the results obtained for car data, again with $x_1 =$ `engine size`, $x_2 =$ `curb weight`, and $y =$ `city distance`, with two forms of graphical representation: perspective and level curves.

Another type of generalization particularly useful for regression splines is based on *tensor products of splines.* The extension to multiple dimensions is obtained by constructing a set of basis functions in $\mathbb{R}^p$, multiplying together the basis of one-dimensional functions for each covariate. If, for example, we consider the two-dimensional case of cubic splines, where $x = (x_1, x_2)^\top \in \mathbb{R}^2$ and we have a basis of functions $h_{1k}(x_1)$ with $k = 1, \ldots, K_1 + 4$, relative to the first covariate

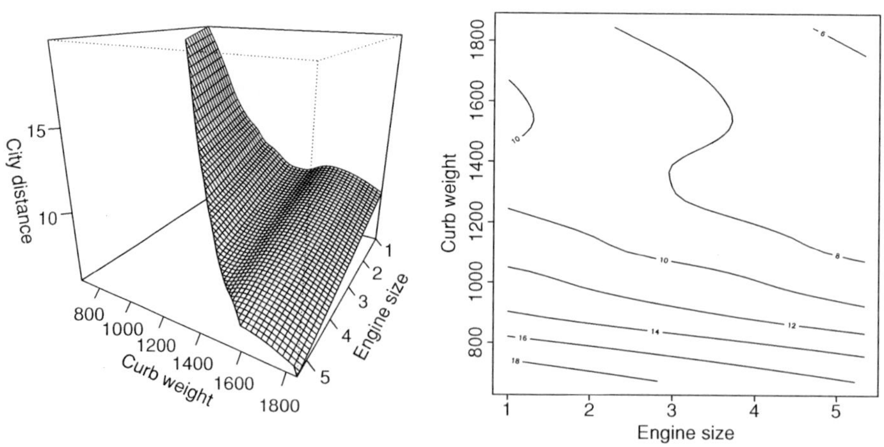

Figure 4.7 Car data: estimation of `city distance` according to `engine size` and `curb weight` by smoothing splines.

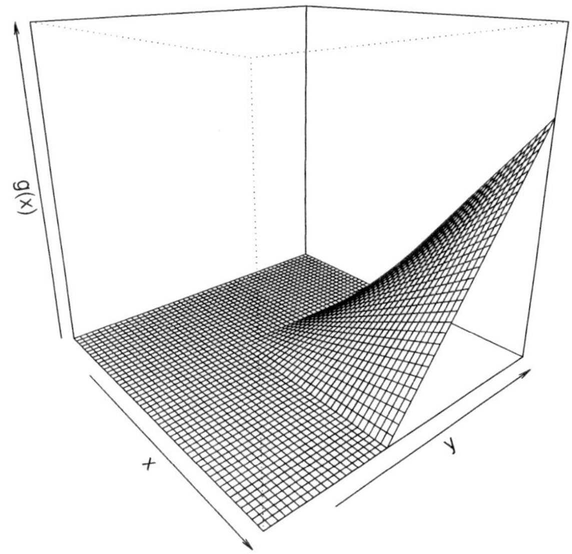

Figure 4.8 Tensor product basis functions, obtained as product of scalar basis functions of type $(x - \xi)_+$, as in figure 4.5.

x_1 and a basis of functions $h_{2k}(x_2)$ with $k = 1, \ldots K_2 + 4$, relative to the second explanatory variable x_2, the *tensor product basis* of dimension $(K_1 + 4) \times (K_2 + 4)$ is defined by

$$g_{jk}(x) = h_{1j}(x_1)h_{2k}(x_2), \qquad j = 1, \ldots, K_1 + 4, \quad k = 1, \ldots, K_2 + 4$$

and can be used to represent a two-dimensional function

$$g(x) = \sum_{j=1}^{K_1+4} \sum_{k=1}^{K_2+4} \theta_{jk} g_{jk}(x).$$

Parameters θ_{jk} can be estimated by the penalized least squares criterion.

Figure 4.8 is an example of tensor product basis functions obtained with one-dimensional components of the type $(x - \xi)_+ = \max(x - \xi, 0)$; see figure 4.5.

4.4.5 MARS

When the number of covariates is high, extension of the previous approach is not easy, due to computational and interpretive difficulties. It is therefore important to use a process that, starting from the information present in the data, allows us to select variables reasonably and provides criteria for the choice of the number of knots necessary for each variable.

Multivariate adaptive regression splines (MARS) represent a particular iterative specification of regression splines (see section 4.4.2), the aim of which is to model problems with many explanatory variables. The basis functions used are pairs of piecewise linear functions, of the type $(x - \xi)_+$ and $(\xi - x)_+$, with a single knot at point ξ, like those of section 4.4.2.

The aim is to find the relationship between a dependent variable y and the p covariates $x = (x_1, \ldots, x_p)^T$. For every explanatory variable x_j, we determine a pair of basis functions with the knot in each observed value x_{ij}, for $i = 1, \ldots, n$ in addition to the linear one. This gives the set of basis functions that are considered as functions on whole space $\mathbb{R}^p$:

$$\mathcal{C} = \Big\{x_j, (x_j - \xi)_+, (\xi - x_j)_+ \; : \quad \xi \in \{x_{i1}, x_{i2}, \ldots, x_{ip}\}, \\ i = 1, 2, \ldots, n, \quad j = 1, \ldots, p\Big\}.$$

We must select a subset of basis functions in $\mathcal{C}$ to combine in a model suitable for fitting the data. Piecewise basis functions are included in the model in pairs of the form $\{(x_j - \xi)_+, (\xi - x_j)_+\}$. The MARS model is therefore of the type:

$$f(x) = \beta_0 + \sum_{k=1}^{2K} \beta_k h_k(x) \tag{4.13}$$

where $h_k(x)$ are either functions belonging to $\mathcal{C}$ or products of two or more such functions, and K is the number of pairs of basis functions to be included in the model.

To select the h_k functions and estimate parameters β, we follow a recursive process.

- Start with $K = 0$. We first introduce constant function $h_0(x) = 1$.
- Generic step K. We presume that the model already has $2(K - 1)$ terms. We consider, as a new pair of basis functions, each of the possible pairs of products of a function h_k, $k \in \{1, \ldots, K\}$, already included in the model, with another pair of functions in $\mathcal{C}$. We then choose the pair of basis functions that adds to (4.13) the terms

$$\hat{\beta}_{2K-1}\, h_m(x)\, (x_j - \xi)_+ \; + \; \hat{\beta}_{2K}\, h_m(x)\, (\xi - x_j)_+$$

  which minimize the least squares criterion. Here, h_m indicates a function that is already included in the model, and $\hat{\beta}_{2K-1}$ and $\hat{\beta}_{2K}$ are parameters that are estimated by least squares together with all the other β parameters of the model.
- The process continues until a predefined maximum K is reached.

This model is generally very large and may overfit the data. It may be appropriate to formulate a backward procedure in which we iteratively select and remove the terms from the model one by one, at each step deleting the terms that make minor contributions to the residual sum of squares. In this backward procedure, single terms are usually deleted, so the final model is not necessarily characterized by a pair of basis functions for each knot.

Model subsets are then compared by means of some fitting criterion. When many data are available, we choose the best model subset by using a different test set, as in section 3.5.1. Alternatively, we can use cross-validation (see section 3.5.2), which, however, requires a considerable computational load.

Another alternative is to use *generalized cross-validation* (GCV). For each model to be compared, GCV is defined as

$$GCV = \frac{\sum_{i=1}^{n}[y_i - \hat{f}(x_i)]^2}{(1 - d/n)^2}$$

where d is an indicator of the effective number of parameters in the model. For the MARS context, d is the sum of the number of terms in the model and the number of knots defined in the basis selection process weighted by a penalty that, after some theoretical and simulation results, is usually fixed at 2 or 3. Another frequently used approximation chooses d proportional to the number of terms in the model. Note that the formula used by GCV approximates the error, based on (3.4), which would be determined by *leave one out* cross-validation for a linear model: this is why it is called *generalized* cross-validation.

The pairs of linear functions chosen as basis functions for MARS have the advantage of operating locally. When these basis functions are multiplied together, they are different from 0 only in that part of the space where all the univariate functions are positive (see figure 4.8), and this allows the model to be fitted to the data with a relatively small number of parameters. These functions also have the advantage that they can be multiplied together simply, with greatly reduced computational complexity.

The constructional logic of the model is clearly hierarchical, in the sense that we can multiply new basis functions that involve new variables only to the basis functions already in the model; therefore, an interaction of a higher order can only be introduced when interactions of a lower order are present. This constraint, introduced for computational reasons, does not necessarily reflect the real behavior of the data, but it often helps in interpreting the results. However, for easier interpretation, we often constrain the model to have only first- or at most second-order interactions.

So far, we have considered the case in which explanatory variables are quantitative, but it is also easy to introduce qualitative predictors in the MARS model. If we consider all the possible binary partitions of the levels of a qualitative explanatory variable, each partition generates a pair of basis step functions that indicates membership to one of the two groups of levels. These basis functions can be inserted into C and used like all the others, to obtain products with functions already included in the model.

For a simple explanation, we again use the car data, this time with only the two covariates `engine size` and `curb weight`. The surface obtained with the MARS model is shown in figure 4.9.

To build a slightly more realistic example, still based on car data, let us now consider as covariates the variables `fuel type`, `intake`, `bodywork type`, `traction`, `motor position`, `width`, `height`, and `length` in addition to `engine size` and `curb weight`. Table 4.2 lists the relevant information used by the final model at the end of the MARS process. Only pairs of basis functions

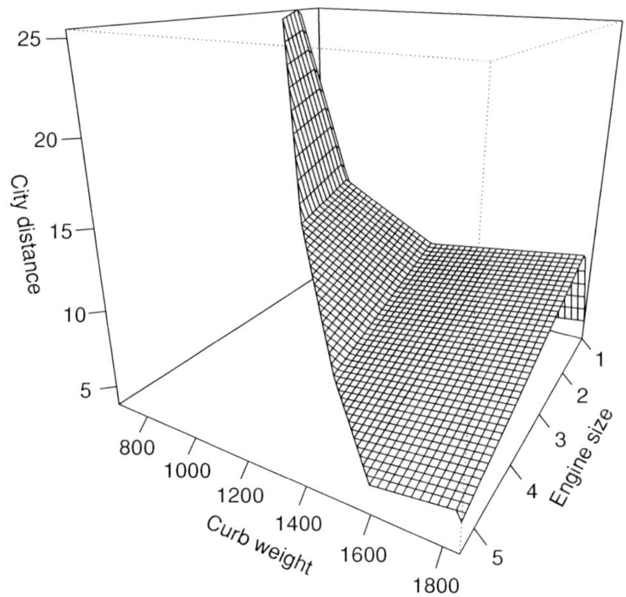

Figure 4.9 Car data: MARS surface, fitted with two quantitative variables.

Table 4.2. Car Data: Parameter Estimates of MARS Model

Variable	Node	Levels	Parameters	SE
`constant`			57.0798	4.4884
`fuel type`		1	−4.0680	0.2768
`intake`		1	1.3412	0.2287
`curb weight`			−0.0639	0.0063
`curb weight`	861.84		0.0510	0.0067
`curb weight`	1149.88		0.0069	0.0013
`engine size`			11.6215	1.7015
`engine size`	1.47		−12.1585	1.7581

based on single variables occur in the final model, so it has no interactions. The table has a line for each pair of basis functions in the final model: the first column shows the explanatory variable linked to the basis, and for basis functions with piecewise linear components, the second column specifies the point at which the knot for that variable is fixed; otherwise, the basis is linear. For qualitative variables, the third column shows the number of levels into which the factor was divided to determine the relative basis. The fourth column lists parameter estimations $\hat{\beta}_k$ relative to each basis, and the last column shows the estimated standard errors of each parameter.

Figure 4.10 shows the one-dimensional plots of the estimates of the response variable for each covariate, where the other explanatory variables are kept constant and equal to their median value in each panel. Figure 4.11 displays a similar plot of

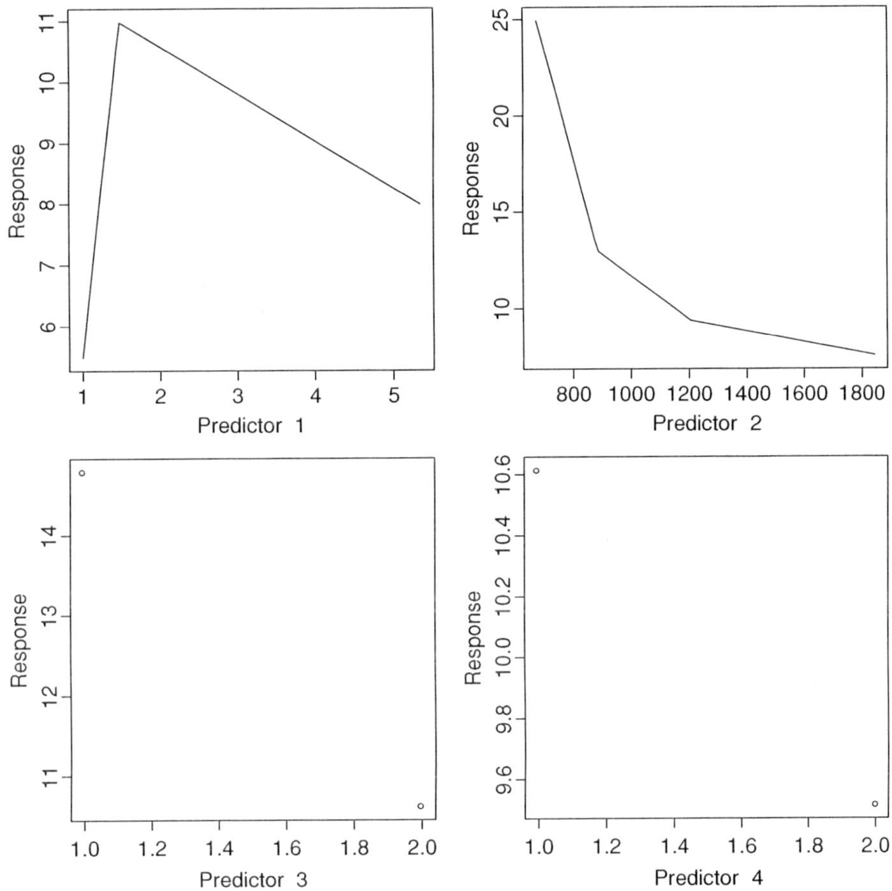

Figure 4.10 Car data: estimates of one-dimensional relationships in MARS model. Other variables are fixed at median values.

the regression function, estimated according to the two variables `engine size` and `curb weight` at the same time.

Bibliographical notes

There are very many other aspects concerning spline functions, for which we refer readers to specialized texts. General coverage of splines and their mathematical properties can be found in the works of de Boor (1978) and Atkinson (1989; section 3.7). Green & Silverman (1994) were among the first to use splines and their variations as thin-plate splines in a statistical environment and were responsible for the spread of this tool in the statistical community. MARS was introduced by Friedman (1991) and is found in many works on data mining (e.g., Hastie et al. 2009; section 9.4). GCV was introduced by Craven & Wahba (1978) and extended to MARS by Friedman (1991).

4.5 Additive Models and GAM

Up to now, we have examined various methods of nonparametric regression estimation, each of which allows us to examine the relationship between a response

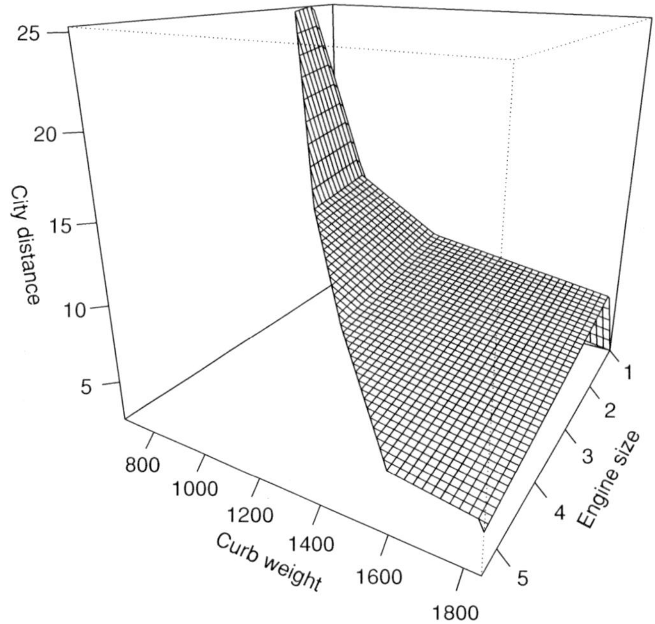

Figure 4.11 Car data: estimates of double relationship in MARS model. Other variables are fixed at median values.

variable y and a certain number p of explanatory variables. All these techniques are valid for the aim, but they also come up against the same problems when p is high: the curse of dimensionality and the other aspects discussed in section 4.2.5.

To overcome this, on one hand we have to introduce some form of "structure," that is, a model of the form of regression function $f(x)$, $x = (x_1, \ldots, x_p) \in \mathbb{R}^p$. On the other hand, for reasons already discussed, we do not want a rigid structure but must maintain ample flexibility.

One option that has been greatly appreciated for its practical usefulness and logical simplicity is the following. Let us presume that a representation of the type

$$f(x) = f(x_1, \ldots, x_p) = \beta_0 + \sum_{j=1}^{p} f_j(x_j) \tag{4.14}$$

holds for $f(x)$, where $f_1, \ldots, f_p$ are functions of one variable, each having smooth behavior, and β_0 is a constant. We say that formulation (4.9) with representation (4.14) of $f(x)$ is an *additive model.*

Note that to avoid what is essentially a problem of model *identifiability,* it is necessary for the various f_j to be centred around 0, that is,

$$\sum_{i=1}^{n} f_j(x_{ij}) = 0, \qquad (j = 1, \ldots, p),$$

where x_{ij} is the jth variable for unit i.

Algorithm 4.1 Backfitting

1. Start: $\hat{\beta}_0 \leftarrow \sum_i y_i/n, \hat{f}_j \leftarrow 0$ for all j.
2. Cycle for $j = 1, 2, \ldots, p, 1, 2, \ldots, p, 1, 2, \ldots$:

$$\text{a. } \hat{f}_j \leftarrow \mathcal{S}\left[\left\{y_i - \hat{\beta}_0 - \sum_{k\neq j} \hat{f}_k(x_{ik})\right\}_1^n\right],$$

$$\text{b. } \hat{f}_j \leftarrow \hat{f}_j - n^{-1}\sum_{i=1}^{n} \hat{f}_j(x_{ij}),$$

until functions $\hat{f}_j$ stabilize.

To fit (4.14) to the data, there is an iterative process based on a nonparametric estimation method of one-variable functions to estimate f_j. This procedure, shown in algorithm 4.1, is called *backfitting* and is essentially a variation of the Gauss-Seidel algorithm.

The specific method for nonparametric estimation is not crucial, and we can even choose different methods for different f_j, but we usually apply a single one, generically indicated by $\mathcal{S}$ in algorithm 4.1, in the sense that $\mathcal{S}(y)$ constitutes the nonparametric estimate, calculated on the observed values $y = (y_1, \ldots, y_n)^\top$, of a scalar function. In many cases, $\mathcal{S}$ is a linear estimator, of type Sy, where S is a suitable smoothing matrix.

A generalization of model (4.14) is of the type

$$f(x_1, \ldots, x_p) = \beta_0 + \sum_{j=1}^{p} f_j(x_j) + \sum_{j=1}^{p}\sum_{k<j} f_{kj}(x_k, x_j) + \sum_{j=1}^{p}\sum_{k<j}\sum_{h<k<j} f_{hkj}(x_h, x_k, x_j) + \cdots$$

which allows us to bear in mind the *interaction effect* between pairs of variables, triplets, or other interactions of a higher order.

Figures 4.12 and 4.13 illustrate how additive models work with reference to the car data, in which the response variable is `city distance` and the covariates are `engine size` and `curb weight`. Figure 4.12 shows the functions indicated in (4.14) by f_1 and f_2, both of which are accompanied by their respective variability bands. Note that the trend of the `engine size` regression function is noticeably modified when the `curb weight` component is introduced with respect to similar graphs in figures 4.1, 4.2, and 4.6, which consider `engine size` alone.

The left panel of figure 4.13 presents the fitted regression surface under the additive hypothesis, combining the two functions shown in figure 4.12; the right

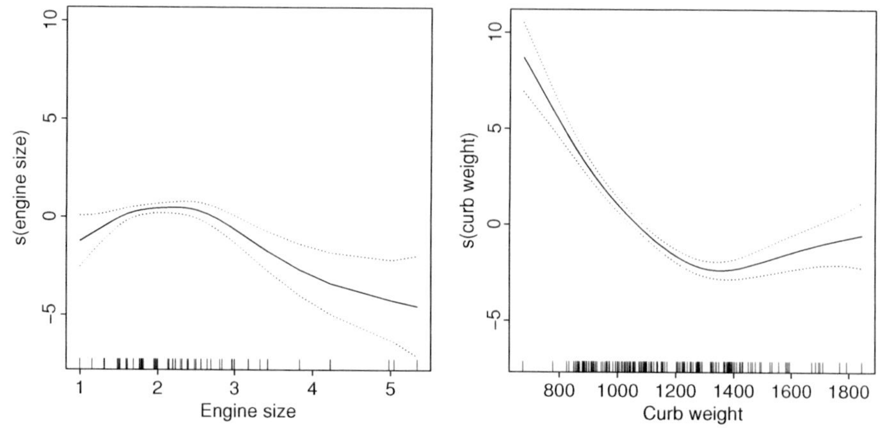

Figure 4.12 Car data: estimate of `city distance` according to `engine size` and `curb weight` by an additive model with a spline smoother.

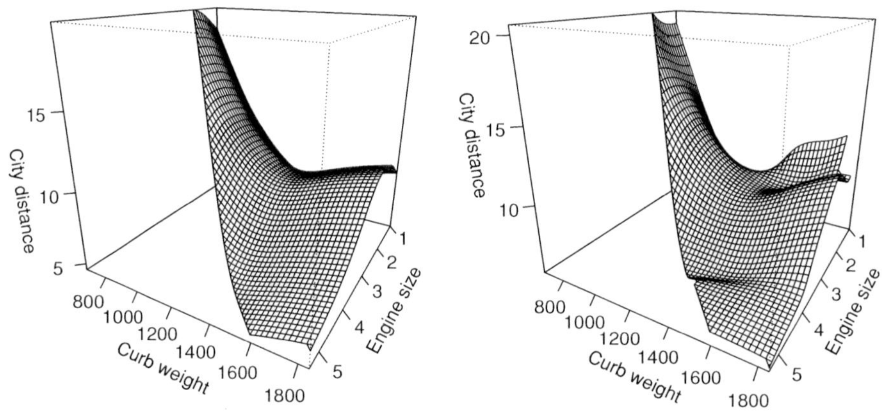

Figure 4.13 Car data: estimate of `city distance` according to `engine size` and `curb weight` by an additive model with a spline smoother (left), and without additive hypothesis by local regression (right).

panel shows the unconstrained estimate, free of the additive hypothesis (see figure 4.4). Comparison between the two plots shows the effect of the additive hypothesis or, rather, the effect of interaction between the variables that cannot be removed from the additive model, which, however, is greatly limited in this example.

Another direction in which model (4.14) is frequently generalized is of the type

$$g\left(\mathbb{E}\{Y|x_1,\ldots,x_p\}\right) = \beta_0 + \sum_{j=1}^{p} f_j(x_j)$$

which follows (2.42), and is called *generalized additive model* (GAM). As in the standard GLM, link function g must be specified. For example, in the case of binomial Y, g is commonly assumed to be logit function (2.43). Instead, the term on the right-hand side is now expressed by an additive form, and consequently the contribution of general variable x_j is no longer linear $\beta_j x_j$ but is of the more general type $f_j(x_j)$.

To estimate functions for a GAM-type model, we use a suitable combination of algorithm 4.1 with that of iterative weighted least squares, applied in the case of GLM.

Bibliographical notes
For complete coverage of additive models and GAM, see Hastie & Tibshirani (1990) and Hastie et al. (2009; section 9.1).

4.6 Projection Pursuit

Additive model (4.14) can also be applied to transformed variables in particular, to projected variables in carefully chosen directions. The model may be written as

$$f(x_1, \dots, x_p) = \beta_0 + \sum_{k=1}^{K} f_k(\beta_k^\top x)$$

and is called the *projection pursuit* regression model, where K is the number of projections that must be chosen, and $\beta_k \in \mathbb{R}^p$ are projection vectors, which must be estimated. Functions $f_k(\cdot)$ are called *ridge functions,* because they are constant in all directions except that defined by vector β_k. Note that unlike additive models, number K of ridge functions does not coincide with number p of variables in the model.

The fitting procedure is based on the least squares criterion, leading to an expression to be minimized by selecting $\beta_1, \dots, \beta_K$ and functions $f_1, \dots, f_K$. The algorithm follows a forward or backward stepwise strategy to select the number of terms K. At each step, it alternates between a Gauss-Newton method to estimate β_k, given f_ks, and a one-dimensional smoothing regression for the f_k, given β_k. After each step, the f_ks from previous steps can be readjusted by backfitting (algorithm 4.1). In a forward stepwise procedure, the number of terms K is selected by stopping the procedure when the next term does not appreciably improve the model fit, but cross-validation can also be used to choose K.

The model is very general, and for large enough K and appropriate choice of f_k, it can arbitrarily approximate any continuous function of the covariates. A class of models with this property is called a *universal approximator.* Note that, for example, additive models do not share this property. Projection pursuit regression is also invariant to nonsingular transformations of covariates, but interpretation of results is usually difficult, because each variable enters the model in different projections.

We illustrate the method by considering the car data, with `city distance` as the response variable and `engine size` and `curb weight` as explanatory variables. We use smoothing splines as smoothing functions and select $K = 3$. Direction vectors β_k are shown in table 4.3 and fitted functions f_k are plotted in figure 4.14. The surface of the fitted `city distance` is shown in figure 4.15.

Table 4.3. Car Data: Direction Vectors for Three-Term Projection Pursuit Regression with Spline Smoother

	Term 1	Term 2	Term 3
`engine size` (rescaled)	0.114	−0.782	0.980
`curb weight` (rescaled)	−0.993	0.623	−0.198

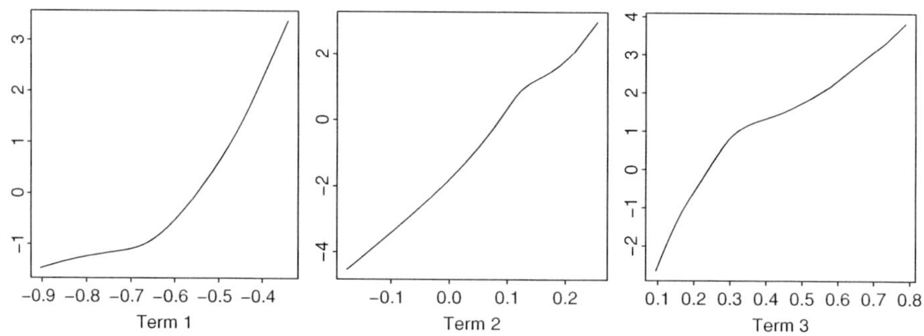

Figure 4.14 Car data: plots of ridge functions for three-term projection pursuit regression with spline smoother.

Bibliographical notes

Projection pursuit was introduced by Friedman & Tukey (1974). A detailed overview is given in Huber (1985), and an introductory account is provided by Hastie et al. (2009; section 11.2). Proof that the projection pursuit model is a universal approximator derives from Kolmogorov's universal approximation theorem (Kolmogorov 1957) and is discussed, for example, by Jones (1992).

4.7 Inferential Aspects

The contents of this chapter so far mostly concern nonparametric estimation of a regression, and we have only marginally considered a statistical inference step, which we now examine in greater depth. In particular, we want to introduce a formulation of *analysis of variance* adapted to the present context to test the hypothesis that a certain covariate does not affect the response variable.

4.7.1 Effective Degrees of Freedom

Let us refer to the general framework (4.9) and to relative estimator $\hat{f}$. We consider the problem of establishing whether a certain explanatory variable, let us call it x_r, is unnecessary and can be removed from the model.

The fact that most of the nonparametric methods described so far are linear forms of the response variable (once the smoothing parameter has been fixed) plays an important role. We can therefore write the vector of fitted values $\hat{y}$ in the form $\hat{y} = Sy$, with S as the $n \times n$ smoothing matrix; the corresponding vector of the residuals is given by $\hat{\varepsilon} = (I_n - S)y$.

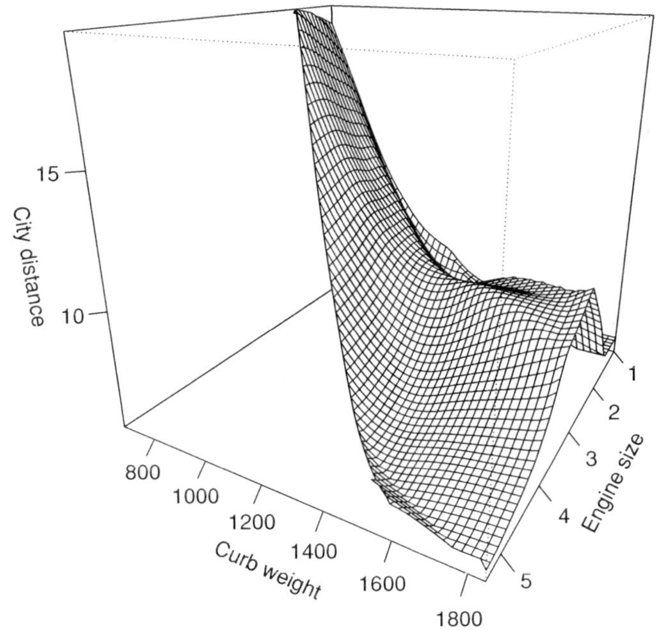

Figure 4.15 Car data: estimate of `city distance` according to `engine size` and `curb weight` through projection pursuit regression with spline smoothers.

To construct a table of analysis of variance, we must introduce some type of "degrees of freedom," even approximately, associated with the quadratic forms connected to an estimator. Consider the residual sum of squares

$$Q = \sum_i \hat{\varepsilon}_i^2 = \hat{\varepsilon}^\top \hat{\varepsilon} = y^\top (I_n - S)^\top (I_n - S) y$$

of which we wish to determine the probability distribution and, in particular, calculate the expected value. Hence, we now consider y as a vector sampled from a multivariate random variable Y.

We take the case of the classic linear model $\hat{\varepsilon} = (I_n - P)y$, where P is projection matrix (2.9) and it is known that $\mathbb{E}\{Q\} = \sigma^2(n-p)$, where $n-p$ are the degrees of freedom of the error component. With the addition of the hypothesis $\varepsilon \sim N_n(0, \sigma^2 I_n)$, we can conclude that $Q \sim \sigma^2\, \chi^2_{n-p}$.

In our case, the residuals are obtained with a formula similar to that of linear models, apart from the fact that projection matrix P is substituted by smoothing matrix S, which does not enjoy the same formal properties. Consequently, even if we assume the normality of ε, the probability distribution of Q is no longer χ^2.

However, we have empirical evidence based on simulations indicating that the shape of the probability density for Q is similar to that of χ^2. The problem now is to find an expression that plays the role of degrees of freedom, and this requires determination of an approximation to $\mathbb{E}\{Q\}$, in view of the correspondence between the average value and degrees of freedom for a χ^2 variable. For this, we write

$$\begin{aligned}\mathbb{E}\{Q\} &= \mathbb{E}\left\{Y^\top (I_n - S)^\top (I_n - S)Y\right\} \\ &= \mu^\top (I_n - S)^\top (I_n - S)\mu + \sigma^2 \mathrm{tr}[(I_n - S)^\top (I_n - S)]\end{aligned}$$

where $\mu = \mathbb{E}\{Y\}$, and we used lemma A.2.4. If we introduce the approximations

$$(I_n - S)\mu \approx 0, \qquad (I_n - S)^\top (I_n - S) \approx (I_n - S),$$

we can then write

$$\mathbb{E}\{Q\} \approx \sigma^2 \{n - \mathrm{tr}(S)\}$$

and $n - \mathrm{tr}(S)$ are called *effective* or *equivalent degrees of freedom* for the error term; correspondingly, $\mathrm{tr}(S)$ are the effective degrees of freedom for the smoother.

Because the foregoing expressions are based on approximations, one implication is that we can introduce slightly different expressions for the same degrees of freedom, based on alternative approximations. For example, forms $\mathrm{tr}(SS^\top)$ or even $\mathrm{tr}(2S - S S^\top)$ have been proposed instead of $\mathrm{tr}(S)$. Dealing with approximations among which there is no clear reason to prefer one form over another, we tend to use the simplest form, $\mathrm{tr}(S)$; in any case, the results do not change radically.

In addition to the role of numerical approximation, it is useful to identify the basic meaning of the idea of effective degrees of freedom. We bear in mind that depending on the choice of smoothing parameter, $\hat{y} = S y$ lies between the linear parametric interpolation and a "totally irregular" fit, which presumes no regularity whatsoever for the underlying function $f(x)$. Choosing the smoothing parameter, and therefore S, between these two extremes corresponds to a form of "partial regularity" of $f(x)$, which is quantified by the degrees of freedom corresponding to the choice of smoothing parameter. In other words, $\mathrm{tr}(S)$ represents the number of effective parameters implied by the model; conversely, $n - \mathrm{tr}(S)$ represents the component of nonregularity and quantifies which fraction of the data is allocated to estimating the error component.

One role played by effective degrees of freedom is that of introducing a uniformly valid smoothing indicator across different types of smoothers.

4.7.2 Analysis of Variance

We now return to the question of evaluating the significance of the individual variables that enter model (4.9).

Like the scheme of analysis of variance for linear models with Gaussian errors, we can establish an extended form of analysis of variance in which total variability is broken down into components that represent the contribution of each covariate.

We can now reproduce (2.35) using two nonparametric estimates for $\hat{y}_0$ and $\hat{y}$, where $\hat{y}_0$ represents the restricted model. Recalling the discussion in the last subsection, we approximate the distribution of test F with a Snedecor F with $(\mathrm{tr}(S) - \mathrm{tr}(S_0), n - \mathrm{tr}(S))$ degrees of freedom.

As an illustration, reconsider the car `city distance` data and examine the effect of `engine size` and `curb weight` using local regression, as in section 4.2, with values of 0.3 and 300 for the smoothing parameters of the two variables, respectively.

As usual, we summarize the essential ingredients in a *table of analysis of variance.*

Component	**Deviance**	**d.f.**	***p*-value**
`engine size`	1169618	12.07	0.000
`curb weight`	729.0	5.40	0.094
`(engine size, curb weight)`	410.2	13.08	

To interpret the elements of this table, we bear in mind that the row headed, for example, `curb weight`, provides the difference of deviance between the complete model, with both terms, `engine size` and `curb weight`, and the restricted model, without the variable `curb weight`—that is, the row reports the contribution made to lowering the deviance due to the variable `curb weight`. In the same way, the row shows the effective degrees of freedom for this component — that is, the difference between the degrees of freedom of the complete model and that without the variable `curb weight`. Last, the p-value is calculated as the complement of the distribution function at the point

$$F = \frac{729.0/5.40}{410.2/(203 - 13.08)} = 1.88$$

of Snedecor's distribution with 5.40 and 203 − 13.08 degrees of freedom, since the sample size is 203.

The values obtained depend to some extent on the choice of smoothing parameters, for example, h. However, we note empirically that the p-values, and therefore the inferential conclusions, are not heavily influenced if the variation of h occurs within a reasonably chosen area. Consequently, the choice of smoothing parameter is not as critical here as we saw in the estimation problem.

Clearly, this form of analysis of variance is used in a particularly natural way within the field of additive models, where the idea of the increase in fit made by each variable is implicit, retracing the logic of classical analysis of variance.

Bibliographical notes

Inferential methods in the context of nonparametric regression are discussed in Bowman & Azzalini (1997; ch. 4). For the introduction of effective degrees of freedom and their various definitions, see, for example, Hastie & Tibshirani (1990; pp. 128–129, and appendix B) and Green & Silverman (1994; pp. 37–38).

4.8 Regression Trees

4.8.1 Approximations via Step Functions

In one sense, the simplest way to approximate a generic function $y = f(x)$, with $x \in \mathbb{R}$, is to use a step function, that is, a piecewise constant function (see figure 4.16).

However, there are various choices to be made: (a) how many subdivisions of the x-axis must be considered? (b) where are the subdivision points to be placed? (c) which value of y must be assigned to each interval?

Of these questions, the easiest to answer is the last one, because it is completely natural to choose value $\int_{R_j} f(x)\,dx/|R_j|$ for any interval R_j, having indicated the length of that interval by $|R_j|$. Regarding positioning the subdivision points of $\mathbb{R}$, and therefore defining the intervals, it is better to choose small intervals where $f(x)$ is steeper. The choice of the number of subdivisions is the most subjective of the three points: intuitively, any increase in the number of steps increases the quality of the approximation, and therefore, in a certain sense, we are led to think of infinite subdivisions. However, this is counter to the requirement to use

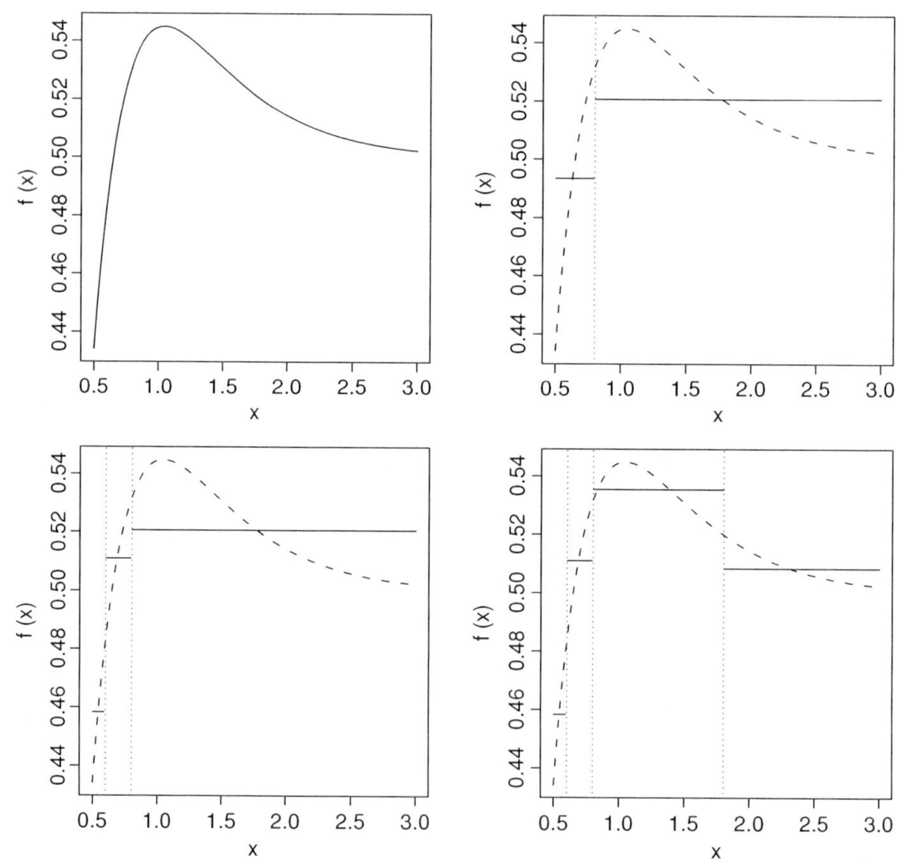

Figure 4.16 A continuous function and some approximations by step functions.

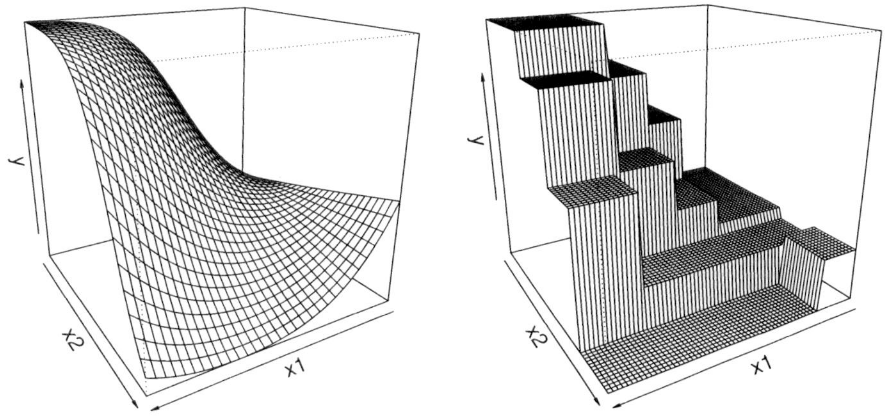

Figure 4.17 A continuous function in $\mathbb{R}^2$ and an approximation via a step function.

a "sparing" approximate representation, and therefore to adopt a finite number of subdivisions.

The scheme can be extended to the case of functions of p variables: we thus write $y = f(x)$ where $x \in \mathbb{R}^p$. There are many ways of extending the idea from the $p = 1$ case to the general p case. Figure 4.17 shows a function in $\mathbb{R}^2$ and its approximation by a step function: the regions with constant values are thus rectangles, the sides of which are parallel to the coordinate axes.

These characteristics of an approximate function, with some additional specifications to be described later, allow it to be represented as a *binary tree*, shown in the top panel of figure 4.18; the bottom panel shows the corresponding partition of the domain of function $f(x)$ and the values of the approximating function in each rectangle.

The components of the tree are inequalities, called *nodes*, relative to any component x of type $x_2 < 1.725$. We begin by examining the inequality of the node at the *root of the tree*, which is at the top. We follow the left branch if the inequality is true and the right branch if it is not. We proceed in the same way, sequentially examining all the inequalities until we reach the terminal nodes, called *leaves*, which give the values of the approximating function.

Graphical representation as a tree is not as visually attractive as that of figure 4.17, but it has important advantages: as the tree is identified by a few numerical elements, it can easily be stored. A second important advantage is that we can move from one approximation to a more accurate one by subdividing one of the components into two subrectangles with the same characteristics as the original. This corresponds to extending a branch of the tree to a further branch level. This characteristic immediately allows us to recursively construct a sequence of approximations that are increasingly accurate, each obtained by refining the previous one, as illustrated in the sequence of three step functions in figure 4.16.

4.8.2 Regression Trees: Growth

We want to use the idea of approximation with a step function to approximate our functions of interest, which are regression functions. Obviously, in our context,

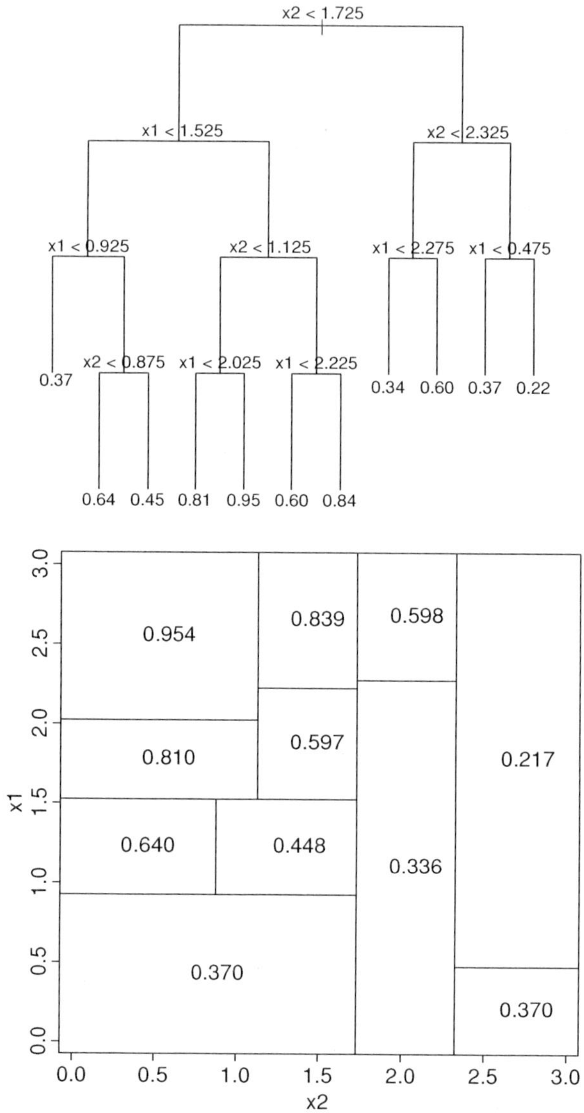

Figure 4.18 Tree corresponding to approximation of bottom panel of figure 4.17 (top), and partition of domain of $f(x)$ induced by tree (bottom).

regression function $f(x)$ is not known, but we can observe it indirectly through n sample observations, generated by model (4.9).

For simplicity, we begin from the case where $p = 1$ and consider the data of figure 4.19, which represents the 60 pieces of data already seen in chapter 2, subdivided into two groups: 'yesterday' and 'tomorrow'. We can estimate regression curve $f(x)$ underlying the data by a step function of the type just described, that is

$$\hat{f}(x) = \sum_{h=1}^{J} c_h I(x \in R_h) \tag{4.15}$$

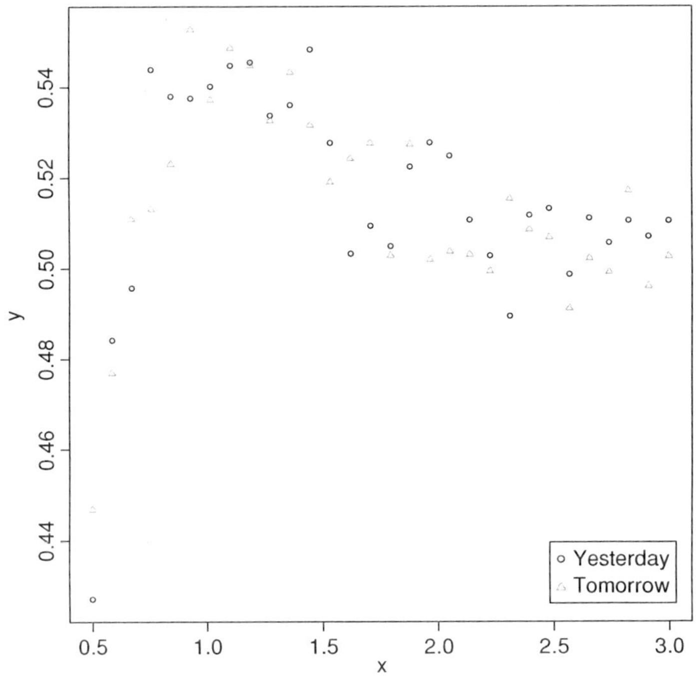

Figure 4.19 Scatterplot with 60 pairs of values.

where $c_1, \ldots, c_J$ are constants and $I(z)$ is the *indicator function* $0 - 1$ of logical predicate z. In general, sets $R_1, \ldots, R_J$ are rectangles, in the p-dimensional sense, with their edges parallel to the coordinate axes. In the specific case where $p = 1$, obviously R_h are reduced to line segments.

We need an objective function to choose R_h and c_h. The reference criterion is deviance, but its minimization, even if we fix step number J, involves very complex computation. Therefore, operatively we follow a suboptimal approach of *step-by-step optimization*, in the sense that we construct a sequence of gradually more refined approximations and to each of these we minimize the deviance relative to the passage from the current approximation to the previous one.

The algorithm starts by splitting the real line associated with one of the variables, for example, x_j, into two parts; which variable is to be considered is discussed later. Each of the subintervals is assigned a value, c_h, given by the arithmetic mean of the observed y_i having component x_j falling in this subinterval, irrespective of the other covariates. Note that this step divides the $\mathbb{R}^p$ space into two regions via a hyperplane parallel to the jth coordinate axis. The subsequent steps of the algorithm proceed similarly, each time splitting one of the existing regions of $\mathbb{R}^p$ into two further regions, again with a split parallel to one of the coordinate axes.

The right panel of figure 4.18 illustrates the outcome of this process in a simple instance with $p = 2$. Figure 4.20 shows three instances of portions that are not compatible with the foregoing process; the fourth one is admissible.

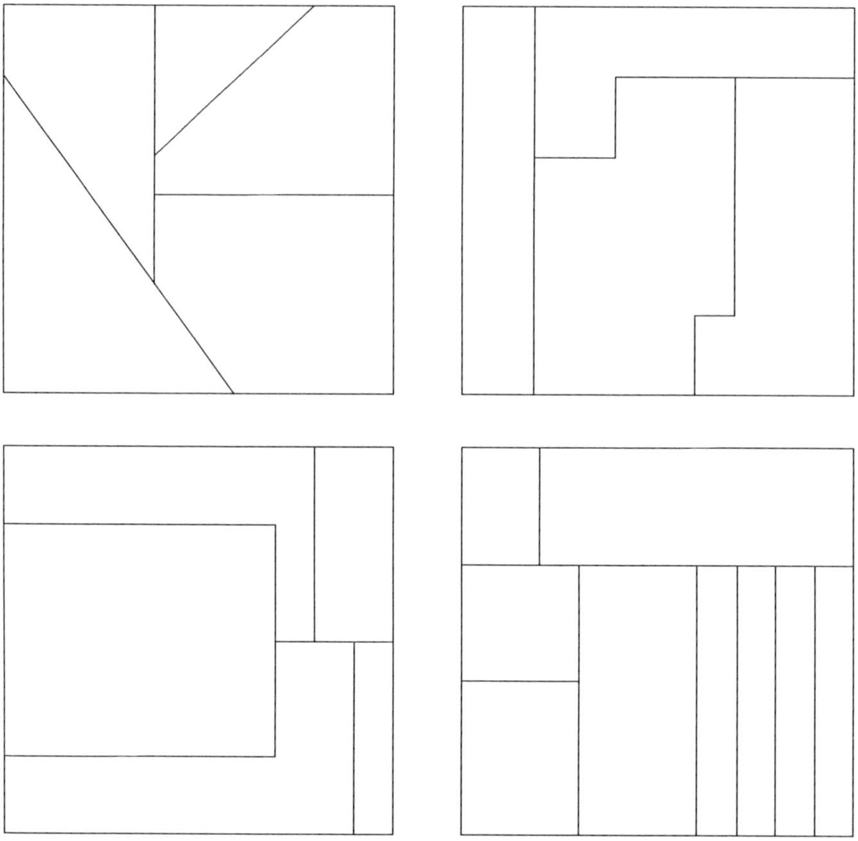

Figure 4.20 Three partitions of domain of $f(x)$, not consistent with a tree, and one partition induced by a tree (lower right).

A crucial aspect is the fact that at each step, one of the already constructed rectangles is divided into two, and so is the portion of data belonging to it; we optimize deviance with respect to this operation. Therefore, this is a *myopic optimization* procedure. Although it does not guarantee global minimization of deviance, it does provide acceptable solutions, maintaining limited *computational complexity*.

At least in principle, this procedure can be applied iteratively through successive subdivisions of $\mathbb{R}^p$ until we can no longer distinguish sets containing a single sampled observation and thus obtain a tree with n leaves. To be useful, the number of leaves must be less than n, preferably much less. Therefore, after the stage of *tree growth*, with the complete or almost complete development of all the leaves, we move to a stage of *tree pruning*. We describe the growth algorithm now and return to the pruning phase later.

To develop the growth algorithm, first note that whatever the division of $\mathbb{R}^p$ into hyper-rectangles, we can break down the deviance as follows

$$D = \sum_{i=1}^{n} \{y_i - \hat{f}(x_i)\}^2 = \sum_{h=1}^{J} \left\{ \sum_{i \in R_h} (y_i - \hat{c}_h)^2 \right\} = \sum_{h} D_h. \tag{4.16}$$

We also bear in mind the general property that the minimum of $\sum_{i=1}^{n}(z_i - a)^2$ with respect to a is obtained for $a = M(z)$, where $M(\cdot)$ is the average operator of the vector.

The growth process starts with $J = 1$, $R_J = \mathbb{R}^p$, $D = \sum_i (y_i - M(y))^2$, and proceeds iteratively for a number of cycles, according to the following scheme:

- once a rectangle R_h is chosen, the appropriate value c_h is the average of the corresponding values

$$\hat{c}_h = M(y_i : x_i \in R_h)$$

- if we subdivide region R_h into two parts, R_h' and R_h'' (therefore moving to $J + 1$ leaves), summand D_h of D is replaced by

$$D_h^* = \sum_{i \in R_h'} (y_i - \hat{c}_h')^2 + \sum_{i \in R_h''} (y_i - \hat{c}_h'')^2$$

  with a "gain" of

$$g_h = D_h - D_h^*$$

- we can inspect all p explanatory variables and, for each of them, all the possible points of subdivision, selecting the variable and its point of subdivision that maximize g_h.

We stop when $J = n$, at least conceptually. Mainly, if n is enormous, we stop earlier — for example, when all the leaves contain a number of sample elements that is less than a preassigned value, or when the relative fall of deviance is less than a prefixed threshold.

4.8.3 Regression Trees: Pruning

A large tree with n leaves is conceptually equivalent to interpolation through a polynomial of order $n - 1$, which passes exactly through all the points; hence, it is not very useful. We have to prune the tree by removing branches of little or no use.

Let us therefore introduce an objective function that incorporates a penalty for the *cost-complexity* of the tree which we assess by dimension J. This objective function is given by

$$C_\alpha(J) = \sum_{h=1}^{J} D_h + \alpha J \tag{4.17}$$

where α is a nonnegative penalty parameter. Breiman et al. (1984) showed that the set of rooted subtrees that minimize the cost-complexity measure is nested. That is, as we increase α we can find the optimal trees by a sequences of pruning operations on the current tree. So for each α, there is a unique smallest tree minimizing $C_\alpha(J)$ (Breiman et al. 1984; proposition 3.7) and we select the tree that minimizes $C_\alpha(J)$ for a fixed α.

To minimize (4.17), we proceed by sequentially eliminating one leaf at a time. At each step, we select the leaf for which elimination causes the smallest increase in $\sum_h D_h$. The question is therefore reduced to choosing α, and for this we can use one of the methods described in section 3.5. We can show that suitable adaptation of the AIC gives $\alpha = 2\hat{\sigma}^2$, where $\hat{\sigma}^2$ is the estimate of residual error variance, but how this can be estimated reliably is not very clear. However, the widespread opinion is that AIC tends to overfit the data in this area. Therefore, the methods of cross-validation and simple subdivision of data into a training set and a test set, as seen in sections 3.5.1–3.5.2, are more widely used.

Predicting $f(x)$ on a new piece of data x_0 is done by allowing the observation to descend from the root of the available tree. Datum x_0 follows one of the branches, according to the components of x_0, which describe it, until it reaches a leaf with a value of $\hat{f}(x_0)$. We repeat this process for the n' components of the test set, (x_i, y_i) for $i = 1, \ldots, n'$. Comparing $\hat{f}(x_i)$ with observed class y_i, we compute the contribution from the ith unit to deviance (4.16), and the sum over the n' terms provides the observed value of the deviance.

For illustration, let us consider the data in figure 4.19, using the subgroups of yesterday's data for growth and tomorrow's data for pruning. The tree developed to fullness using only yesterday's data is shown in the first panel of figure 4.21, where the length of the vertical lines is proportional to the reduction of the deviance obtained by subdividing the node. Clearly, after some ramifications, there is no substantial gain due to the lower branches.

The top-right panel represents function $\sum_h D_h$ calculated from tomorrow's data. The graph indicates choice $J = 4$, associated with $\alpha = 4.33 \times 10^{-4}$. The suitably pruned tree appears in the lower-left panel, and function $\hat{f}(x)$ is found lower-right, overlapping the points.

In this case, the small sample size allows us to use cross-validation.

Note that pruning is often very radical and can easily lead to a tree with a small number of nodes with respect to the numbers of variables and their levels, if they are categorical. This fact automatically leads to a choice of the useful variables, regarding the variables that remain excluded. In reverse, it is not easy to rank importance for those that remain in the tree, as the reduction in deviance associated with each node is not directly interpretable. This difficulty is due to at least three aspects: (a) the reduction of the deviance due to the node quantifies the gain of that particular dichotomization of the variable and not the whole variable; (b) the logic of myopic optimization used to make the tree grow makes it difficult to attribute global significance a posteriori to local aspects and (c) each variable may be included in more than one node.

To overcome these problems, specific measures of the relative importance of each covariate in predicting the response have been proposed. For example, a simple measure of the contribution of a variable, like x_k, is based on improvements g_h in lowering the deviance at each step, involving x_k as splitting variable. The sum of squared g_h^2 over all the internal nodes for which x_k was chosen as splitting variable is a squared relative measure of the importance of that variable.

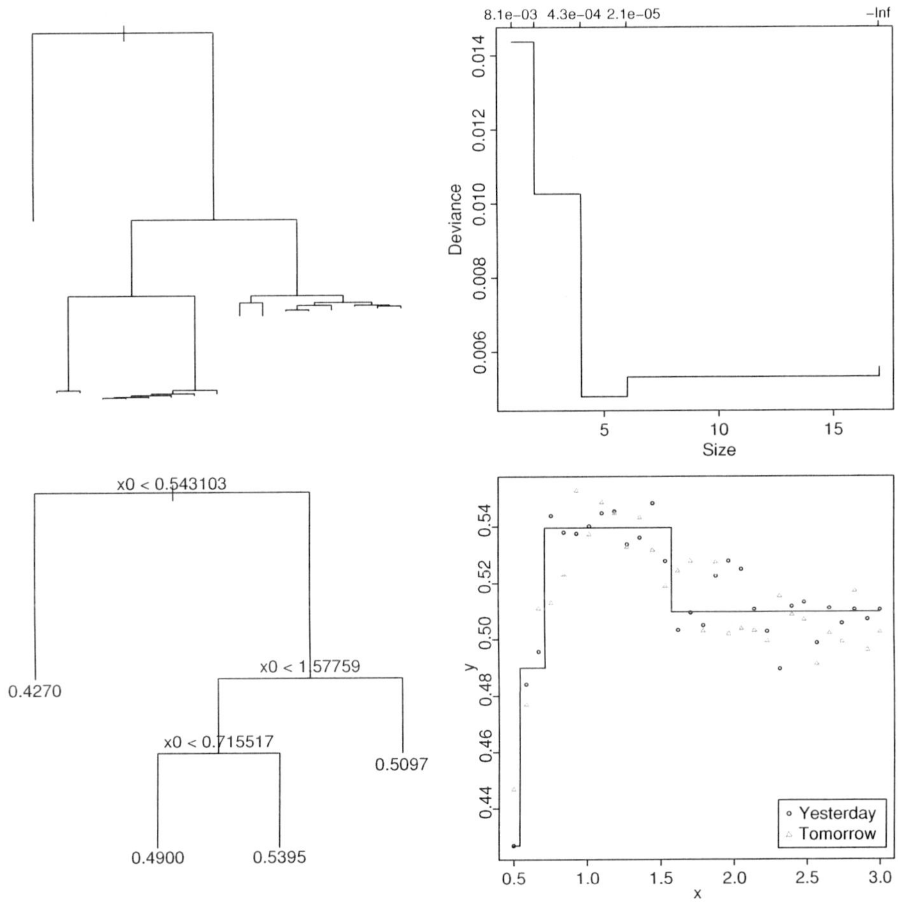

Figure 4.21 For data of figure 4.19, top-left panel displays a nearly fully grown tree. Top-right panel: deviance function from tomorrow's data, which selects four-node tree in lower-left panel. Lower-right panel: data with overlapping selected four-level function.

4.8.4 Discussion

Because trees are very frequently used in practice, we note their advantages and disadvantages.

Advantages

- Logical simplicity and ease of "communication," above all with those who have a nonquantitative background. Trees are logical structures usually used by many people in decision-making processes, for example, by physicians and businesspeople, perhaps not consciously.
- The step function has a simple, compact mathematical formulation in terms of information to be stored.
- Speed of computation: the process is not very taxing from this point of view, and it can also take advantage of the potential of *parallel calculation*.

- Use of discrete and categorical variables: although the previous description referred to continuous covariates, there is no specific reason to limit oneself to them, and the method can proceed in the same way if some of the variables are discrete or qualitative.
- Robust forms of deviance: clearly, having seen the construction, we immediately can substitute deviance with another criterion and the average with the corresponding operator, thus allowing the use of criteria based on robustness considerations.
- Missing data: not particularly complicated variations can be introduced, which allow for missing values, in both tree construction and prediction.
- Variable selection: the method automatically selects the important variables.

Disadvantages

- Instability of results: a tree is often very sensitive to the insertion of new data or changes in existing data.
- Difficulty in upgrading: if more data arrive, they cannot be added to the already constructed tree; it is necessary to start again from the beginning.
- Difficulty of approximating some mathematically simple functions, particularly if they are steep, and a straight line or other simple function would approximate them very well.
- Statistical inference: formal procedures of statistical inference such as hypothesis testing, confidence intervals, and others are not available.
- Selection of variables: it is not simple to evaluate the order of importance of variables remaining in the pruned tree.

Bibliographical notes

Breiman et al. (1984) introduced not only the idea of regression trees and classification trees (see later discussion) but also the acronym CART, which then became synonymous with the same method. This work was among the first to promote a particular philosophy of data analysis and examine issues that later to became the characteristic elements of data mining. Venables & Ripley (2002) describe the practical usage of trees.

4.9 Neural Networks

The term *neural network* encompasses a wide family of techniques developed in *machine learning*. We describe only the simplest version here.

Figure 4.22 shows p explanatory variables (*input*) in a relationship with q response variables, or *output*. The most characteristic aspect is the *layer* of r *latent variables*, which is not observable (hidden) and comes between the two previous groups in the sense that the covariates influence the latent variables; these in turn influence the response variables. The number of *input* and *output* variables is determined by the problem, but the number r of latent variables is something we can choose, because they are only conceptual entities. In figure 4.22, we have $p = 4$, $r = 3$, and $q = 2$, and some additional "constant variables," identical to 1, are also shown.

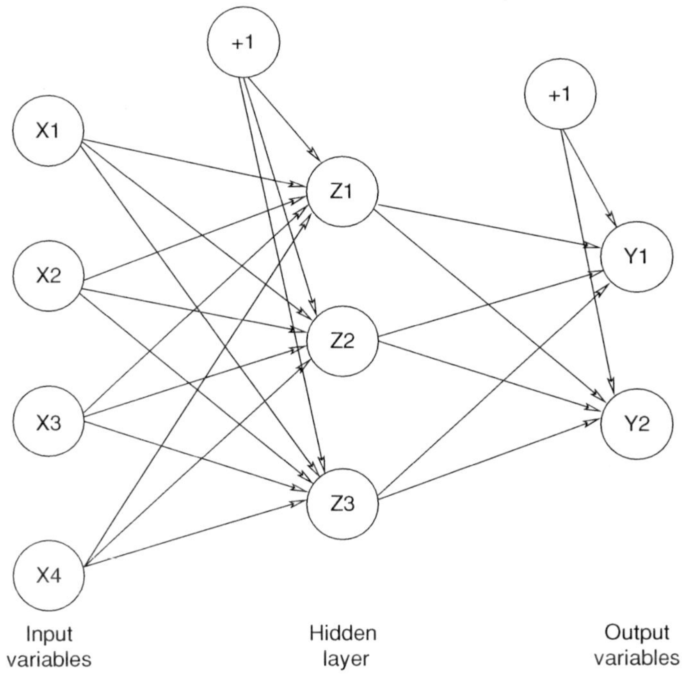

Figure 4.22 A simple neural network.

The term *neural network* originated as a mathematical model that in the past was believed to be the mechanism controlling the working of the animal brain: every node of the graph represented a neuron, and the arcs represented the synapses. We now know that the animal brain is much more complex, but the term *neural network* survives.

A neural network is essentially a two-stage regression scheme, generally of nonlinear or at least partially nonlinear type. We indicate the generic *input*, latent, and *output* variables by x_h, z_j, and y_k, respectively, and add constant variables x_0 and z_0 equal to 1. The previous scheme can now be expressed as

$$z_j = f_0\left(\sum_{h\to j} \alpha_{hj}\, x_h\right), \qquad y_k = f_1\left(\sum_{j\to k} \beta_{jk}\, z_j\right), \tag{4.18}$$

where α_{hj} and β_{jk} are parameters to be estimated, and the sums are over the indices of the variables for which a dependence relation is predicted. Figure 4.22 shows these dependencies by arrows and involves all the compatible variables, although this is not necessarily the case. We can therefore see that the resulting structure is an acyclic *graph* with directed edges and *weights* associated with coefficients α and β.

To complete the picture, we must specify *activation functions* f_0 and f_1. In regression problems, where the y_k are generally nonlimited, we presume

$$f_0(u) = \frac{e^u}{1 + e^u}, \qquad f_1(u) = u, \tag{4.19}$$

where the choice of f_0 is the logistic function, as seen in section 2.4. We note, however, that at least one of the two functions f_0 and f_1 must be nonlinear to avoid reducing the whole network to a set of linear relations, effectively eliminating the latent layer.

There are mathematical results that give rise to interesting properties for the framework. In particular, we can show that a neural network with linear *output* units can approximate any continuous function f uniformly on compact sets, by appropriately increasing the number of units of the latent layer; see Ripley (1996; p. 147 and 174).

Extensions are possible in various directions. One of the most common is to consider several layers of latent variables. Another is to introduce edges that skip a layer: in the case of the single latent layer considered here, this means inserting an edge directly between some variables of the *input* layer and some of the *output*. Two elements must be specified: the number r of units in the hidden layer and the set of coefficients α and β of (4.18). For the choice of r, there are no criteria that are easy to use in practice, apart from experimenting with various ones and comparing the results.

Therefore assume that r has been fixed and we want to estimate coefficients α and β according to sample observations. This is done by minimizing the usual objective function

$$D = \sum_i \|y_i - f(x_i)\|^2$$

where y_i now indicates the q-dimensional vector of the response variables of the ith observation. Analogously, x_i is the corresponding p-dimensional vector of the covariates, and $f(x)$ is the vector, whose kth component is

$$f(x)_k = f_1\left\{\sum_{j\to k} \beta_{jk} f_0\left(\sum_{h\to j} \alpha_{hj}\, x_h\right)\right\}, \qquad (k = 1, \ldots, q).$$

More elaborate versions of this objective function can be obtained by including a penalty term to avoid overfitting problems, for example, functions of the type

$$D_0 = D + \lambda J(\alpha, \beta). \tag{4.20}$$

Here λ is a positive tuning parameter and $J(\alpha, \beta)$ is a penalty function, according to a path already seen previously, for example in section 4.4.3. Among the most common penalty forms there are

$$J(\alpha, \beta) = \int \sum_{h,k} \frac{\partial^2 y_k}{\partial x_h^2}\, \mathrm{d}x \approx \frac{1}{n} \sum_i \sum_{h,k} \frac{\partial^2 y_{ki}}{\partial x_{hi}\, \partial x_{hi}}, \qquad J(\alpha, \beta) = \|\alpha\|^2 + \|\beta\|^2 \tag{4.21}$$

of which the first form penalizes the amplitude of the second derivative, and the second tends to shrink the parameters toward 0; the latter is called *weight decay*. Here y_{ki} denotes the kth component of y_i.

These formulations, both D and penalty function J, make sense if the variables are measured on the same scale. As a preliminary operation, it is therefore better to normalize them—for example, by rescaling all the variables between 0 and 1 (at least approximately). For regulation parameter λ, Venables & Ripley (2002; p. 339) advise choosing a value between 10^{-4} and 10^{-2}.

Clearly, minimization of D_0 requires a numerical optimization process. Much effort has been invested in developing such algorithms. The most common method is called *back-propagation*, which has interesting properties. One of the most important aspects in this context is that there exists a variant of the back-propagation algorithm, which allows for later updating of parameter estimates in an incremental way as new data become available.

It must be stressed that practical experience has provided extensive evidence that objective function D_0 often has many points of local minima, and it is therefore wise to start the optimization algorithm from several initial points. This difficulty in turn affects something else: in choosing λ it is difficult to take advantage of techniques like cross-validation, as the algorithm varies widely in locating the minimum.

To illustrate the method, let us consider the engine size and curb weight from our car data to predict city distance. We consider a neural network with f_0 and f_1 as in (4.19) and one latent layer with $r = 3$ nodes. We minimize function D_0 with penalty $J(\alpha, \beta)$ in the second form, and $\lambda = 10^{-3}$. After various executions of the minimization algorithm, starting from different initial points of the parameters, we reach what would seem to be an acceptable minimum point. The results are shown in figure 4.23, in which the top diagram is a graph of the neural network with estimated weights and the lower one is a prospective representation of $f(x)$.

In conclusion, we review of the advantages and disadvantages of this approach.

Advantages

- Flexibility: the method allows for good approximation of practically any regression function $f(x)$, that is, the model is a *universal approximator*.
- Compactness of representation: the estimated regression function is identified by a limited number of components.
- Sequential upgrading: coefficients α and β can be updated sequentially as new data arrive by means of a suitable variation of the back-propagation algorithm.

Disadvantages

- Arbitrariness: there are no strong criteria with which to choose the number r of latent nodes; in addition, we only have rough indications for the choice of λ.
- Instability in the estimation stage: the nature of objective function D, or its variations, implies that its properties are difficult to identify, especially the existence of a single minimum point. Instead, there is empirical evidence

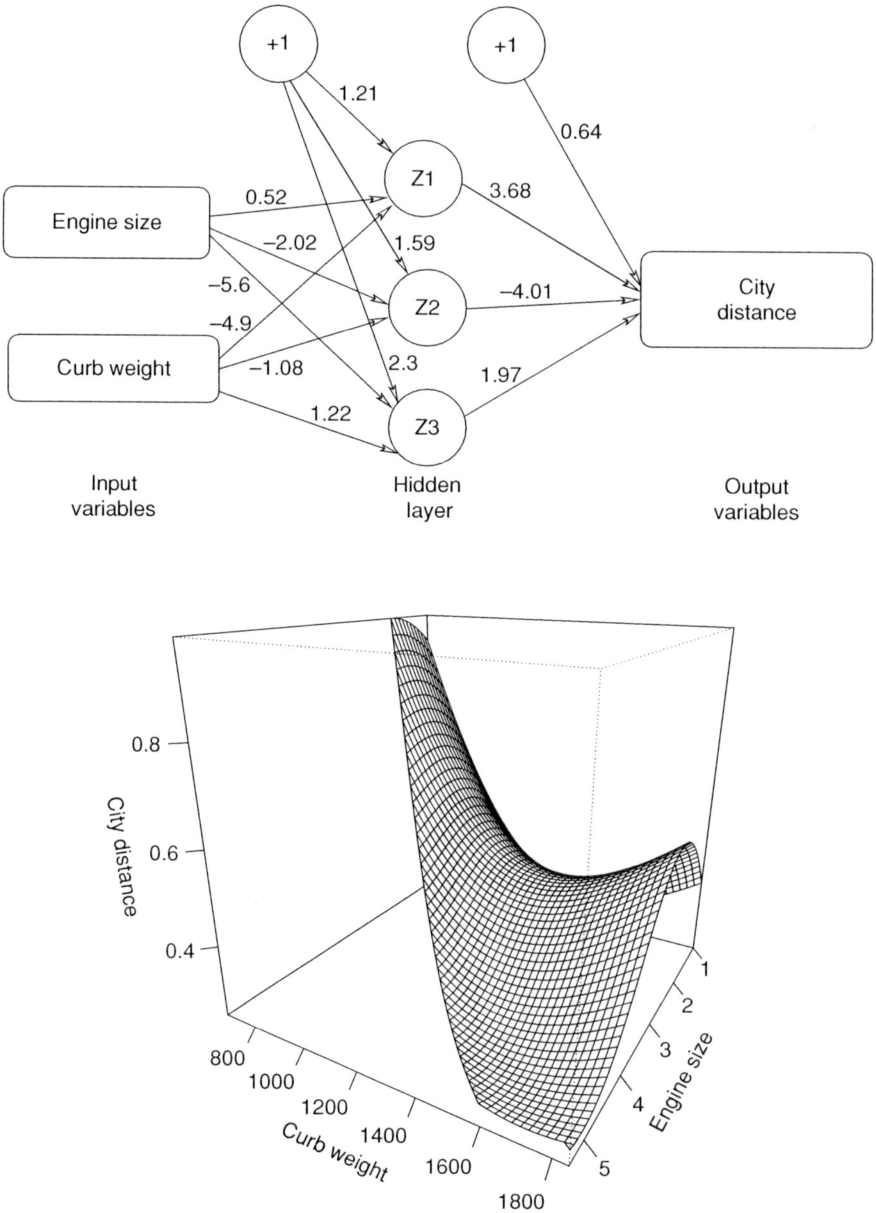

Figure 4.23 Car data. Top: neural network 2–3–1 with `engine size` and `curb weight` to predict `city distance`, bottom: surface of estimated function.

of the frequent presence of local minima, and different results may be obtained if the optimization algorithm is started from different points.

- Inference: there are no standard errors associated with the coefficients or other inferential procedures—for example, to reduce the number of coefficients.

- Interpretation: there are major problems in interpreting results, particularly when r increases.

Bibliographical notes

The literature on neural networks is extremely ample, and ranges from very technical presentations to very informal ones. Among the latter, from the viewpoint of readers with a statistical background, we mention the works by Ripley (1996; ch. 5) and Hastie et al. (2009; ch. 11). Fine (1999) provides a more mathematical account.

4.10 Case Studies

The data used up to now to illustrate the various methods, although obtained from real cases, were suitably simplified to avoid over-specific details in applied problems that interfere with presentation. Here we focus on operational aspects and treat a couple of real cases in their original complexity.

4.10.1 Traffic Prediction in Telecommunications

The first problem presented here was handled by a group of marketing analysts in a telecommunications company. Our aim is not to analyze the associated marketing themes in detail but to present the use of data mining methods as a tool for business choices.

The data and the background problem

The group within the marketing section of a telecommunications company managing customer relationships (*customer base management*) is interested in analyzing customer behavior regarding telephone traffic. Of the many types of analysis the group uses to study customer traffic characteristics, identifying a tool to predict the traffic of every single customer in the coming months is often extremely useful. Not only can appropriate estimations of overall traffic provide necessary elements for predicting the company's budget, but tools can be supplied to evaluate each client's *value* to the company. Marketing actions can be organized to incentivize the use of company services to those whose traffic could potentially increase and to avoid doing the same to those who do not need them. Traffic predictions are also used to note possibly anomalous behavior by customers, particularly those who are more valuable to the company, for early identification of possible dissatisfaction, problems in using the main services the company offers, or even fraudulent situations.

In this context, let us consider a set of customers who possess a SIM (subscriber identity module) card with a call plan that is of particular interest to the company. We tackle the problem of predicting traffic for the coming month using data available so far. Therefore, as response variable, we choose the total number of seconds of outgoing calls made in a given month.

A typical way of proceeding in these cases is based on the idea that in essence, customers' traffic behavior can be considered as stable in time if reduced time intervals are considered. Under such a hypothesis, traffic in month t using data for months $t-1, t-2, \ldots, t-k$, can be predicted as a first approximation, irrespective of specific month t. Therefore, in the first search we do not keep count of seasonal components or cycles in the model, but, following common usage in this field, we consider a model constructed with data for month t as a good prediction tool for each successive month.

The approximation thus introduced may seem excessive, and it is in fact possible to consider components that gather effects due to the specific months in constructing the model itself. Or, in contexts in which prediction models are often updated, we can validate the model on the basis of test sets extracted from data that refer to t months differing from those used for the estimate. For the sake of simplicity, we concentrate only on the stable hypothesis.

We must now choose our covariates. First, we determine for how many months it would be useful and suitable to "go backwards" in time to continue to find meaningful relationships with the response variable and then identify the variables to be observed for each customer. Some of these are observed for all months, for example, the number of text messages (SMS) sent or the number of calls to the customer services helpline (*customer care*), but others do not depend on the time interval in question, for example, gender or the day of activation of the service.

The company's DWH (see section 1.1.3) yields a data mart for 30,619 customers, for whom information on a total of 99 variables is available. Some of these are intrinsic customer characteristics (e.g., gender and age) or have to do with the specific relationship between customer and company (e.g., activation data or any value-added services), and some have to do with information on traffic in each of the consecutive nine months previous to the month of interest. Last, there is the variable relative to the total duration of outgoing calls in the tenth month, which is our response variable. The data are presented in greater detail in section B.4.

The high number of customers allows us to divide the data set into two parts, one for the estimate and the other for validation of results and comparisons. We subdivide the available data into two equal sets, composed of 15,310 and 15,309 customers, respectively.

We now examine the training set. An initial descriptive graphical analysis shows that the distribution of the response variable is highly asymmetric and concentrated around 0. In particular, the training set contains 5,131 customers who have not made outgoing calls. This data characteristic involves some difficulty in automatically using the models proposed here. Clearly, the response variable cannot be treated as a continuous variable, because it has the characteristics of a mixed variable: it is the combination of a continuous component for some of the observations and a discrete, binary component for the other group of customers who did not make calls in that month.

It is therefore reasonable to take advantage of this information to construct our prediction model of the duration of outgoing calls. One possibility is to organize the process into two stages. First, we fit a model for the probability that the duration is not 0, and, conditionally on this event, we then construct a model for positive duration values.

To construct a model that predicts an indicator variable, we still need to introduce more elements (see chapter 5 for the analysis of this aspect). In the

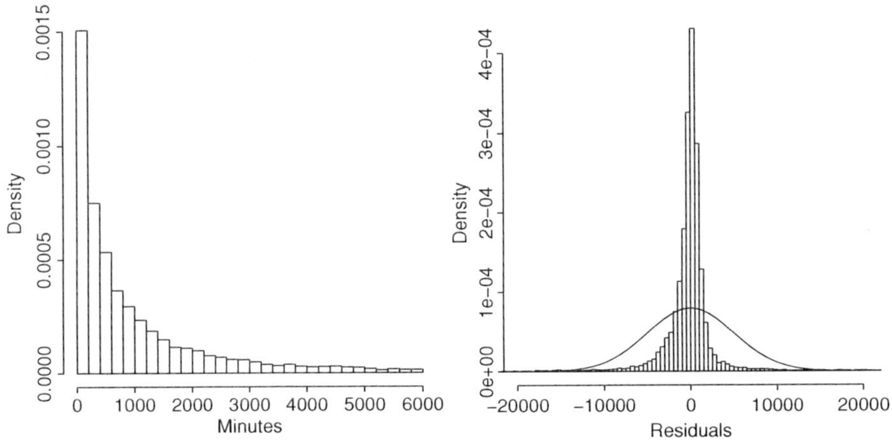

Figure 4.24 Telecommunications customers. Left: distributions of outgoing call duration for month of interest; right: residuals of stepwise linear model with density of Gaussian distribution of average 0 and variance equal to variance of residuals.

next section, we describe some models to predict the total duration of calls on condition that they were made.

Some prediction models

We now consider, among the customers selected for the estimate, the set containing the 10,179 customers who had positive total call durations in the month of interest. The left part of figure 4.24 is a histogram of the response variable for this set.

The first prediction method is a linear model obtained with all available variables. The fitted model with 98 covariates gives $R^2 = 0.613$. Clearly the estimate of many parameters indicates that the variables in question do not significantly influence the response variable. Therefore, a *stepwise* procedure is formulated to select the relevant variables. After much computer work, the final model contains 56 covariates and gives $R^2 = 0.612$. Note that in situations like these, in which we have an extremely high number of observations, it is not useful to carry out formal hypothesis testing with test F to verify the combined influence of all the eliminated variables on the response variable.

The histogram of the residuals of the model with 56 variables appears in the right part of figure 4.24. The quantities, obtained by estimating the parameters in R, are listed below:

```
Residuals:
     Min        1Q    Median       3Q       Max
-69152.5    -790.7      66.8    663.4 148323.2

Coefficients:
                      Estimate Std. Error t value Pr(>|t|)
(Intercept)           4.04e+03   2.97e+02   13.64  < 2e-16 ***
tariff.plan4          1.50e+04   4.92e+02   30.57  < 2e-16 ***
```

```
tariff.plan6          -3.78e+03   2.52e+02  -15.00  < 2e-16 ***
tariff.plan7          -4.02e+03   1.99e+02  -20.24  < 2e-16 ***
tariff.plan8          -3.78e+03   1.97e+02  -19.21  < 2e-16 ***
etac1                 -2.92e+01   6.24e+00   -4.68  2.9e-06 ***
activ.zone2           -4.64e+01   1.24e+02   -0.37  0.70829
activ.zone3            4.87e+02   1.32e+02    3.70  0.00022 ***
activ.zone4           -2.87e+01   1.92e+02   -0.15  0.88146
vas1Y                  3.93e+02   1.13e+02    3.46  0.00053 ***
q01.out.ch.peak       -4.26e+00   1.58e+00   -2.70  0.00698 **
q01.out.dur.peak       3.01e-02   1.26e-02    2.40  0.01635 *
q01.out.ch.offpeak     1.67e+01   5.91e+00    2.82  0.00481 **
q01.out.dur.offpeak    1.92e-01   4.45e-02    4.31  1.7e-05 ***
q01.out.val.offpeak   -6.45e+01   1.30e+01   -4.98  6.4e-07 ***
q01.in.ch.tot          3.85e+00   1.33e+00    2.90  0.00370 **
q01.ch.cc             -6.54e+01   4.16e+01   -1.57  0.11609
q02.out.dur.peak      -4.37e-02   2.04e-02   -2.15  0.03180 *
q02.out.val.peak       1.81e+01   4.47e+00    4.05  5.1e-05 ***
q02.out.ch.offpeak     1.11e+01   6.85e+00    1.62  0.10539
q02.out.dur.offpeak   -2.13e-01   4.24e-02   -5.03  5.1e-07 ***
q02.out.val.offpeak   -1.28e+01   6.91e+00   -1.85  0.06398 .
q02.in.ch.tot         -3.82e+00   1.37e+00   -2.79  0.00525 **
q02.ch.cc             -1.08e+02   4.03e+01   -2.68  0.00736 **
q03.out.val.peak       4.94e+00   1.62e+00    3.05  0.00232 **
q03.out.dur.offpeak    1.20e-01   3.70e-02    3.25  0.00115 **
q03.out.val.offpeak    2.03e+01   8.81e+00    2.30  0.02129 *
q03.in.dur.tot        -3.06e-02   8.19e-03   -3.73  0.00019 ***
q04.out.ch.peak       -3.59e+00   1.27e+00   -2.82  0.00485 **
q04.out.dur.peak      -3.62e-02   1.90e-02   -1.90  0.05713 .
q04.out.val.peak       1.19e+01   4.29e+00    2.77  0.00568 **
q04.out.ch.offpeak    -3.71e+01   5.00e+00   -7.42  1.3e-13 ***
q04.in.dur.tot         2.60e-02   9.58e-03    2.71  0.00678 **
q05.out.dur.peak       5.44e-02   1.66e-02    3.27  0.00108 **
q05.out.val.peak      -1.46e+01   3.37e+00   -4.34  1.4e-05 ***
q05.out.ch.offpeak     3.35e+01   6.69e+00    5.00  5.9e-07 ***
q05.out.val.offpeak    1.46e+01   9.44e+00    1.55  0.12220
q05.ch.cc              6.74e+01   3.93e+01    1.72  0.08637 .
q06.out.dur.peak      -4.48e-02   1.77e-02   -2.53  0.01134 *
q06.out.val.peak       1.14e+01   3.88e+00    2.93  0.00342 **
q06.out.ch.offpeak    -5.43e+01   8.54e+00   -6.35  2.2e-10 ***
q06.out.dur.offpeak   -1.11e-01   7.23e-02   -1.54  0.12357
q06.out.val.offpeak    2.04e+02   2.61e+01    7.82  5.8e-15 ***
q06.in.dur.tot         1.59e-02   9.45e-03    1.68  0.09219 .
q06.ch.sms            -4.29e+00   1.86e+00   -2.30  0.02139 *
q07.out.dur.peak      -3.59e-02   1.37e-02   -2.62  0.00893 **
q07.out.val.peak       1.26e+01   3.06e+00    4.12  3.8e-05 ***
q07.out.ch.offpeak    -2.34e+01   8.74e+00   -2.68  0.00728 **
q07.out.dur.offpeak   -1.12e-01   7.72e-02   -1.45  0.14819
q07.out.val.offpeak    4.01e+01   2.66e+01    1.51  0.13233
q07.in.dur.tot        -1.86e-02   9.48e-03   -1.96  0.04975 *
q07.ch.cc             -3.23e+01   1.84e+01   -1.76  0.07900 .
q08.out.ch.peak       -2.71e+00   1.34e+00   -2.03  0.04280 *
q08.out.dur.peak       4.69e-02   1.36e-02    3.46  0.00055 ***
q08.out.val.peak      -1.37e+01   3.11e+00   -4.41  1.1e-05 ***
```

```
q08.out.ch.offpeak   -2.18e+01   9.03e+00   -2.42  0.01569 *
q08.out.dur.offpeak   2.48e-01   6.35e-02    3.90  9.5e-05 ***
q08.in.ch.tot         3.43e+00   1.19e+00    2.89  0.00389 **
q09.out.val.peak      1.34e+01   9.95e-01   13.51  < 2e-16 ***
q09.out.ch.offpeak    1.27e+02   8.67e+00   14.63  < 2e-16 ***
q09.out.dur.offpeak   1.47e+00   6.31e-02   23.35  < 2e-16 ***
q09.out.val.offpeak -1.99e+02    1.88e+01  -10.53  < 2e-16 ***
---
Signif. codes:  0 '***' 0.001 '**' 0.01 '*' 0.05 '.' 0.1 ' ' 1

Residual standard error: 5020 on 10117 degrees of freedom
Multiple R-Squared: 0.612,      Adjusted R-squared: 0.61
F-statistic:  262 on 61 and 10117 DF,  p-value: <2e-16
```

Did you stop to look at the individual numbers in the list? They seem to be useless and too many. In fact, they contain much useful information for analysts looking for reasons customers increase their traffic. So if we analyze the parameter values in more detail, we see that they offer interesting suggestions for marketing choices.

Let us make a single example of simple interpretation: the high value of the parameter of the first value-added service (`vas1`) tells us that, taking into account the linear effect of all the other variables in the model, subscription to such a service is a strong incentive to use the phone. This result may give rise to marketing choices aimed at increasing the use of this value-added service—for example, a targeted marketing campaign or sending personalized letters presenting the service to customers who have not yet subscribed to it.

However, predictions obtained by applying this model to new data may also give rise to negative values for total call duration. To avoid this annoying problem, the prediction for all these customers to whom the model would assign negative duration at 0.5 was fixed. The choice of 0.5 is reasonable here because it is lower than any other value in the training set for total call duration, but it is not too small to have any great influence on the results of the following analysis.

To evaluate the quality of these two prediction models more completely, we measure performance on the test set. The resulting squared prediction errors are 257.74×10^9 and 258.52×10^9 for the complete and restricted models, respectively.

The models minimize objective function (2.3), which assigns the same importance to each observed entity. However, as we note that total monthly call duration (the variable we are predicting) is certainly a positive quantity and we expect that it frequently has low or medium values and only rarely high ones, we also expect that it will have a skew shape. This consideration, also corroborated by the right panel of figure 4.24, which clearly shows that the residuals of the linear model do not have Gaussian distribution, leads us to consider different objective functions for estimate evaluation.

A simple, widespread choice in these cases is to consider as new output the logarithmic transform of the response variable, leading to a deviance on the logarithmic scale:

$$D(\beta) = \sum_i \{\log(y_i) - f(x_i;\, \beta)\}^2 \tag{4.22}$$

where, as usual, y_i indicates the response variable observations, x_i the corresponding covariate observations, β the vector of the unknown parameters to be estimated, and f the function identified by the model.

To evaluate these two linear models in terms of this new loss function, we can calculate the prediction error on a logarithmic scale on the test set, calculating the function

$$\sum_i \{\log(y_i) - \log(g(x_i;\, \hat{\beta}))\}^2$$

where g is the linear predictor on the original scale. On the logarithmic scale, deviance is 113,472 for the complete model and 112,061 for the restricted one, respectively, confirming that the model with fewer covariates produces a slightly better prediction.

The linear model can also be directly fitted so as to minimize (4.22). This time, too, we fit the model with all the covariates to the data and then select the most important ones with a stepwise procedure. The prediction errors on the validation set of all the models fitted using the original and logarithmic scales are listed in table 4.5.

In figure 4.25, the left panel shows the histogram of the logarithm of the outgoing call duration in the month of interest; the right panel shows the histogram of the model residuals on the logarithmic scale obtained by stepwise selection of the variables. These histograms support the hypothesis that the loss function on the logarithmic scale is a reasonable choice for the problem in question.

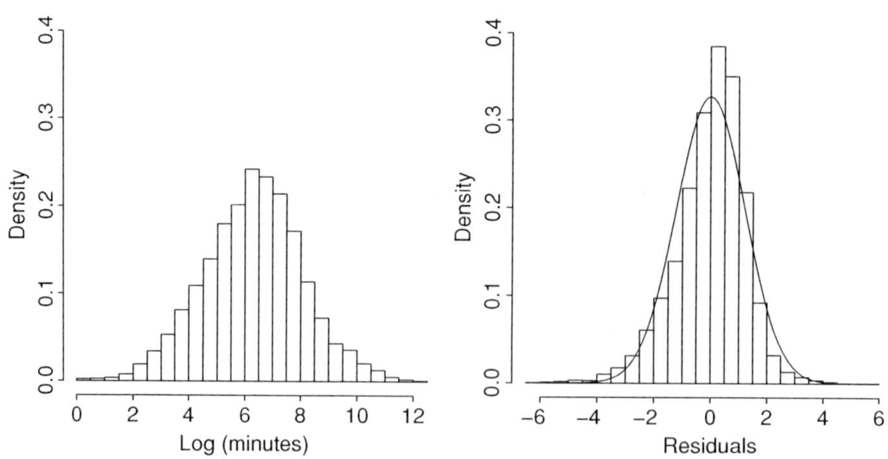

Figure 4.25 Telecommunications customers. Left: distribution of logarithms of outgoing call duration for month of interest; right: histogram of residuals of linear model on logarithmic scale with stepwise selection of variables with Gaussian distribution density (average 0 and variance equal to variance of residuals).

A second group of models fitted to these data is based on GAM models (see section 4.5). In this case, a first model with all available variables was fitted and then a second, with only the variables resulting significant in the first model, determined through analysis of variance (section 4.7.2). A GAM model was also fitted, with only those variables for the last observed month, as well as customer characteristics, which do not vary in time. Also in this case, to estimate the functions of the additive models, both logarithmic and original scales were used. Table 4.5 lists the prediction errors of the six additive models obtained by selecting the covariates (all of them, only significant ones, only those relative to the last month) and using the two estimation criteria (original and logarithmic scales).

For all continuous variables, smoothing splines (see section 4.4.3) were used as nonparametric estimators, and the number of effective degrees of freedom was fixed at 4 for each univariate function as a choice of spline smoothing parameter. The estimates of the functions of variables significant for the model on the logarithmic scale are shown in figure 4.26.

Looking at all the coefficients of the linear model may give rise to feelings of confusion or uselessness. However, also in this case, each figure may have useful consequences for company policy. Simple examples are that value-added services (`vas1` and `vas2`) cause an increase in net traffic, other estimated elements being fixed. Regarding traffic variables, note the narrowness of the variability bands of the function for off-peak call duration in the ninth month, identified as a very important predictor for increased traffic in the tenth month, and the nonmonotone trend of the curve for the same variable in the sixth month.

The other family of models based on splines used here is MARS (see section 4.4.5). In this case, because the procedure automatically chooses variables useful for predictions, one model was used for the original scale and one for the logarithmic scale. Table 4.4 lists the information used by the final model on the original scale. The prediction errors of these models are also listed in table 4.5 to aid comparison with other predictions.

A neural network (see section 4.9) was also fitted on both original and logarithmic scales. Three nodes were used for the hidden layer and to control overfitting. We selected $\lambda = 10^{-3}$ as the *weight decay* parameter. Prediction errors are listed in table 4.5.

Last, two regression trees (see section 4.8) were "grown" on the two scales. The trees, like MARS, automatically select the variables that most influence the response variable, taking advantage of pruning phases; it is not necessary here to carry out any preliminary operations to reduce the models. The training set was divided into two parts: one of 5,089 customers, used to grow the tree, and the other of 5,090 customers, for pruning. Figure 4.27 shows the deviance plot versus number of tree nodes for the two models on the original and logarithmic scales, respectively, and figure 4.28 shows the two final trees.

For the first tree, the function that describes deviance with respect to number of nodes (top panel, figure 4.27) shows two local minima, and the absolute minimum attained with deviance on the pruning set of 119.52×10^9 refers to the tree with 44 leaves, which is obviously a tree with many branches.

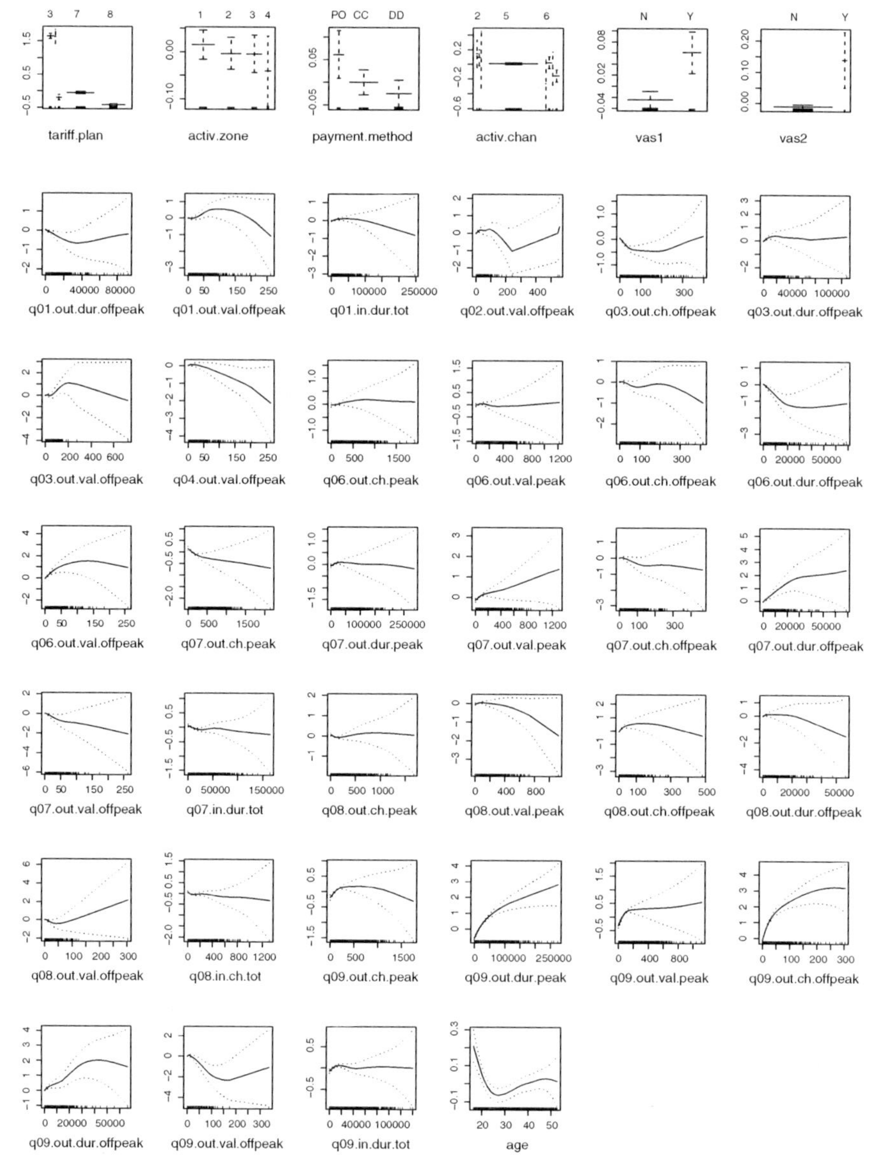

Figure 4.26 Telecommunications customers: GAM model on logarithmic scale, with significant covariates only.

It seems unreasonable in this case to apply the algorithm automatically, which suggests choosing the tree that minimizes deviance on the pruning set. A more careful analysis indicates we should consider both models proposed by the deviance curve, which therefore also correspond to the local minimum of 121.32×10^9—a tree with seven leaves. In fact, as well as finding the optimal prediction criterion, it is always useful to look for a model which also responds to simplicity criteria. If, as in this case, deviance on the pruning set is very similar in the two models, we often prefer the simpler one, apart from other considerations regarding

Table 4.4. TELECOMMUNICATIONS CUSTOMERS: ESTIMATES FOR MARS MODEL

First Variable	First Node	Second Variable	Second Node	Parameters	SE
constant				988.29	251.77
q09.out.dur.offpeak				−2.84	0.85
q09.out.dur.offpeak	365.00			4.42	0.85
q09.out.val.peak				40.80	3.27
q09.out.val.peak		q09.out.dur.offpeak		0.34	0.01
q09.out.val.peak		q09.out.dur.offpeak	365.00	−0.34	0.01
plan.tariff				−119.08	56.43
plan.tariff		q09.out.val.peak		−8.77	0.78
q09.out.val.offpeak				−48.60	39.37
q05.out.val.offpeak				342.32	25.45
plan.tariff		q05.out.val.offpeak		−70.93	4.92
q05.out.val.offpeak		q09.out.val.peak		−0.44	0.13
q05.out.val.offpeak	48.07			−396.81	32.59
plan.tariff		q05.out.val.offpeak	48.07	183.68	12.05
q05.out.val.offpeak	48.07	q09.out.val.peak		−3.13	0.18
q05.out.val.offpeak		q09.out.val.offpeak		−2.53	1.03
q05.out.val.offpeak		q09.out.dur.offpeak		−0.001	0.01
q09.out.val.offpeak	29.22			−515.14	46.37
q09.out.val.peak	189.57			4.91	1.47
q05.out.val.offpeak		q09.out.val.peak	189.57	3.35	0.25
q05.out.val.offpeak		q09.out.val.offpeak	29.22	11.44	1.30

model interpretation. It was for this reason that in figure 4.28 we preferred to design the final tree with seven leaves. Table 4.5 lists the prediction errors of the trees with both 7 and 44 leaves, so as to highlight the real differences between the models on the test set.

Comparisons and discussion

Examination of table 4.5 and other elements prompt some reflections on both models and estimates.

The choice of the objective function is obviously linked to the marketing problem in question. In our case requests vary; on one hand the most precise prediction possible of the traffic of every customer for the month of interest is required—to be able, for example, to run budget predictions, measure the value of each customer, or redesign the network. On the other hand, we can study which operative tools can incentivize customers with medium or low traffic to increase their use of company services. Both objective functions are therefore used to provide useful suggestions for these types of requirements.

After choosing one of the two objective functions, the optimized models are clearly better than those obtained by minimizing the other function. Table 4.5 shows that similar models obtained with different optimizations can also differ greately from each other (see, for example, results on linear models). In the following, therefore, we compare the models obtained by maximizing each of the two objective functions separately.

Regarding the original scale—that is, analysis of the measure of total traffic, and therefore of the gain obtained by the company—we focus on the fact that

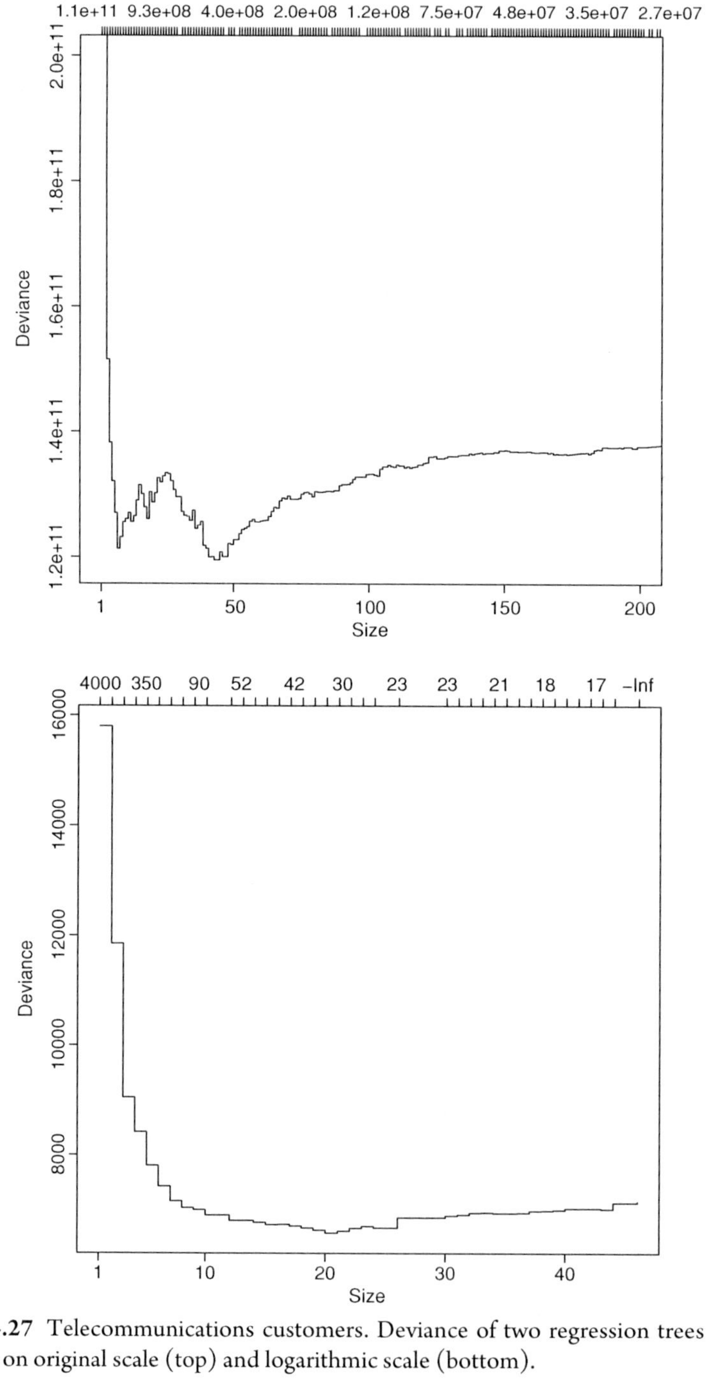

Figure 4.27 Telecommunications customers. Deviance of two regression trees with call duration on original scale (top) and logarithmic scale (bottom).

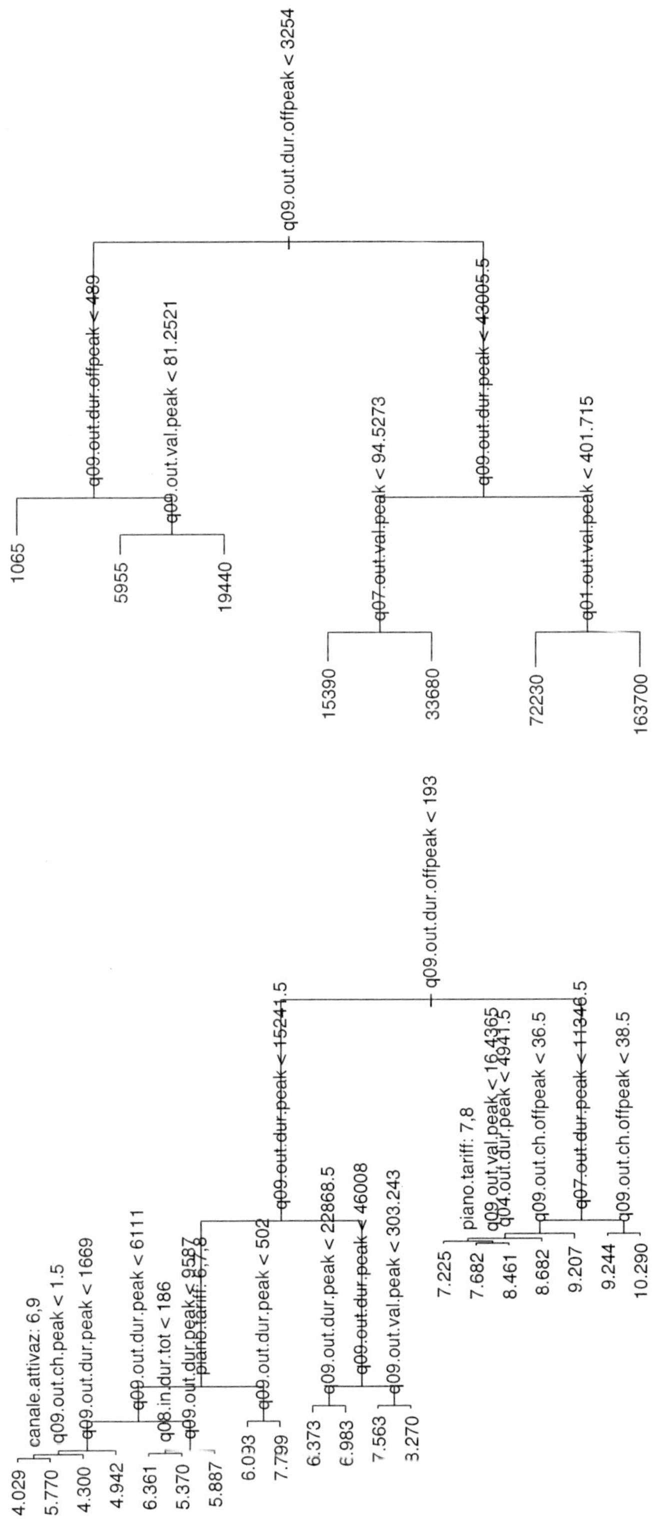

Figure 4.28 Telecommunications customers. Final regression trees, fitted to data with call duration on original scale (top) and logarithmic scale (bottom).

Table 4.5. TELECOMMUNICATIONS CUSTOMERS: PREDICTION ERRORS IN ORIGINAL AND LOGARITHMIC SCALES FOR MODELS FITTED TO DATA

Model	Variables	Optimization Scale	Squared Error Original Scale	Squared Error Logarithmic Scale
Linear	all	original	257,736,193,454	113,472
Linear	only signif.	original	258,524,520,314	112,061
Linear	all	logarithmic	79,407,475,570,006,224	15,838
Linear	only signif.	logarithmic	41,140,000,000,000,000	15,853
GAM	all	original	392,419,005,268	94,137
GAM	only signif.	original	391,286,779,004	98,628
GAM	prev. month	original	299,872,658,074	109,552
GAM	all	logarithmic	666,446,084,652	229,497
GAM	only signif.	logarithmic	1,190,485,331,659	13,989
GAM	prev. month	logarithmic	1,668,869,970,637	13,779
MARS		original	211,786,338,287	35,151
MARS		logarithmic	276,868,390,512	13,317
Neural network		original	604,876,539,104	35,151
Neural network		logarithmic	601,084,507,392	36,875
Tree	44 leaves	original	324,547,309,675	20,252
Tree	7 leaves	original	252,187,094,517	32,852
Tree		logarithmic	344,247,954,167	13,796

customers with high traffic are special and are considered much more important than customers with medium or low traffic. We therefore note the following.

- All the models are essentially equivalent as for prediction error with the only exception of the neural network.
- The tree that the pruning set suggested was "optimal" (i.e., 44 final nodes) performs worse than the tree selected as "sub-optimal," with 7 leaves, which combines simplicity and precision. This example indicates that trees may provide very interesting results if they are studied with care and evaluated in all their relevant aspects.
- The preferable model in terms of prediction error is MARS, which is used to make a precise prediction of total call duration in the following month.
- After precise prediction of total call duration, it is very important for marketing experts to have a precise description of the characteristics of those customers who make large numbers of calls with respect to those who do not.

    - Table 4.4 gives us a first, albeit preliminary, idea of the mechanism that allows MARS to predict total duration.
    - Other models are of more help in interpreting results. In this case, the regression tree does not only perform well in terms of

prediction error, but is also extremely easy to interpret and, as seen in figure 4.28, offers simple but useful cues for marketing actions.
   - Other than trees, linear models and GAM are also simple to interpret, through the table of coefficients presented at the beginning of this section for the former, and graphs like those of figure 4.26 for the latter.
   - Neural networks present greater difficulties in interpreting relationships. In this specific case, they also appear to perform worse than the other models.

Similar reasoning can be made for the last column of table 4.5, showing the squared errors on the logarithmic scale. In this case, the objective is to reduce the effect on estimates of best customers and search for the levers on which the company's marketing department can act to increase the traffic of customers with low value.

Again, the best model seems to be MARS, followed by GAM and the regression tree. The linear model behaves slightly worse than the others, probably due to the nonlinearity induced by the logarithmic transformation, which the other, more flexible models can handle.

When working on the logarithmic scale, the need to interpret the results is greater than that of simply being able to suggest actions to carry out on the customer base. It therefore favors a GAM-type model, which, as we have seen, offers not only good performance in terms of prediction error but also easy interpretation (see figure 4.26).

Summary

- We need a model to predict the traffic of each customer in a fixed month, using information on customers and services to them in the previous months.
- There are at least two types of aims: (1) to predict total traffic in the month of interest with the greatest possible precision, (2) to identify lines of action to persuade customers with less traffic to increase it.
- The model chosen for the first aim was MARS, with the smallest prediction on the original scale, which is appropriate for problems of type (1).
- The model chosen for the second aim, for which we used the logarithmic transform, was GAM, which, although not having the best prediction error on the logarithmic scale, is appropriate for problems of type (2). It also makes quite good predictions, offers easy interpretation of the model, and indicates possible up-sell actions.

4.10.2 Insurance Pricing

The problem described next was handled by the *nonlife* actuarial office of an insurance company. However, our objective is not to analyze actuarial issues in detail but to present data mining methods as tools for business choices. Much literature is available on statistical models for insurance pricing problems (e.g., Ohlsson & Johansson 2010; Tse 2009). A specific characteristic of this type of problem is that the value of the *premium* must be defined before a *claim* is made, that is, before its cost is known. Large numbers of predictive actuarial models have been developed to price different types of insurance.

The marketing managers of insurance companies are also interested in the combinations of products customers have. They want to "segment" the base by grouping customers with similar behavior and identify the characteristics of customers having similar premium value for some specific product. In this section, we examine the problem of predicting the amount of the *pure premium* (the total claim amount, divided by the duration of exposure to risk) for private car third-party liability insurance by considering subscription to other products offered by the same company, as well as other policyholder variables such as age, gender, and residential area. In particular, the company is interested in which insurance products are bought by customers who buy third-party liability insurance.

The data

The data are described in detail in section B.5 and refer to a random sample of 5,000 policyholders in a given year. Available variables include the number and total amount of premiums paid in the current year. It is easy to obtain the average pure premium for third-party liability policies, which is the response variable in our analysis. Clearly, a different strategy would be to predict the single variables, number of policies, and total sum paid for premiums by using two different models and then obtain the predicted ratio of the single-model predictions. There are circumstances in which either of these strategies is preferable (see exercise 4.10). Data on other customer policies are also available and are used as covariates.

We divide the original data set into two parts: 4,000 policyholders are used for training the models and the remaining 1,000 are reserved to validate results.

Simple descriptive analysis of the training set shows that 26.70% of customers do not hold any third-party liability insurance, and only 12.12% subscribe to more than one policy.

The first four panels of figure 4.29 show some characteristics of customers in the training set. The marginal distribution of the average pure premium paid by customers holding at least one policy is quite skewed, as shown by the bottom-left panel of figure 4.29. The last panel of figure 4.29 shows the histogram of the logarithm of the average pure premium added to 1, so that customers not subscribing to third-party liability policies are all included in the bar at 0.

Table 4.6 shows the probabilities that in the year considered, a customer subscribes to products of one or two types, usually called "lines," at the same time. The diagonal elements represent the probabilities of subscribing to a product of every single line, and all other elements represent the probabilities that a customer subscribes both to a product of the type indicated by the row and to one of a type indicated by the column in the same year, so that the matrix is symmetric.

The probability of subscribing to a third-party liability policy and a policy of type 1 at the same time is clearly higher than in other groups. Policy type 4 also shows a relatively high association with the policy of interest.

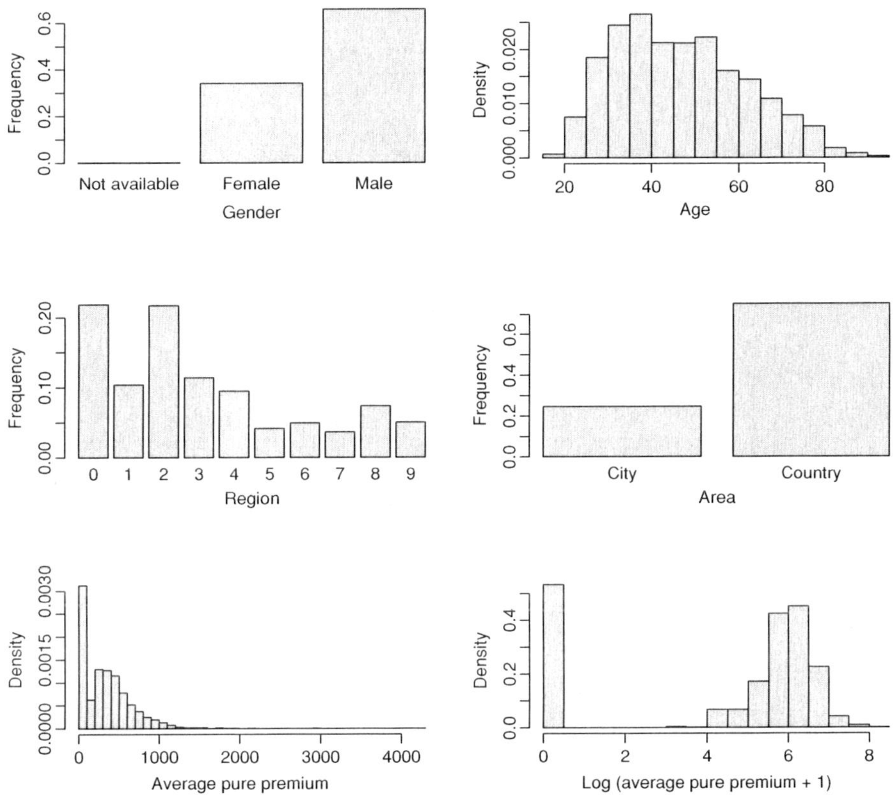

Figure 4.29 Insurance customers: Plots of distribution of a few selected variables.

Some prediction models

To better understand the characteristics of third-party liability customers, we formulate a number of regression models by considering either the original average pure premium or its logarithm.

We can organize the analysis into two phases, as we did for telecommunications customer prediction (section 4.10.1): first fitting a model for the probability that the premium is not 0, and then, conditionally on this event, fitting a model for the amount of the premium when it assumes positive values. We leave this implementation as an exercise (exercise 4.11), and prefer here to predict the average pure premium directly, including customers with 0 amount paid in the response variable.

A linear model to predict the average pure premium with all available covariates gives $R^2 = 0.27$. The same value is obtained by selecting the most important 30 variables in a stepwise procedure based on AIC. We also fit a linear model with lasso. The LARS algorithm allows us to estimate the entire set of models. We then select the one producing the smallest squared error on the evaluation set. Figure 4.30 shows the whole path of the coefficients for all lasso models obtained by changing the value of s in (3.11). Coefficients are plotted versus $t = s / \sum_{j=1}^{p} |\hat{\beta}_j|$. A vertical line is drawn at $t = 0.14$, the value chosen according to the test set.

Table 4.6. INSURANCE CUSTOMERS. CUSTOMERS SUBSCRIBE TO PRODUCTS OF ONE OR TWO GROUPS AT THE SAME TIME: OBSERVED FREQUENCIES

Policy	Third-Party Liability	Policy 1	Policy 2	Policy 3	Policy 4	Policy 5	Policy 6	Policy 7	Policy 9	Policy 10
Third-party liability	0.7330	0.1137	0.0130	0.0027	0.0515	0.0065	0.0160	0.0022	0.0002	0.0065
1	0.1137	0.1867	0.0287	0.0012	0.0245	0.0045	0.0072	0.0017	0.0000	0.0027
2	0.0130	0.0287	0.0455	0.0005	0.0080	0.0020	0.0032	0.0002	0.0000	0.0015
3	0.0027	0.0012	0.0005	0.0165	0.0030	0.0010	0.0017	0.0000	0.0000	0.0002
4	0.0515	0.0245	0.0080	0.0030	0.1340	0.0130	0.0050	0.0007	0.0000	0.0027
5	0.0065	0.0045	0.0020	0.0010	0.0130	0.0165	0.0015	0.0000	0.0000	0.0005
6	0.0160	0.0072	0.0032	0.0017	0.0050	0.0015	0.0372	0.0002	0.0000	0.0002
7	0.0022	0.0017	0.0002	0.0000	0.0007	0.0000	0.0001	0.0047	0.0000	0.0000
9	0.0002	0.0000	0.0000	0.0000	0.0000	0.0000	0.0000	0.0000	0.0002	0.0002
10	0.0065	0.0027	0.0015	0.0002	0.0027	0.0005	0.0002	0.0000	0.0002	0.0120

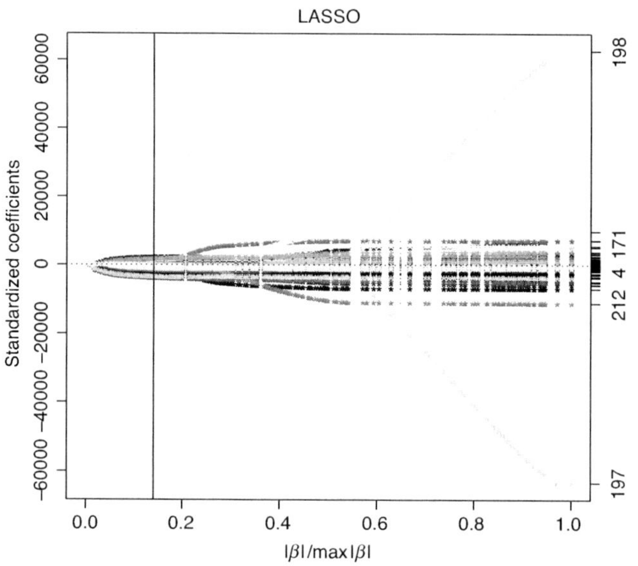

Figure 4.30 Insurance customers: profiles of lasso coefficients as tuning parameter s is varied. Standardized coefficients plotted versus $t = s/\sum_1^p |\hat{\beta}_j|$.

Lasso shrinks parameters and gives a model with only 9 variables (15 parameters). The estimated coefficients obtained by lasso with R are listed in table 4.7.

We also fit some nonlinear models to the data. To choose a suitable neural network for our problem, we divide the training set into two subsets of equal size and fit a number of different networks on the first subset by modifying the number of nodes in the hidden layer and the weight decay. Networks with 10 to 19 hidden nodes and 10 values for weight decay between 0.001 and 0.1 are evaluated, and we select the one with the smallest squared prediction error on the second subset of the training set. The best model has 12 nodes and weight decay of 0.1.

Table 4.7. INSURANCE CUSTOMERS: ESTIMATE OF COEFFICIENTS FOR BEST LASSO NONZERO ESTIMATE OF LINEAR MODEL

Variable	Level	Coeff.
`occupation.1`	9	6.94
	99	26.57
`occupation.2`	5	25.54
	12	7.63
	13	20.47
`area`	1082	73.15
	1191	17.27
	1542	94.19
`number.claims.3`		100.18
`amount.claims.last`		0.0041
`prem.non-life.5`		0.1098
`prem.payed.life.1b`		0.00007
`number.life.2b`		102.57
`region`	2	61.84
	8	10.24

A regression tree is also fitted to data. We let the tree grow by using data on the first subset of the training set and then prune it by using the second subset. The top panel of figure 4.31 shows deviance versus number of nodes. The global minimum is observed for size 2 and the pruned tree splits the pure premium once: if there were no claims in the last three years, the predicted pure premium is €306.00; otherwise, it is €497.50. However, this tree cannot describe the various characteristics of customers, particularly if we are interested in connections with other lines of products. Thus, we consider a second tree, the one corresponding to the local minimum for size equal 11, plotted in the bottom panel of the same figure 4.31.

Last, we fit a MARS model, a projection pursuit regression model, and an additive model by selecting variables by a stepwise procedure based on AIC.

We then consider the logarithm of the average pure premium and fit different models predicting this transformed variable. Linear models with different selection strategies, GAM, MARS, projection pursuit, neural network, and trees are fitted to the data to predict the transformed response variable, choosing appropriate tuning parameters.

The top part of figure 4.32 shows the pruned tree resulting from prediction of the transformed variable with the same growing and pruning subsets as for the original-scale tree.

Table 4.8 shows the prediction errors obtained on the validation set of some of the models used (the better-fitting ones). For each, we present the squared prediction error on original and logarithmic scales.

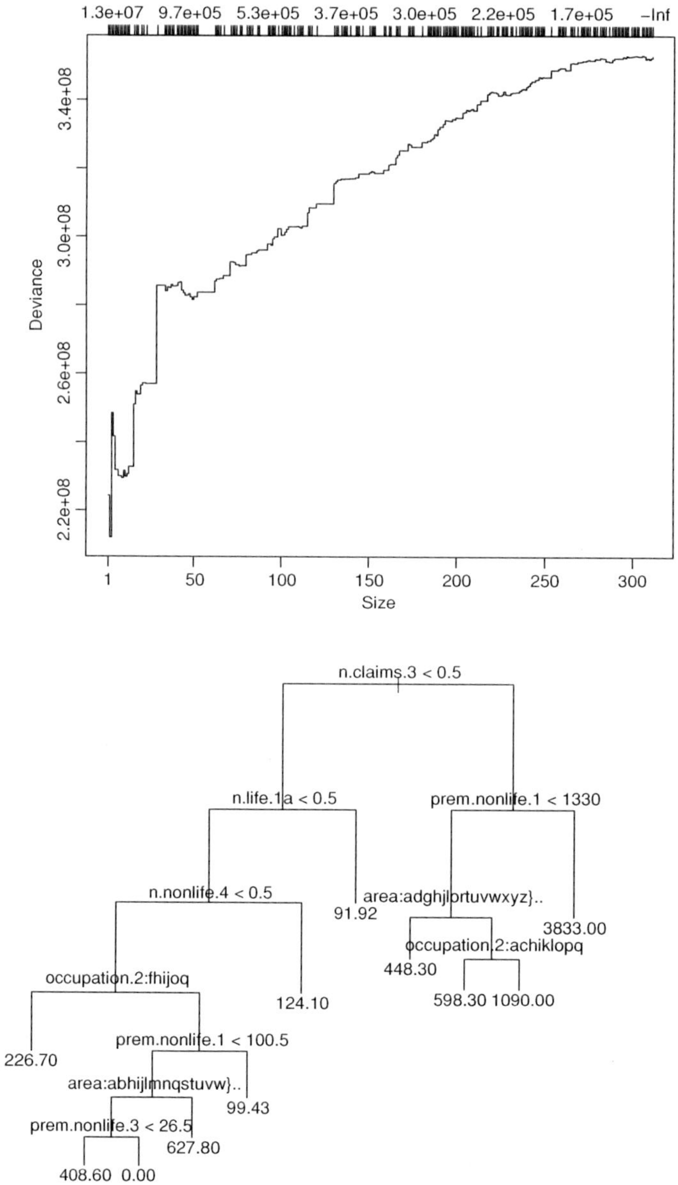

Figure 4.31 Insurance customers. Top: deviance of regression tree with pure premium on original scale; bottom: second-best regression tree fitted to data with average pure premium on original scale.

Comparisons and discussion

Analysis of table 4.8 provides ingredients for comparing the various models and allows for different choices according to marketing managers' aims.

In the original scale, the best prediction is that obtained by lasso estimates of the linear model, and GAM and MARS predictions give a squared error on the original

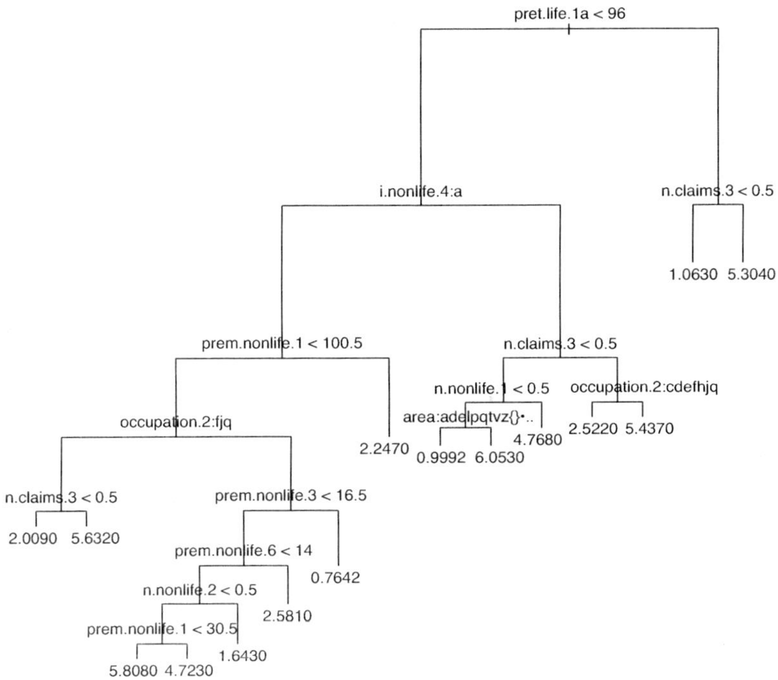

Figure 4.32 Insurance customers. Regression tree with pure premium on logarithmic scale.

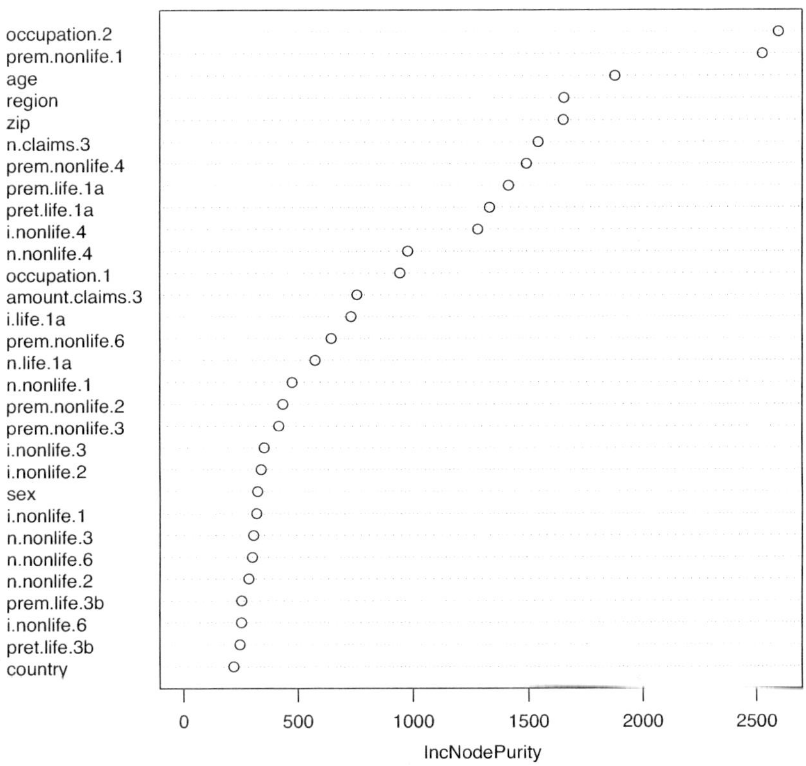

Figure 4.33 Insurance customers. Measure of importance of each variable of random forest for pure premium on same scale.

Table 4.8. Insurance Customers: Prediction Errors in Original and Logarithmic Scales for Models Fitted to Data

Model	Optimization Scale	Squared Error Original Scale	Squared Error Logarithmic Scale
Linear (all variables)	original	1,924,095,198	7,918
Linear (lasso)	original	74,270,206	7,967
Linear (stepwise)	original	1,870,139,302	7,744
GAM (stepwise)	original	73,171,214	7,622
Neural network	original	94,077,740	9,002
Neural network	logarithmic	192,971,718	4,483
Tree (2 leaves)	original	92,697,508	9,503
Tree (9 leaves)	original	94,077,740	7,795
Tree	logarithmic	94,425,522	4,036
Projection pursuit	original	85,829,668	7,454
Projection pursuit	logarithmic	3,607,884,000,000	4,258
MARS	original	74,288,572	6,732
MARS	logarithmic	620,993,900,000,000	4,201

scale that is only slightly larger. These models, in addition to good predictions, allow easier interpretation of the characteristics of various types of customers.

For example, from table 4.7 it is clear that customers paying large premiums for third-party liability insurance live in certain regions (in particular region 2 and, to a limited extent, region 8) and geographic areas (area 1542 shows premium increases of about €94, area 1082 about €73, and area 1191 about €17, when compared with other regions) and they work in specific sectors. In addition, there are some characteristics more related to customer behavior: subscription to one or more life insurance policies of type 2b and the total amount of the premium paid for type 1b life insurance policies both increase the level of the pure premium for third-party liability. Moreover, the amount of premiums paid for nonlife insurance of type 5 increases the pure premium, and customers who made claims in the past three years, particularly those who spent more in the last year, have larger premiums.

The model that best predicts the squared prediction error on the logarithmic scale is the tree. Looking at figure 4.32, we see that the regression tree for the logarithm of the average pure premium mainly contains splits related to nonlife products.

Summary

- We want a model to predict the average pure premium of customers for private car third-party liability insurance, using information on the number of policies and amount of premiums paid by customers for other lines of business of the same insurance company, in addition to some sociodemographic data.

Table 4.9. INSURANCE CUSTOMERS: PREDICTION ERRORS IN ORIGINAL AND LOGARITHMIC SCALES FOR SOME MODELS DESCRIBED IN CHAPTER 5 FITTED TO DATA

Model	Optimization Scale	Squared Error Original Scale	Squared Error Logarithmic Scale
SVM (radial kernel)	original	68,795,451	5,790
Random forest (40 variables for each split)	original	76,643,407	6,820
Random forest (40 variables for each split)	logarithm	88,553,872	3,599

- There are two objectives: (1) to predict the average pure premium with the greatest possible precision; (2) to predict low and medium levels of premiums more carefully, for the moment neglecting precision for high premiums.
- For the first target, our choice is a linear model fitted with a lasso procedure, which shows good prediction error on the original scale, is appropriate for problems of type (1), and has an easily interpretable output in terms of characteristics of customers profiled with respect to the average pure premium.
- For the second objective, we choose a regression tree, which presents the best prediction error on the logarithmic scale.

Back from the future

Some methods, which are modifications of classification methods discussed in the next chapter, that is, support vector machines (SVM), bagging, boosting, and random forests, are also fitted to this data, considering both original and logarithmic scales. Table 4.9 shows the prediction errors obtained on the validation set of some of these models (the better-fitting ones).

In the original scale, SVM shows better prediction than the lasso linear model, but as we see in section 5.8.2, it does not allow for easy interpretation of results.

In logarithmic scale the random forest (see section 5.9.3) over-perform the previously chosen tree. In this case, although easy interpretation of the tree is lost, a useful plot can give some information about the importance of the variables. Figure 4.33 shows an importance measure of each variable for random forests. As discussed in section 5.9.3, this measure is the average over all the trees in the forest of the measures of importance of a variable for each single tree, introduced at the end of section 4.8.3. This plot shows that when we consider logarithmic scale errors (that is, when we want to limit the effect of errors for very large premiums) sociodemographic characteristics such as age, occupation, and region of residence are still relevant variables, and important variables regarding customer behavior are more related to nonlife products: policies of types 1 and 4 are more important than life products, which were more important in predicting the original scale.

Exercises

4.1 Prove (4.3).

4.2 Prove (4.5).

4.3 Every nonparametric regression model involves a smoothing parameter. For example, consider parameter h of local regression. Why is it not estimated by a standard method such as maximum likelihood?

4.4 Given n points $(x_1, y_1), \ldots, (x_n, y_n)$, with all x_i distinct and increasing, show that the function that minimizes

$$\int_{x_1}^{x_n} (f''(t))^2 \, dt$$

under the constraint that $f(x_i) = y_i$ $(i = 1, \ldots, n)$ is a natural cubic spline with nodes at points $x_1, \ldots, x_n$. This function is called an *interpolation spline*.

4.5 Show that function (4.10) satisfies the following three conditions, which characterize cubic splines

1. f is a cubic function in each subinterval $[\xi_j, \xi_{j+1})$, for $j = 1, \ldots, K - 1$;
2. f has two continuous derivatives;
3. f has a third derivative that is a step function with jumps at points $\xi_1, \ldots, \xi_K$.

4.6 Prove (4.12).

4.7 Consider non-parametric model $Y_i = f(x_{i1}, \ldots, f(x_{ip}) + \varepsilon_i$, where $\mathbb{E}\{\varepsilon_i\} = 0$, $\mathrm{var}(\varepsilon_i) = \sigma^2$, for $i = 1, \ldots, n$, and assume that all error terms ε_i are independent of each other. Under the assumption of smoothness of f, consider linear smoother $\hat{Y} = SY$ evaluated at the observed covariate points, i.e., S is a $n \times n$ smoothing matrix and $Y = (Y_1, \ldots, Y_n)^\top$. Prove that:

$$\sum_{i=1}^{n} \mathrm{cov}(\hat{Y}_i, Y_i) = \mathrm{tr}(S)\sigma^2$$

4.8 In the tree growth algorithm, show that $D_j - D_j^* > 0$, apart from a degenerate case (which one?).

4.9 Consider the step of the tree growth algorithm when examining a generic variable x_r. How can the value of the point of subdivision of its range be determined efficiently?

4.10 In the case study on prediction of third-party liability insurance premiums in section 4.10.2, we directly predicted the average pure premium per customer. A different strategy would be to predict, separately with two different models, the number of policies per customer and the total amount of premiums paid by each customer, and then obtain the ratio between the two predictions. Follow this strategy and compare results with those presented in section 4.10.2.

4.11 Analyze the insurance data in two steps: first fit a model for the probability that the premium is not 0, and then, conditionally on this, fit a model for the amount of the premium when it assumes positive values. Compare the results with those presented in section 4.10.2 and those obtained in this exercise.

5
Methods of Classification

5.1 Prediction of Categorical Variables

One of the most frequent practical problems in statistics is allocating a unit to a *category* or a *class* among K possible alternatives, using observations about its variables. The examples that follow illustrate various situations of this type, focusing on a business context, an area where this kind of problem arises.

- A bank must decide on the degree of solvency of a customer who is asking for a loan. The problem is to assign the customer to the category of "solvent" or "insolvent" borrowers, which are two mutually exclusive and exhaustive categories—presuming, that is, the bank conventionally allocates its customers to one of the two categories. To make such a classification, various pieces of information, both personal and historical, about the customer are available to the bank. In the credit sector, this type of problem is associated with the terms *credit scoring* and *credit rating*.
- An insurance company must evaluate whether a motorist who takes out a third-party liability policy will have 0, 1, 2, or more accidents in the next year. The available information here is customer's personal information, vehicle characteristics, and data on insurance history. In the business sector, this type of problem is associated (indirectly) with *pricing*.
- An airline wants to predict which of its customers, among those in possession of a loyalty card, will make an intercontinental flight to a holiday destination within the next 12 months. To avoid contacting people who

are not interested, the airline sends a catalog of promotional deals to those customers with a high inclination to do so. In this case, customers are divided into two groups: those who will and those who will not make holiday flights, and the available information for predictions is recorded in the airline loyalty card database. In the business sector, this type of problem is associated with terms like *up-sell* and *cross-sell*.

- A car company wants to identify customers who, within the next six months, intend to purchase a new car of the type "luxury car," so that a presentation brochure of the latest model can be sent to them. It therefore needs to turn to a specialized company for lists of potential customers. These are created from extremely large collections of data from diverse sources, which all contribute toward the formation of individual economic behavior profiles . In the business sector, this type of problem is associated with the management of *prospects*.

The number K and the nature of the classes in each problem are well defined, in the sense that the allocation criterion must be able to decide the membership of each unit to one and only one class in a nonambiguous way. In all the previous examples, we had $K = 2$ (apart from the second example, where $K = 3$). The predominance of examples with $K = 2$ corresponds to a predominance in real situations.

The objective, therefore, is to construct a rule to arrange available observations on the variables relative to an individual and allocate that person to one of the classes. The following is based on the hypothesis that we use a certain set of n cases for which membership class is known, in addition to observed variables. In this case, we use this information to construct the classification rule.

The problem is similar to that considered in chapter 4, with the difference that response variable y is categorical with K levels, which represent membership class. We indicate by $y_1, \ldots, y_n$ the membership classes of elements in the sample, and by n_k the number of units belonging to the kth class, for $k = 0, \ldots, K - 1$. We denote by Y the parent random variable from which the y_i are sampled.

Therefore, apart from methods specifically developed in this context, many of the techniques presented here go back to the contents of chapter 4. However, there are some necessary adaptations, one of which concerns the usual discrepancy measure (2.10) between observed values and estimates, which is not adequate here. Another aspect is that we have $K(K - 1)$ possible forms of *misclassification error*, and the adequacy measures of various methods are constructed in this context.

5.2 An Introduction Based on a Marketing Problem

5.2.1 Prediction via Logistic Regression

We have already seen one of the methods used to overcome classification problems when $K = 2$, that is, logistic regression (section 2.4). This model predicts a categorical response variable with two levels, usually indicated by 0 and 1 so that an appropriate transformation of the probability of result 1 is expressed as a linear combination of the covariates. We can use this tool to face a first example of classification and examine further aspects of the problem in greater detail.

Consider data on the preferences of consumers of two brands of fruit juice in some American supermarkets, considering $n = 1070$ purchases that included fruit juice. The source and other information on the same data are reported in section B.6. To predict the customer choice between the two brands, CH and MM, other available variables are used: the prices of the two brands, `priceCH` and `price MM`; the discounts applied, `discountCH` and `discountMM`; a loyalty indicator for MM, `loyaltyMM`; an identifier of the week, and the store where the purchase was made. Indicator `loyaltyMM` reflects the fraction of preference given in previous purchases to brand MM; the similar indicator `loyaltyCH` is also available, so that their sum is constantly 1, and therefore only one of the two needs to be considered.

According to the procedure already introduced in section 3.5.1, we select a random portion of 75% of the total set, to be used for fitting and other operations. The remaining 25% is then used to evaluate the results.

Figure 5.1 shows the behavior of the variables, taken individually. The first six panels are box-plots of the continuous variables, stratified with respect to the response variable. The last panel shows a bar plot of the percentage of cases in which MM was preferred, stratified by store.

As a first classification tool of customers with respect to their purchase preferences, we fit a logistic regression model for probability π of choosing MM, using the covariates indicated above. The model takes the form

$$\begin{aligned}\text{logit}(\pi) = \beta_0 &+ \beta_1\,\texttt{week} + \beta_2\,\texttt{priceCH} + \beta_3\,\texttt{priceMM} \\ &+ \beta_4\,\texttt{discountCH} + \beta_5\,\texttt{discountMM} \\ &+ \beta_6\,\texttt{loyaltyMM} + \beta_7^\top I_{\texttt{store}}\end{aligned} \tag{5.1}$$

where the notation $I_{\texttt{factor}}$ represents a set of indicator variables equal in number to the levels of the qualitative variable `factor`, decreased by 1; in this case, the corresponding parameter β_j is a vector with a matching dimension. Here we adopt the so-called *corner-parameterization* for the qualitative variable, for which the first level is taken as reference and the parameters for other levels represent deviation from it. The parameter estimates and related quantities are listed in table 5.1.

We remove the term `week` from (5.1) in light of the p-values of table 5.1. After refitting the model, the parameter estimates and other relevant quantities are as listed in table 5.2. The appropriateness of the reduction is confirmed by the likelihood ratio test $D_2 - D_1$, which is virtually 0 on the scale of reference distribution χ_1^2, bearing in mind (2.33).

5.2.2 Misclassification Tables and Adequacy Measures

We now apply the fitted model to the portion of data not yet used to classify the remaining units and examine the prediction ability of the identified model. To allocate a new unit, we evaluate the probability of choosing MM according to the chosen model and assign the customer to one category or the other, according to

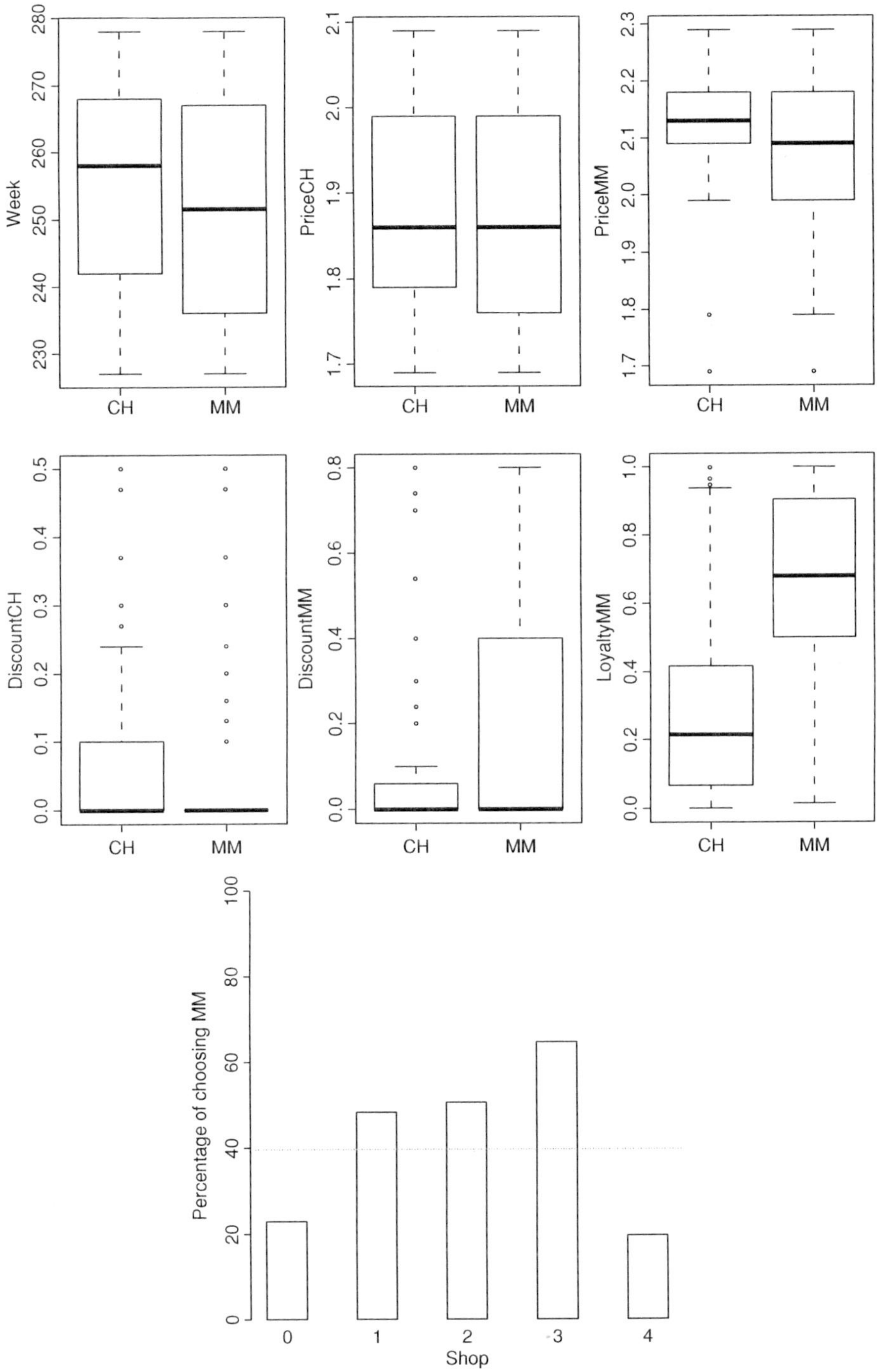

Figure 5.1 Fruit juice data: Preliminary graphical representations.

Table 5.1. FRUIT JUICE DATA: SUMMARY OF FITTED LOGISTIC REGRESSION MODEL (5.1)

	Estimate	**SE**	***t*-value**	***p*-value**
`(intercept)`	−3.816	2.059	−1.85	0.064
`week`	−0.002	0.013	−0.13	0.895
`priceCH`	4.435	2.114	2.10	0.036
`priceMM`	−3.706	1.006	−3.68	0.000
`discountCH`	−3.648	1.140	−3.20	0.001
`discountMM`	2.095	0.500	4.18	0.000
`loyaltyMM`	5.864	0.448	13.09	0.000
`store1`	0.551	0.315	1.75	0.080
`store2`	0.656	0.285	2.30	0.021
`store3`	0.574	0.368	1.56	0.119
`store4`	0.039	0.419	0.09	0.927

$D = 631.63$ with 791 d.f.

Table 5.2. FRUIT JUICE DATA: SUMMARY OF FITTED LOGISTIC REGRESSION MODEL WITHOUT TERM `week`

	Estimate	**SE**	***t*-value**	***p*-value**
`(intercept)`	2.056	2.015	1.02	0.308
`priceCH`	4.241	1.520	2.79	0.005
`priceMM`	−3.744	0.963	−3.89	0.000
`discountCH`	−3.695	1.084	−3.41	0.001
`discountMM`	2.082	0.491	4.24	0.000
`loyaltyMM`	5.868	0.447	13.12	0.000
`store1`	0.543	0.309	1.76	0.079
`store2`	0.651	0.283	2.30	0.021
`store3`	0.593	0.338	1.75	0.079
`store4`	0.055	0.401	0.14	0.892

$D = 631.64$ with 792 d.f.

whether this probability is greater or less than $\frac{1}{2}$. We thus construct a cross-table that counts the number of correctly or incorrectly predicted cases, for each of the two levels (table 5.3). This is called a *misclassification table* or a *confusion matrix.*

Because we want to compare various classification procedures, we look for a summarizing index of the quality of the result, and therefore introduce some *adequacy measures* of prediction. The first of these is simply constructed from the fraction of cases correctly classified or, conversely, those wrongly classified. In this case, we obtain

$$(150 + 76)/268 = 0.843 \qquad \text{and} \qquad (19 + 23)/268 = 0.157$$

Because these two quantities provide equivalent information, one of them suffices and, by convention, we call this error frequency *misclassification error.*

Table 5.3. FRUIT JUICE DATA: MISCLASSIFICATION TABLE OF MODEL WITHOUT TERM week IN TEST SET

Prediction	Actual response CH	MM	Total
CH	150	23	173
MM	19	76	95
Total	169	99	268

Table 5.4. CONFUSION MATRIX AND TABLE OF PROBABILITY ERRORS

Prediction	Actual response −	+	Total
−	n_{00}	n_{01}	$n_{0\cdot}$
+	n_{10}	n_{11}	$n_{1\cdot}$
Total	$n_{\cdot 0}$	$n_{\cdot 1}$	n

Prediction	Actual response −	+
−	$1-\alpha$	β
+	α	$1-\beta$
Total	1	1

However, this method of proceeding is somewhat simplistic: there are various reasons to be aware of the two separate types of error. If the "positive" event is the purchase of MM, MM customers classified as CH purchasers are called *false negatives,* taking a term originally used in medical context. In reverse, CH customers classified as MM purchasers are called *false positives.* The situation is shown in table 5.4, where the terms n_{ij} on the left correspond to the absolute frequencies of the four possible results; therefore, n_{01} is the count of the false negatives and n_{10} that of false positives.

There is a similarity between this set-up and that of hypothesis testing in the sense that false positives are analoguous to type I error and false negatives to type II error, as listed in the right side of table 5.4. According to the terminology of hypothesis testing,

$$\alpha = \mathbb{P}\{\text{false positive}\}, \qquad \beta = \mathbb{P}\{\text{false negative}\}.$$

These two probabilities are unknown and not fixed by us, but they can be estimated by

$$\hat{\alpha} = n_{10}/n_{\cdot 0}, \qquad \hat{\beta} = n_{01}/n_{\cdot 1}.$$

A first remark in this regard is that the *cost* of a classification error—that is, damage caused by an error—is not the same in the two situations. Depending on the problem, we can give more weight to one type of error or the other. For example, if we are interested in identifying customers who choose MM, we want to minimize the error in identifying them.

We therefore consider the error fractions for each observed subpopulation, thus distinguishing between false positives and false negatives. Recalling the previous comments, we use the selected model in a slightly different way: the logistic model itself provides a set of probabilities, and $\frac{1}{2}$ need not be used as the threshold value to allocate the units. We can grade the weight assigned to each category by moving the threshold value.

5.2.3 ROC Curve

Although from some points of view it is convenient to use very concise and comprehensive indicators of the performance of a classification procedure, such as the simple fractions just considered, it is useful to evaluate the predictive ability of various models more analytically.

One tool to evaluate the adequacy of a classification criterion is provided by the *ROC curve* (receiver operating characteristic). This was introduced during World War II in the context of communication theory, specifically radar signal detection, and was then extensively used elsewere, especially in quality control and medical statistics.

We return to table 5.3 to quantify the proportion of false positives with respect to the total of positive individuals, here 19/169, and the proportion of false negatives, here 23/99. However, these values are linked by the threshold value, which is $\frac{1}{2}$ for this table. We now move this threshold between 0 and 1, and calculate the corresponding proportion of false positives and negatives. We call these proportions:

- *specificity* for the proportion of predicted negatives with respect to the number of actual negatives, that is, $1 - \alpha$.
- *sensitivity* for the proportion of predicted positives with respect to the number of actual positives, that is, $1 - \beta$.

These quantities are naturally estimated by

$$\text{specificity} \approx \frac{n_{00}}{n_{00} + n_{10}}, \qquad \text{sensitivity} \approx \frac{n_{11}}{n_{01} + n_{11}}.$$

The ROC curve is made up of the coordinate points (1 − specificity, sensitivity) from these fractions for each of the possible threshold values.

For the fruit juice data, the results, are shown in the left panel of figure 5.2; the right panel shows a smooth version of the same points. This smoothing was done by regrouping the data into portions with one-tenth of the points each.

To interpret this curve, we bear in mind that the bisector of the origin corresponds to random classification of subjects. We are searching for a classification rule in which the ROC curve is as high as possible above the diagonal.

5.2.4 Lift Curve

Another frequently used tool to evaluate the performance of a classification procedure more analytically is the *lift function*, which provides a measure of the improvement gained by the model with respect to random classification, with uniform probability equal to the observed fraction in the test set.

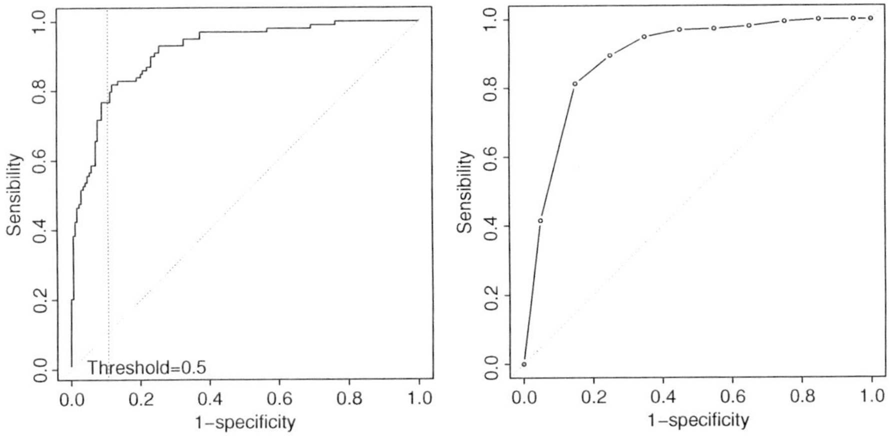

Figure 5.2 Fruit juice data: ROC curve for logistic model. Vertical dotted line in left panel: threshold $\frac{1}{2}$ in allocation rule.

One way of introducing this tool refers to the previous question on the threshold at which to discriminate customers. Let us imagine that company CH wants to acquire new customers, and a prediction error of MM customers is very worrying; we want to highlight the predictive ability of this set. We return to the response provided by the model in terms of the values of estimated probability. These fall in the interval $(0, 1)$, and we simplify it by dichotomizing it with respect to a threshold. One way of scaling such a threshold is to order units according to the probability assigned by the model and then verify whether the parts of the units with a greater predicted probability are those that do correspond to greater frequency of events—in this example, by choosing MM.

Figure 5.3 shows the results of such an operation, with two variants. In both panels, the left-most points of the line correspond to sets of customers for whom the estimated probability is higher, and the y-axis represents the proportion of observed purchasers of MM of those customers, divided by the average proportion calculated on all the data. The left panel shows the calculation made for every possible fraction of subjects, ordered according to estimated probabilities; the right panel shows a smooth form of the same curve, in which the calculated points refer to fractions of 10%, 20%, ..., 100% of the data. The smooth variant is more commonly used, both to obtain a more regular trend and for computational simplicity.

Both panels of figure 5.3 also show a vertical dotted line, which corresponds to the classification of subjects with the probability of the indicated value as a threshold. For every fixed value of this threshold, a misclassification table is identified, of the type presented in table 5.4, from which, we can extract the y-axis of the lift curve for event $+$, represented by

$$\frac{n_{11}/n_{1\cdot}}{n_{\cdot 1}/n}$$

In table 5.3 constructed with threshold $\frac{1}{2}$, the y-axis of the lift curve for "purchase of MM" is $(76/95)/(99/268) = 2.17$, which is the observed value on figure 5.3

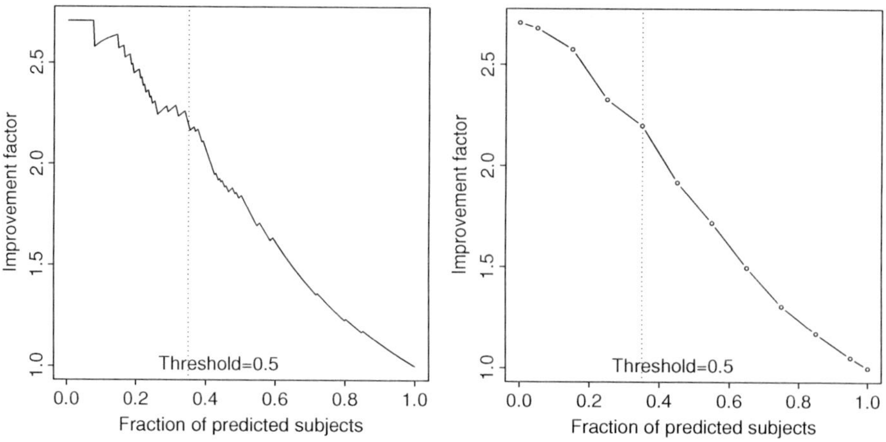

Figure 5.3 Fruit juice data. Lift curve for logistic model. Left: curve calculated for every fraction of subjects; right: curve calculated for grouped data.

where the vertical dotted line crosses the lift curve. In the right panel, the value of the y-axis is subject to approximation because we constructed the lift curve by reorganizing the data into 10 groups.

To better appreciate the value of the information in this type of graph, and also the reason for the term *lift*, we refer to an example of a different type. Imagine that a company wants to promote a product to already known customers who are contacted individually—for example, by mail. For cost reasons, the company decides to send a limited number N of letters, and therefore the problem arises of which customers to send them to.

The trivial option, without taking advantage of any information about customers, is to send letters to N customers chosen at random. Instead, let us use a logistic regression model for the probability of responding positively to the promotion, constructed according to available data, following analogous promotional actions in the past. Clearly, if we take advantage of the indications of the model, we send letters to those N subjects who have a higher probability of responding positively.

The lift curve of this model allows us to quantify the *expected improvement* of the logistic model with respect to random choice. At x, that is, the proportion between N and the size of the customer base, point y of the lift curve represents the ratio between the probability of success in reaching the customers selected by the model and a randomly chosen set.

A further observation regarding the asymmetry of the behavior of lift with respect to the choice of "favorable" or "unfavorable" events: different graphs are obtained if we invert the choice of the event in question.

5.3 Extension to Several Categories

5.3.1 Multivariate Logit and Multinomial Regression

The case $K > 2$ may be treated by extending the previous method as follows. If we call the K classes $0, 1, \ldots, K-1$, and denote by $\pi_k(x)$ the probability that

$Y = k$ by the fixed value of x, with $\sum_{k=0}^{K-1} \pi_k(x) = 1$, we assume that

$$\log \frac{\pi_k(x)}{\pi_0(x)} = \eta_k(x) \tag{5.2}$$

holds, where $\eta_k(x)$ is a linear combination of the covariates, of type $\beta_0 + x^\top \beta$, where the components of vector β vary with k, for $k = 1, \ldots, K-1$. A simple algebraic manipulation leads us to write

$$\frac{1}{\pi_0(x)} \sum_{k=1}^{K-1} \pi_k(x) = \sum_{k=1}^{K-1} e^{\eta_k(x)}$$

and, therefore, adding 1 to both sides,

$$\begin{aligned} \pi_k(x) &= \frac{e^{\eta_k(x)}}{1 + \sum_r e^{\eta_r(x)}}, \qquad \text{for } k = 1, \ldots, K-1, \\ \pi_0(x) &= \frac{1}{1 + \sum_r e^{\eta_r(x)}}. \end{aligned} \tag{5.3}$$

These relations extend (2.41), and the derived model is called a *multivariate logistic regression model*. In principle, each of functions $\eta_k(x)$ may use different covariates, but conceptually the substance does not change.

The $p\,(K-1)$ parameters of this model may be estimated by fitting $K-1$ logistic regression models. Each of these is applied to compare classes 0 and k, conditional on the fact that the subject belongs to one of these two classes. Because

$$\log \frac{\mathbb{P}\{Y = k | Y = 0 \cup Y = k\}}{\mathbb{P}\{Y = 0 | Y = 0 \cup Y = k\}} = \log \frac{\pi_k(x)/(\pi_0(x) + \pi_k(x))}{\pi_0(x)/(\pi_0(x) + \pi_k(x))} = \eta_k(x)$$

it is immediately verified that the parameters estimated in this way are those of interest for the multivariate model.

A different estimation strategy is based on the assumption that $\pi_0(x)$, $\pi_1(x), \ldots, \pi_{K-1}(x)$ in (5.3) are the parameters of *multinomial* distribution, which specifies the probability of each way of allocating n observations in K categories. The estimates are obtained by numerically maximizing the log-likelihood function, which is proportional to

$$\sum_{k=0}^{K-1} y_k \log \pi_k(x), \tag{5.4}$$

where $y_0, \ldots, y_{K-1}$ represent the number of observed events for each category. In this case, the model is called *multinomial* (or *polytomous*) *logit*.

Table 5.5. BANK DATA: SUMMARY OF MULTINOMIAL LOGIT MODEL, WITH LINEAR (TOP) AND QUADRATIC (BOTTOM) EFFECTS OF AGE. STANDARD ERROR VALUES IN BRACKETS

Model with linear age effect

Logit	`(Intercept)`	`Age`	`Car possession`
$\log(\pi_1/\pi_0)$	−0.973 (0.744)	0.0358 (0.0153)	−0.916 (0.483)
$\log(\pi_2/\pi_0)$	0.373 (0.617)	0.0099 (0.0124)	0.047 (0.439)
$\log(\pi_3/\pi_0)$	−0.737 (0.597)	0.0510 (0.0122)	−0.795 (0.406)

$D = 1164.493$ with 9 d.f.

Model with quadratic age effect

Logit	`(Intercept)`	`Age`	`Age`^{2}	`Car possession`
$\log(\pi_1/\pi_0)$	3.07 (0.0009)	−0.171 (0.0198)	0.0024 (0.0004)	−0.8939 (0.043)
$\log(\pi_2/\pi_0)$	2.97 (0.0031)	−0.125 (0.0170)	0.0016 (0.0003)	0.0523 (0.110)
$\log(\pi_3/\pi_0)$	3.98 (0.0030)	−0.177 (0.0169)	0.0027 (0.0003)	−0.7620 (0.168)

$D = 1156.30$ with 12 d.f.

Note that in (5.2) the choice of 0 as reference class, called baseline category, is arbitrary but irrelevant in that we could use any other class for this aim, and the probabilities resulting from (5.3) would remain unchanged.

For a numerical illustration, we analyze the Brazilian bank data (described in section B.3 and already used in section 2.3.3), examining the satisfaction of the bank's customers as a categorical variable with four categories, modeled as a function of customer age and an indicator of car ownership. Table 5.5 lists the estimate operations of a multinomial logit model, with satisfaction level 4 as baseline category.

For a given age, the estimated odds that customers not possessing a car have a satisfaction level of 1 instead of 4 are $\exp(-0.92) = 0.40$ times the estimated odds for customers possessing a car; the Wald 90% confidence interval is $\exp(-0.92 \pm 1.64 \times 0.483) = (0.18, 0.88)$. For example, the age effect indicates that the estimated odds that satisfaction level is 1 instead of 4 are relatively higher for older customers. The left part of figure 5.4 plots the estimated probabilities that satisfaction level is 1, 2, 3, or 4 as a function of age, for customers owning a car.

We also consider a model in which a quadratic component for age is added. The lower part of table 5.5 lists these estimates and the right part of figure 5.4 plots the estimated probabilities for the four satisfaction levels as a function of age. Figure 5.5 plots the distributions of the predicted probabilities that the response variable falls in each category with the model that includes the quadratic component of age.

5.3.2 Ordinal Categorical Variables and Cumulative Logit Models

Sometimes, as in the case of customer satisfaction in the Brazilian bank example, the categorical response variable is ordinal but the multinomial logit model does

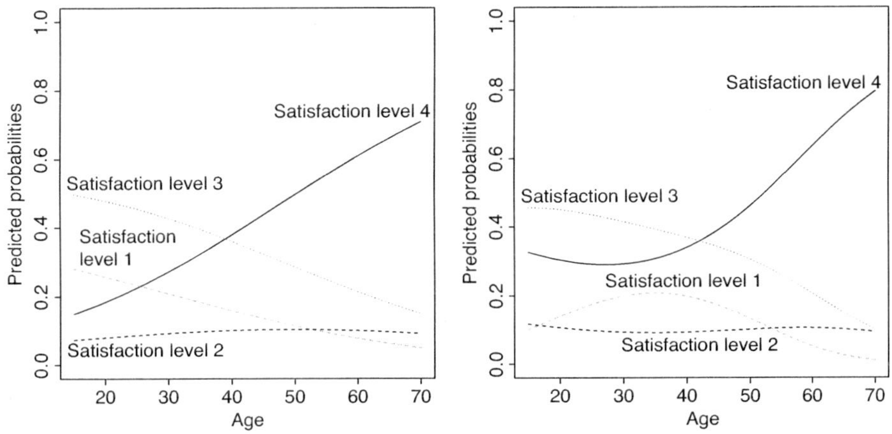

Figure 5.4 Bank data: Estimated probabilities with a multilnomial logit model. Left: satisfaction levels with linear age effect; right: with quadratic effect.

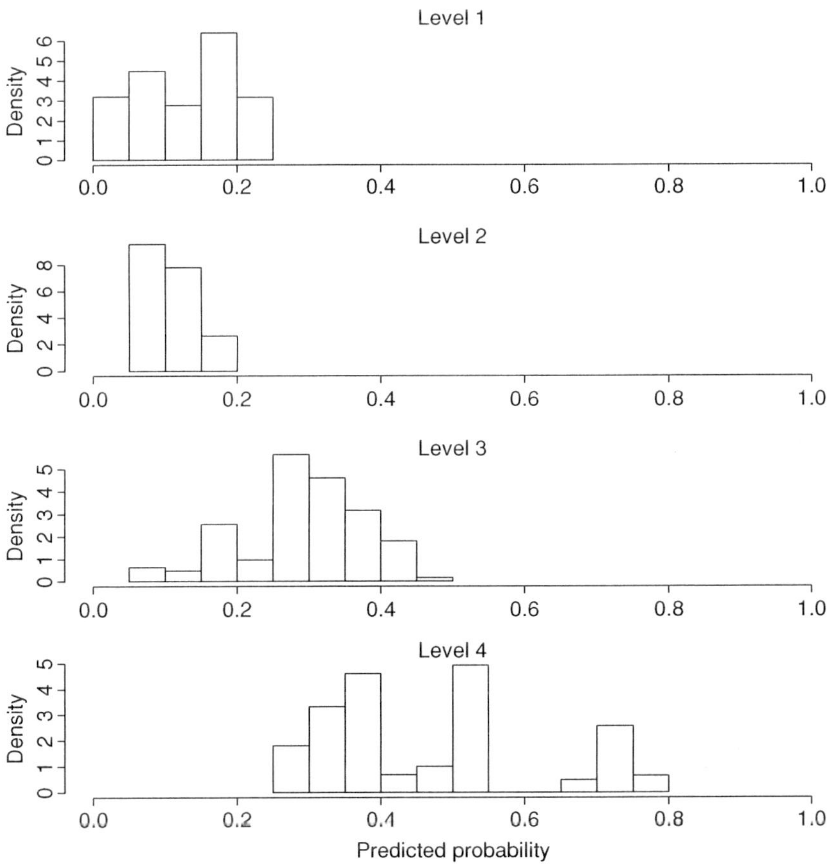

Figure 5.5 Bank data: Distribution of predicted probabilities for each satisfaction level with a multinomial logit model with quadratic age effect.

not take this information into account. Models for ordinal responses may be introduced for simpler interpretation and potentially greater precision.

Considering an ordered response variable, for each category we define as *cumulative probabilities* the probabilities that response variable Y belongs to a class not higher than the nominated category

$$\mathbb{P}\{Y \leq k\} = \pi_0 + \ldots + \pi_k \qquad k = 0, \ldots, K-1.$$

A model for cumulative logits

$$\log \frac{\mathbb{P}\{Y \leq k\}}{1 - \mathbb{P}\{Y \leq k\}} = \log \frac{\pi_0 + \ldots + \pi_k}{\pi_{k+1} + \ldots + \pi_{K-1}},$$

automatically incorporates category order. The simplest model of this type is the *proportional odds model*, in which an identical effect of the explanatory variable is assumed for all $K-1$ cumulative probabilities

$$\log \frac{\mathbb{P}\{Y \leq k\}}{1 - \mathbb{P}\{Y \leq k\}} = \beta_{0k} - \beta_1 x_1 - \ldots - \beta_p x_p = \eta(x, k),$$
$$k = 1, \ldots, K-1, \tag{5.5}$$

where $\eta(x, k) = \beta_{0k} - \beta_1 x_1 - \ldots - \beta_p x_p$. The choice of the negative sign preceding β_j is conventional and is adopted for easier interpretation of the parameters, as will be made clear shortly. Here, each cumulative logit has its own intercept β_{0k}, but effects β_j of the jth covariate, for $j = 1, \ldots, p$, are the same for all categories.

Model (5.5) satisfies the property

$$\begin{aligned} &\text{logit}(\mathbb{P}\{Y \leq k|x'\}) - \text{logit}(\mathbb{P}\{Y \leq k|x''\}) \\ &= \log \frac{\mathbb{P}\{Y \leq k|x'\}/(1 - \mathbb{P}\{Y \leq k|x'\})}{\mathbb{P}\{Y \leq k|x''\}/(1 - \mathbb{P}\{Y \leq k|x''\})} \\ &= \beta_1(x_1'' - x_1') + \ldots + \beta_p(x_p'' - x_p'), \end{aligned}$$

where $x' = (x_1', \ldots, x_p')$ and $x'' = (x_1'', \ldots, x_p'')$ are two points of covariate space and, in this case, a notation of the type $\mathbb{P}\{Y \leq k|x'\})$ makes explicit the dependence on x that was implicit previously. This means that the odds of making response $Y \leq k$ when the covariates assume value x' are $\exp\{\beta_1(x_1'' - x_1') + \ldots + \beta_p(x_p'' - x_p')\}$ times the odds at $x = x''$. The log cumulative odds ratio is therefore proportional to the distance between the two points x' and x''. This is why the model is called *proportional odds model*.

The equivalent model expression for cumulative probabilities is

$$\mathbb{P}\{Y \leq k|x\} = \frac{\exp\{\eta(x, k)\}}{1 + \exp\{\eta(x, k)\}}, \qquad k = 1, \ldots, K-1$$

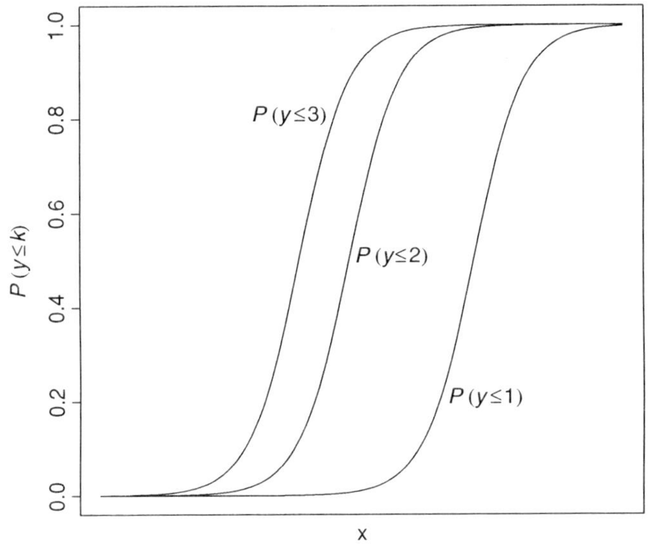

Figure 5.6 Cumulative probabilities in a proportional odds model. Each curve corresponds to a category.

and the single category probabilities are

$$\mathbb{P}\{Y = k|x\} = \frac{\exp\{\eta(x, k)\}}{1 + \exp\{\eta(x, k)\}} - \frac{\exp\{\eta(x, k-1)\}}{1 + \exp\{\eta(x, k-1)\}}.$$

Figure 5.6 shows an example of the trend of $\mathbb{P}\{Y \leq k\}$ for a proportional odds model versus one covariate, all other explanatory variables being given.

By hypothesizing multinomial distribution for independent observations, estimates are obtained by maximizing the log-likelihood

$$\log L(\beta_{01}, \ldots, \beta_{0K-1}, \beta_1, \ldots, \beta_p)$$
$$= \sum_{i=1}^{n} \sum_{k=1}^{K-1} y_{ik} \log \left(\frac{\exp\{\eta(x_i, k)\}}{1 + \exp\{\eta(x_i, k)\}} - \frac{\exp\{\eta(x_i, k-1)\}}{1 + \exp\{\eta(x_i, k-1)\}} \right).$$

Note that models based on cumulative probabilities can use a link function other than a logit. Models belonging to this broad family are usually called "cumulative link models." They are characterized by an interesting interpretation, which considers the response categorical variable as a discretization of an underlying continuous variable. If Y^* denotes such a latent variable, we consider the model for Y^* as a function of x

$$Y^* = \beta_1 x_1 + \ldots + \beta_p x_p + \varepsilon, \tag{5.6}$$

where we assume that ε has a distribution function $G(\cdot)$, with $\mathbb{E}\{\varepsilon\} = 0$. If $-\infty = \beta_{00} < \beta_{01} < \cdots < \beta_{0K-1} < \beta_{0K} = \infty$ are *cut-points* or *thresholds* on a continuous scale, we assume that

$$Y = k \qquad \text{if} \quad \beta_{0k} < Y^* \leq \beta_{0k+1}, \quad k = 0, \ldots, K-1.$$

This means that

$$\mathbb{P}\{Y \leq k|x\} = \mathbb{P}\left\{Y^* \leq \beta_{0k}|x\right\} = G(\beta_{0k} - \beta_1 x_1 - \ldots - \beta_p x_p)$$

and, equivalently,

$$G^{-1}\{\mathbb{P}\{Y \leq k|x\}\} = \beta_{0k} - \beta_1 x_1 - \ldots - \beta_p x_p.$$

It is now clear why we adopted the negative sign for the β_j in (5.5). Negative signs in (5.5) correspond to positive signs in (5.6), so that the parameters have the usual directional interpretation—that is, if β_j is positive, then Y is more likely to assume high values as x_j increases.

If $G(\varepsilon) = e^{\varepsilon}/(1 + e^{\varepsilon}) = \ell(\varepsilon)$—that is, if G is the standard logistic distribution—G^{-1} is the logit link function and the model is a proportional odds model (5.5). Other latent distributions are implied by different link functions: for example, if ε is Gaussian, G^{-1} is the *probit* link function, the inverse of the normal distribution function.

To illustrate the proportional odds model, we analyze the Brazilian bank data. We consider the simple model with only age and car possession as predictors. Table 5.6 lists the estimates for the proportional odds model. For each parameter, the 95% level Wald confidence interval is adopted. Analyzing these confidence intervals, we observe that the interval for car possession includes 0, so we are led to test the hypothesis that this parameter is null. The likelihood ratio test statistic is $2(\log L_1 - \log L_0)$, where L_0 is the maximized log-likelihood function under the null hypothesis constraint that $\beta_{\texttt{car}} = 0$, and L_1 is the maximized log-likelihood function without that constraint. The observed test statistic, $1185.64 - 1182.31 = 3.33$ on 1 degree of freedom, leads to an observed significance level of 0.068, suggesting that we could eliminate the variable car possession from the model.

The lower part of table 5.6 lists the proportional odds model with only age as a covariate. The top part of figure 5.7 plots the estimated probabilities for the 4 satisfaction levels as functions of age, and the bottom part shows the distributions of predicted probabilities for each category.

Bibliographical notes

Fahrmeir & Tutz (2001) deal with GLM in the multidimensional case, including extension of logistic regression to the multivariate case, to treat categorical variables with more than two levels.

Multinomial classification models and cumulative odds models are also discussed in many works on categorical data analysis, for example, Agresti (2002) and, specifically for ordinal categorical data, Agresti (2010). For a discussion of GLM, including a presentation of proportional odds models, see McCullagh & Nelder (1989).

Table 5.6. BANK DATA: PROPORTIONAL ODDS VERSION OF CUMULATIVE LOGIT MODEL WITH LINEAR EFFECT OF AGE. `Intercept 1|2`, `Intercept 2|3`, `Intercept 3|4`: PARAMETERS $\beta_{01}, \beta_{02}, \beta_{03}$

	Estimate	SE	*t*-value	Wald 95% conf. limits	
Model with age and car possession					
`(Intercept 1|2)`	−0.5803	0.3569	−1.6256	−1.2798	0.1193
`(Intercept 2|3)`	0.1778	0.3499	0.5081	−0.5080	0.8636
`(Intercept 3|4)`	1.5289	0.3560	4.2951	0.8312	2.2265
`age`	0.0386	0.0068	5.6450	0.0252	0.0519
`car possession`	−0.4080	0.2259	−1.8060	−0.8508	0.0348

$D = 1182.31$ with 5 d.f.

Model with age only					
`(Intercept 1|2)`	−0.2901	0.3187	−0.9105	−0.9147	0.3344
`(Intercept 2|3)`	0.4678	0.3110	1.5041	−0.1418	1.0774
`(Intercept 3|4)`	1.8137	0.3198	5.6719	1.1870	2.4405
`age`	0.0374	0.0068	5.5240	0.0242	0.0573

$D = 1185.64$ with 4 d.f.

5.4 Classification via Linear Regression

We tackled our first problem of classification with a fairly simple and familiar method: logistic regression. There are more sophisticated methods, but we now move on to another, which is even simpler and more familiar: linear regression. After all, simple methods often give very good results.

5.4.1 Case with Two Categories

We start by considering the case with $K = 2$ classes, 0 and 1. We introduce a linear regression model in which response variable y is formed exactly of labels 0 and 1 of the two classes, and value $\hat{y} = \frac{1}{2}$ is the discriminatory threshold for predicting the two categories.

To illustrate the method, consider the artificial data of the two parts of figure 5.8. Here, we have two continuous covariates z_1 and z_2, and membership of the points to the two groups is distinguished by the symbol used. There are 120 points in one category and 80 in the other.

The simplest form of linear regression we can consider is

$$y = \beta_0 + \beta_1 z_1 + \beta_2 z_2 + \varepsilon. \tag{5.7}$$

It is important to note that the nature of ε as implied here is truly original, in the sense that it must be a random variable, so that its value added to the deterministic part gives 0 or 1. However, the really crucial assumption for the least squares criterion to provide reasonable results is that $\mathbb{E}\{\varepsilon\} = 0$, but in fact

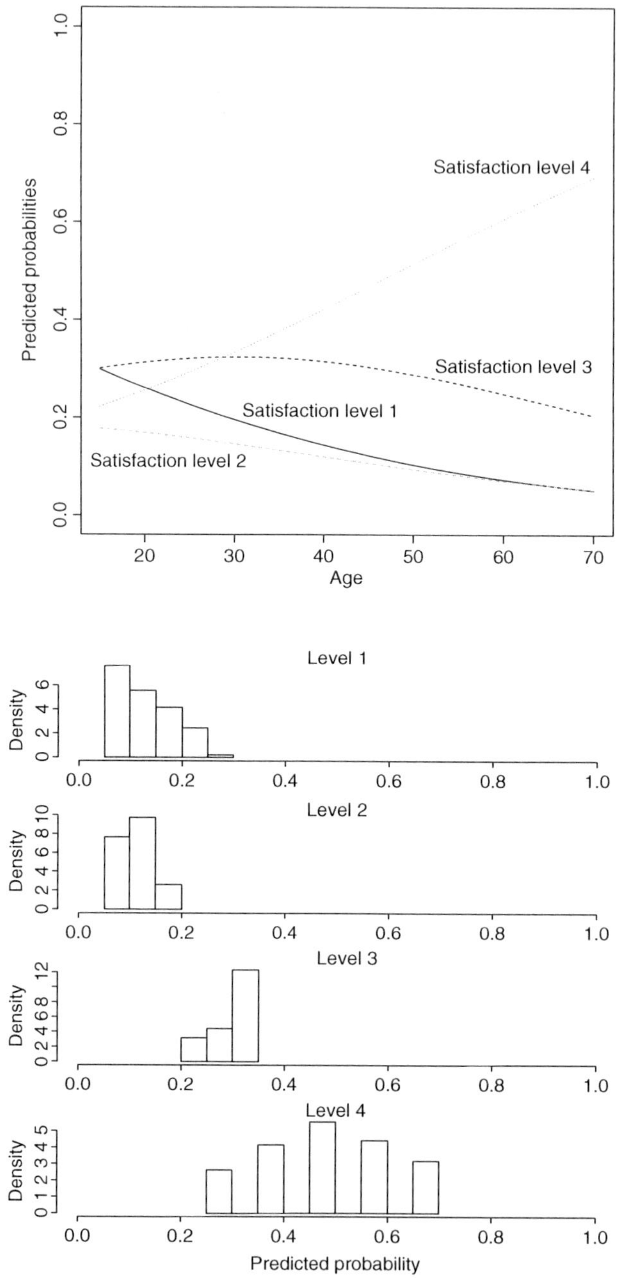

Figure 5.7 Bank data: Proportional odds version of cumulative logit model. Top: estimated probabilities of satisfaction levels with linear effect of age; bottom: distributions of predicted probabilities for each satisfaction level.

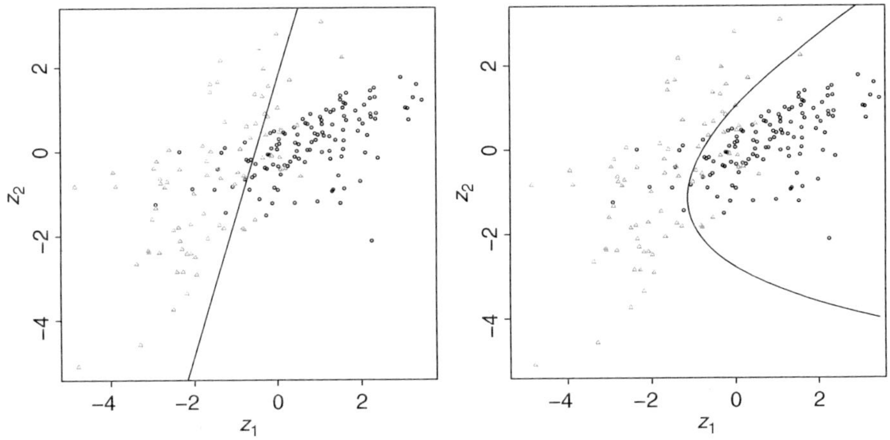

Figure 5.8 Simulated data with two groups. Left: classification with simple regression; right: classification with quadratic regression.

this requirement is automatically satisfied when the model includes an intercept, because any nonzero value can be included in β_0.

After least squares estimation, the $\mathbb{R}^2$ plane is divided into two parts by the line

$$\hat{y} = \hat{\beta}_0 + \hat{\beta}_1 z_1 + \hat{\beta}_2 z_2 = \tfrac{1}{2} \tag{5.8}$$

where self-explanatory notation is used. This is the line plotted in the left part of figure 5.8.

Elaborating on this formulation, we can extend the process by inserting nonlinear functions of z_1 and z_2 into the linear predictor. The simplest choice is that of polynomial functions—for example, the quadratic form

$$\beta_0 + \beta_1 z_1 + \beta_2 z_2 + \beta_3 z_1^2 + \beta_4 z_1 z_2 + \beta_5 z_2^2.$$

After estimation of the parameters, equating the resulting function to $\frac{1}{2}$ leads to subdivision of $\mathbb{R}^2$, indicated by the separation curve in the right part of figure 5.8.

We now apply this procedure to the fruit juice data, using the variables already shown in figure 5.1. Obviously, with many variables in play, it is not possible to produce a plot like that of figure 5.8. The misclassification table in the test sample is identical to that in table 5.3, and so are the error percentages. The lift and ROC curves are practically indistinguishable from those of the logistic model and are therefore not shown. However, figure 5.9 shows the scatterplot of logit$(\hat{\pi})$ of the fitted logistic model, with respect to the predicted values according to the linear model. This reveals astonishing agreement between the two classification rules, at least for threshold value $\frac{1}{2}$, which corresponds to 0 on the logit scale. The essential equivalence of the two methods is quite common, although not an absolute rule.

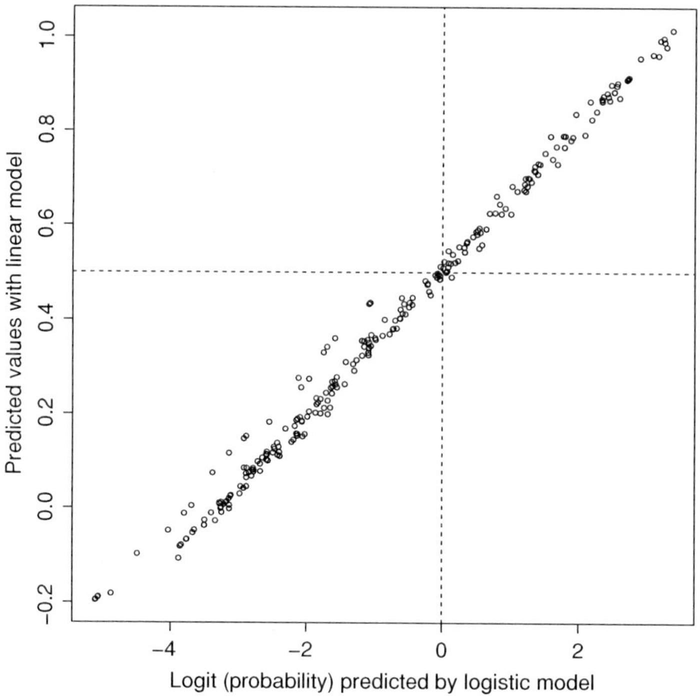

Figure 5.9 Fruit juice data: Scatterplot of logit$(\hat{\pi})$ predicted by fitted logistic model and values predicted by linear model.

5.4.2 Case with Several Categories

The case of $K > 2$ can be tackled with an extension of the previous process combined with the multivariate linear model described in section 2.1.3. We construct the $n \times K$ dimension matrix Y made by the indicator variables of the levels of y. The columns of Y are linearly dependent, in the sense that the sums of each row are identically equal to 1, but in this case it is convenient not to eliminate a column. We can therefore arrange multivariate multiple linear regression of the type (2.20),

$$Y = XB + E,$$

where X represents design matrix $n \times p$ and B the $p \times K$ of the parameters. For the columns of error matrix E, the comments made for ε in (5.7) hold.

Once matrix B has been estimated by (2.21), we can allocate a new point x_0 ($x_0 \in \mathbb{R}^p$) to one of the classes, calculating

$$\hat{y}_0 = \hat{B}^\top x_0$$

and assigning x_0 to the class for which component $\hat{y}_0$ is greater ($\hat{y}_0 \in \mathbb{R}^K$).

For numerical illustration, we can refer to figure 5.10, showing three groups of simulated data, of which two coincide with those in figure 5.8 and the new set has 100 points. The two panels of figure 5.10 correspond to those of figure 5.8 in the

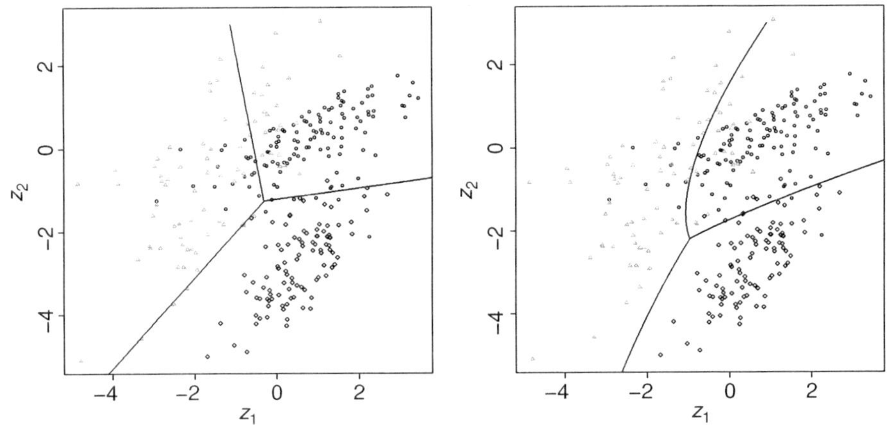

Figure 5.10 Simulated data with three groups. Left: classification with linear regression; right: classification with quadratic regression.

sense that a regression plan is used for the former and a second-degree polynomial for the latter.

5.4.3 Discussion

Using linear models for classification purposes is somewhat unnatural. The domain of y is $\{0, 1\}$, which does not fit the logical set-up of least squares because a linear regression function does not remain constrained within this set. One consequence of this was already mentioned when the nature of error term ε of (5.7) was discussed. In turn, this nature causes difficulty in using inferential methods: the usual hypothesis of homoscedasticity is not guaranteed in this case. Therefore, the usual standard errors and other inferential procedures are not fully sustained by a fixed theory, although some numerical tests give comforting indications in the sense that approximate standard errors are essentially valid.

It is appropriate here to remark on the interpretation of the model parameters. Because labels 0 and 1 are conventional, parameters vary if we choose other labels. As for the nonconstant terms of the linear predictor, the estimate of the parameter and the corresponding standard error vary in proportion, so the overall interpretation is not modified. However, the intercept has a purely arbitrary value: it changes simply if other values are used for the classes, for example, -1 and 1, the associated standard errors change appreciably, and so do the observed significance levels of the parameters. However, the constant term in the linear model is needed to guarantee that $\mathbb{E}\{\varepsilon\} = 0$ for every label choice, and it must therefore be maintained.

A particular problem of this approach comes from the possible "masking" of a class, in the sense that we can construct a classification rule for which a new individual will never be allocated to a certain class: this class is masked by the others. For a more detailed illustration of the problem, see Hastie et al. (2009, p. 105). The remedy is to consider polynomial expressions of the explanatory variables in the linear predictor up to order $K - 1$, which involves $O(p^{K-1})$ terms.

To conclude, we list the advantages and disadvantages of this approach.

Advantages

- Familiarity of the method: linear regression is one of the most widespread and familiar statistical tools.
- Computational simplicity: the computational side is noniterative, with minimal computational complexity. The recursive updating formulas of algorithm 2.2 can be used, thus allowing real-time applications.
- Effectiveness: in spite of its simplicity, the method produces satisfactory results, competitive with more sophisticated ones.

Disadvantages

- Improper use of the linear model: the domain of y is in no way similar to the set of values of a linear function.
- Masking problems: if we are not careful, we risk masking a class.
- Difficulties with inferential aspects: there is no completely satisfactory theoretical basis to support inferential processes.

Apart from these aspects, there are the standard considerations about using a parametric method, in both positive and negative senses.

5.5 Discriminant Analysis

5.5.1 General Remarks

Linear regression and logistic regression are not really tools specifically designed for classification. "Proper" treatment of the problem follows the procedure shown next, in which we refer to a p-dimensional random variable X, assumed for the moment to be continuous, and a random categorical variable Y, which represents the class to which a subject belongs.

The total population is made up of K subpopulations (classes), having probability density functions $p_0(x), \ldots, p_{K-1}(x)$ for the conditional distribution of X, and weights $\pi_0, \ldots, \pi_{K-1}$ with respect to the total population ($\sum_k \pi_k = 1$). Therefore, marginal density in the total population is

$$p(x) = \sum_{k=0}^{K-1} \pi_k \, p_k(x), \tag{5.9}$$

For the moment, we argue as if the various ingredients of $p(x)$ were known. A priori, the probability that a still unclassified subject belongs to the kth subpopulation is given by π_k. For this subject, if the observed value of X is x_0, then by Bayes theorem the a posteriori probability that this subject belongs to group k is given by

$$\mathbb{P}\{Y = k|X = x_0\} = \frac{\pi_k p_k(x_0)}{p(x_0)}$$

or, equivalently, comparison of probability between class k and class m takes place according to

$$\log \frac{\mathbb{P}\{Y = k|X = x_0\}}{\mathbb{P}\{Y = m|X = x_0\}} = \log \frac{\pi_k}{\pi_m} + \log \frac{p_k(x_0)}{p_m(x_0)}.$$

We therefore compare the various classes through the *discriminant function*

$$d_k(x_0) = \log \pi_k + \log p_k(x_0)$$

linked to the posterior probability of the classes. The value of k that maximizes the discriminant function selects the group to which we assign the new subject.

This constitutes the framework of *discriminant analysis*. However, to make the process operative, we must know and therefore estimate from the data the ingredients of (5.9). Regarding π_k, it is natural to estimate it as $\hat{\pi}_k = n_k/n$, unless we have further information. However, there are various approaches we can take for $p_k(x)$: parametric or nonparametric; the former includes various options relating to the family of density functions to be considered, and the latter various alternatives among estimation methods.

From now on, we develop the more classical procedure, that of Fisher (1936), which are placed within the parametric environment. We do not deal with the nonparametric approach, as it has not yet found widespread application, both because it does not lend itself easily to combining quantitative and qualitative variables and because it falls quite rapidly into the curse of dimensionality, and therefore is not suitable for dealing with the problems that interest us here.

5.5.2 Linear Discriminant Analysis

For discriminant analysis, the simplest parametric hypothesis is that in which each density $p_k(x)$ is multivariate normal with parameters dependent on k, say, $N_p(\mu_k, \Sigma_k)$, which results in

$$p_k(x) = \frac{1}{(2\pi)^{p/2} \det(\Sigma_k)^{1/2}} \exp\left\{-\tfrac{1}{2}(x - \mu_k)^\top \Sigma_k^{-1}(x - \mu_k)\right\} \quad (5.10)$$

for $k = 0, \ldots, K - 1$. For a brief recap of multivariate normal distribution, see appendix A.2.3.

In the simplified case, in which all the variance matrices are equal to the same Σ, the discriminant function takes the form

$$d_k(x) = \log \pi_k - \tfrac{1}{2}\mu_k^\top \Sigma^{-1} \mu_k + x^\top \Sigma^{-1} \mu_k$$

which is a linear function of x, leading to its name, *linear discriminant analysis* (LDA).

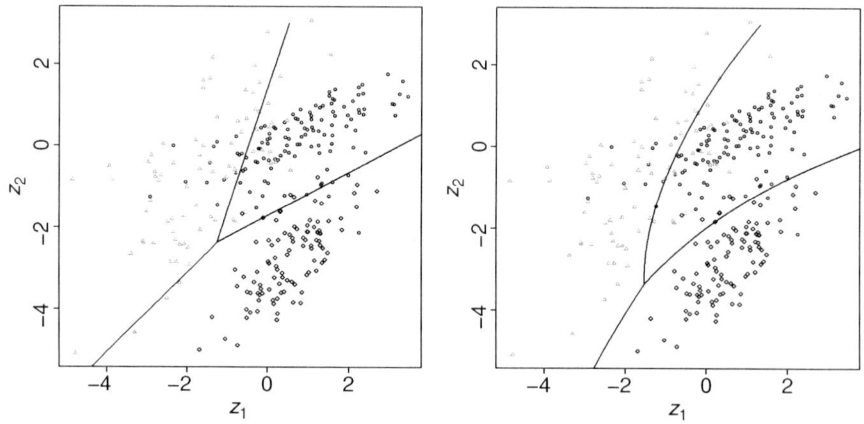

Figure 5.11 Simulated data with three groups: Classification with linear discriminant analysis for two sets of variables given in (5.11).

Parameter estimation poses no difficulties because it is immediate to set

$$\hat{\mu}_k = \frac{1}{n_k} \sum_{i:y_i=k} x_i, \qquad \hat{\Sigma} = \frac{1}{n-K} \sum_{k=0}^{K-1} \sum_{i:y_i=k} (x_i - \hat{\mu}_k)(x_i - \hat{\mu}_k)^\top$$

where denominator $n - K$ follows from the same logic of the denominator of (2.11), and x_i denotes the value of X taken on the ith sample unit. Therefore, the total number of estimated parameters is $pK + p(p+1)/2$.

Consider the simulated data of figure 5.10 and use them as in section 5.4 adopting the same linear predictor. Figure 5.11 was made with the general term x_i of the type

$$x_i = (z_{i1},\ z_{i2})^\top, \qquad x_i = (z_{i1},\ z_{i2},\ z_{i1}^2,\ z_{i1}\,z_{i2},\ z_{i2}^2)^\top \tag{5.11}$$

for the left and right panels, respectively, and therefore they have $p = 2$ and $p = 5$ components. Here, z_{i1} indicates the ith observation of z_1, and analogously for z_{i2}.

However, we can reach the linear discriminant function just indicated without the multivariate normality assumption simply by using second-order assumptions. This justifies using the technique even when X is not a multivariate normal variable, and it may in fact have noncontinuous components. The development of LDA through the second-order hypothesis was the original path followed by Fisher (1936), but for simplicity of explanation, it is easier to follow the framework based on normal distribution.

5.5.3 Quadratic Discriminant Analysis

If we remove the condition that the K variance matrices in (5.10) are equal, we obtain the discriminant function

$$\delta_k(x) = \log \pi_k - \tfrac{1}{2}(x - \mu_k)^\top \Sigma_k^{-1}(x - \mu_k) - \tfrac{1}{2}\log|\Sigma_k|$$

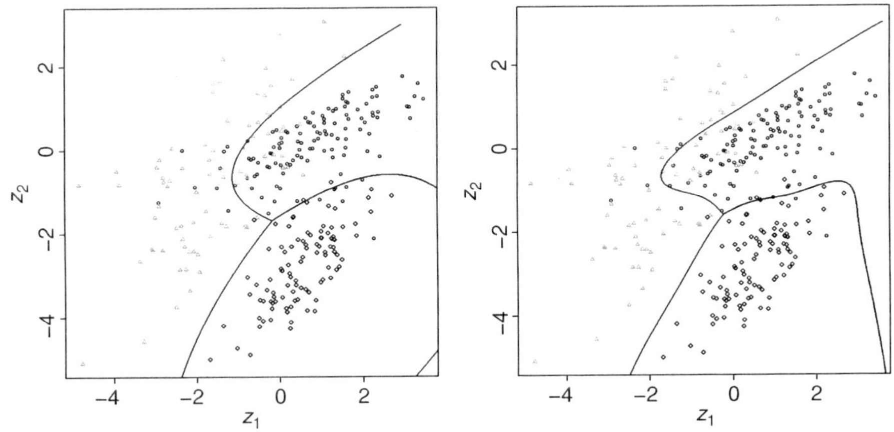

Figure 5.12 Simulated data with three groups: classification with quadratic discriminant analysis for two sets of variables given in (5.11).

which is a quadratic function in x, and the corresponding procedure is therefore called *quadratic discriminant analysis* (QDA).

The estimate of average vectors μ_k is the same as that given in the previous section, whereas the estimate of Σ_k is given by

$$\hat{\Sigma}_k = \frac{1}{n_k - 1} \sum_{i:y_i=k} (x_i - \hat{\mu}_k)(x_i - \hat{\mu}_k)^\top$$

and there are therefore a total of $Kp + Kp(p+1)/2$ distinct estimated parameters.

Applying this procedure to the data used earlier and employing the same transformations of variables z_1 and z_2 in the components of x, we obtain the classification regions shown in figure 5.12. Figure 5.13 displays the lift and ROC curves of the LDA and QDA.

It is important to emphasize that unlike LDA, QDA is closely linked to the Gaussian distributive hypothesis. However, the second diagram was produced by violating this assumption, as it cannot be true that z_1^2 and z_2^2 have normal distribution, not even approximately, because z_1 and z_2 assume values around 0. In spite of this, the regions have reasonable shapes.

Let us now apply these two variants of discriminant analysis to the fruit juice data, in both cases using the linear predictor of (5.1). From table 5.7 we obtain the total misclassification percentages, which are $42/268 = 0.157$ and $46/268 = 0.172$, for the linear and quadratic variants, respectively. Note that in this case, the misclassification error of QDA is larger than that of LDA, highlighting the fact that a more complicated model does not always give better results.

5.5.4 Discussion

Advantages

- Appropriateness of the method: the method was specifically developed for the classification problem; it is not an adaptation of a procedure designed for a different aim.

Table 5.7. FRUIT JUICE DATA: MISCLASSIFICATION TABLE OF LINEAR AND QUADRATIC DISCRIMINANT ANALYSIS ON TEST SET

	Actual response		
	CH	MM	Total
Prediction with LDA			
CH	147	20	167
MM	2	79	101
Total	169	99	268
Prediction with QDA			
CH	145	22	167
MM	24	77	101
Total	169	99	268

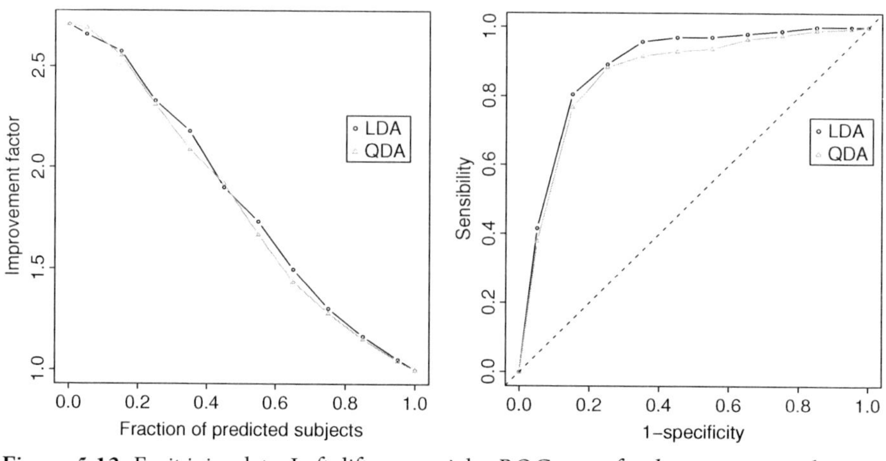

Figure 5.13 Fruit juice data. Left: lift curve; right: ROC curve for discriminant analysis.

- A priori information: if available, this can easily be included in the prior probability of the subpopulation.
- Simplicity of calculation: both parameter estimates and calculation of discriminant functions are extremely simple from a computational viewpoint, and the procedure lends itself well to real-time applications.
- Quality and stability of results: years of accumulated experience on discriminant analysis have shown that the method is highly reliable and produces results that are valid in a large number of cases and stable with respect to new data inputs.
- Robustness with respect to the hypotheses: even when the assumptions of the method are not satisfied, the method tends to produce valid results.

Disadvantages

- Restrictive hypotheses: the method is constructed under quite detailed hypotheses.
- Selection and grading of variables: there are no simple techniques to examine whether a certain variable can be removed without much loss, apart from the universal method of testing on a test set. The same applies to the similar problem of identifying an order of importance among variables.
- Number of parameters: when p and/or K are not small, QDA brings about a rapid increase in the number of parameters. In particular, when n_k is small, some covariance matrices Σ_k may not be identifiable—that is, their estimates may turn out to be singular.
- Nonrobustness of estimates: the estimates of the required parameters are very quickly calculated with the method of moments, but for this very reason they are not robust when outlying observations occur. However, forms of robust estimates exist.

Bibliographical notes

Discriminant analysis was introduced by Fisher (1936). Classic works on classification problems, with a presentation of discriminant analysis, are those of Mardia et al. (1979, ch. 11), Hand (1981, 1982), and McLachlan (1992). A work with a statistical approach but with emphasis on the area of machine learning is that of Ripley (1996).

5.6 Some Nonparametric Methods

Up to now, we have only dealt with parametric methods, but it is also worth exploring nonparametric ones. In the remaining part of this chapter, we consider some of the options, which mainly consist of adapting the matching procedures discussed in chapter 4 to classification problem.

The k-nearest-neighbor estimator (section 4.2.4) is easily generalized to the classification framework by considering, for every fixed point x_0, neighborhood $N_k(x_0)$, including the k points closest in distance to x_0 and classifying x_0 according to a majority vote among k neighbors.

As in the case of regression, number k is a tuning parameter related to the "complexity" of the model. Figure 5.14 shows the classification results obtained by applying k-nearest-neighbors with $k = 1$ and $k = 50$ to the simulated data already used in figure 5.8.

Several nonparametric regression techniques can be adapted to the classification problem, considering the indicator variable that identifies the membership class of each unit as a response variable. To relate this response variable to covariates, exactly as in GLM, we use a link function that transforms the scale of the nonparametric predictor into that of the response variable. For example, where $K = 2$, the link function is again logit function (2.43), and the nonparametric predictor is an unknown function f yielding

$$\text{logit}\left(\mathbb{E}\{Y|x_1, \ldots, x_p\}\right) = f(x)$$

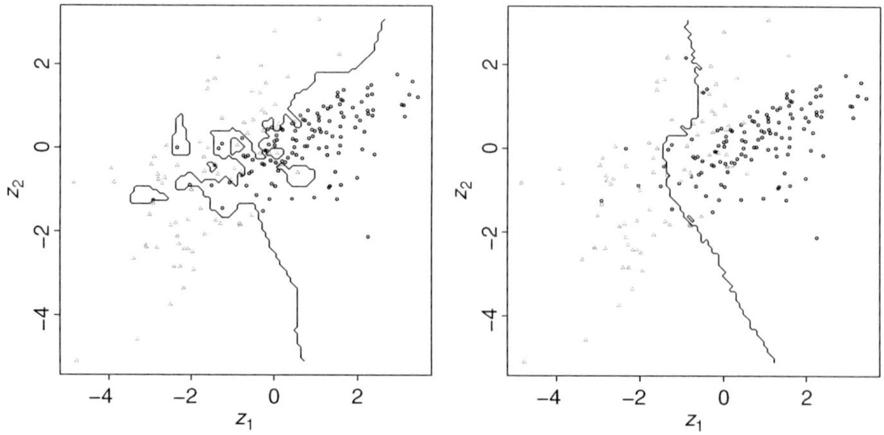

Figure 5.14 Simulated data with two groups classification with k-nearest-neighbors. Left: $k = 1$; right: $k = 50$.

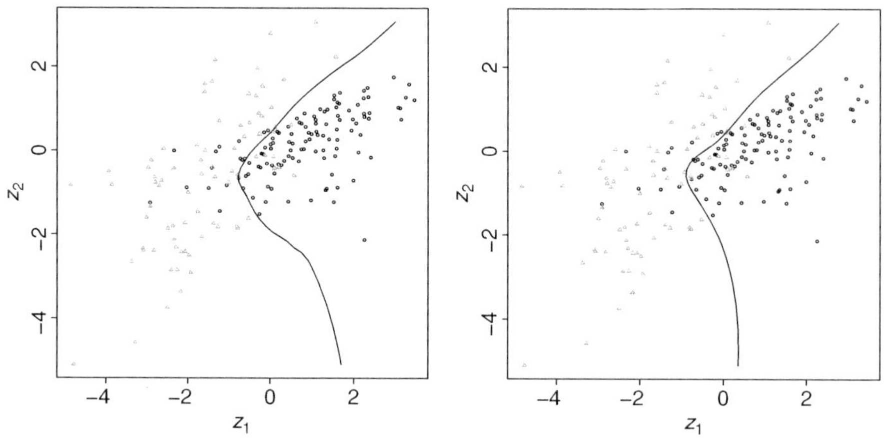

Figure 5.15 Simulated data with two groups: Classification with loess and a thin plate spline.

If one or two covariates are available, then the regression function can be estimated in a nonparametric way by one of the techniques described in section 4.2 and section 4.4. Figure 5.15 shows the classification results applying `loess` and a thin plate spline to the simulated data of figure 5.8.

Extension to the case of K categories is possible, following the scheme of section 5.3 and using the multilogit function (5.3) as the link function.

When many covariates are available, as in the regression case, the models must have an albeit weak structure to reduce both conceptual and computational complexity. The generalized additive models, introduced in section 4.5, can be used for classification, with a suitable distribution of the response variable and an appropriate link function. In the case of $K = 2$, we also use the logit function,

Algorithm 5.1 Local scoring for additive logistic model.

1. Initialization:

$$\hat{f}_j \leftarrow 0, \qquad j = 1, \ldots, p;$$

   a. if the y_i are all 0 or 1, put $\hat{\beta}_0 \leftarrow 0$ or 1, and the algorithm terminates;
   b. otherwise, set

$$\hat{\beta}_0 \leftarrow \log(\bar{y}/(1-\bar{y})),$$

$$\hat{\eta}_i \leftarrow \hat{\beta}_0 + \sum_{j=1}^{p} \hat{f}_j(x_{ij}),$$

$$\hat{p}_i \leftarrow \frac{1}{1+\exp(-\hat{\eta}_i)}$$

   where $\bar{y}$ is the average of y_i.

2. Cycle for $j = 1, 2, \ldots, p, 1, 2, \ldots, p, 1, 2, \ldots$:

   a. set:

$$z_i \leftarrow \hat{\eta}_i + \frac{y_i - \hat{p}_i}{\hat{p}_i(1-\hat{p}_i)},$$

$$w_i \leftarrow \hat{p}_i(1-\hat{p}_i),$$

   b. fit an additive model to variable z_i with weights w_i, using the weighted backfitting algorithm and obtaining new estimates for $\hat{\beta}_0$ and $\hat{f}_j$,

   until functions $\hat{f}_j$ stabilize.

which yields

$$\text{logit}(\pi) = \beta_0 + \sum_{j=1}^{p} f_j(x_j) \tag{5.12}$$

where π is the probability of belonging to class 1. To obtain nonparametric estimates of $f_j(x_j)$, we use a modification of the backfitting algorithm, which in this context is called *local scoring* and is shown in algorithm 5.1.

The result of applying the GAM model to the fruit juice data is shown in figure 5.16, where the estimated functions are represented with the smoothing

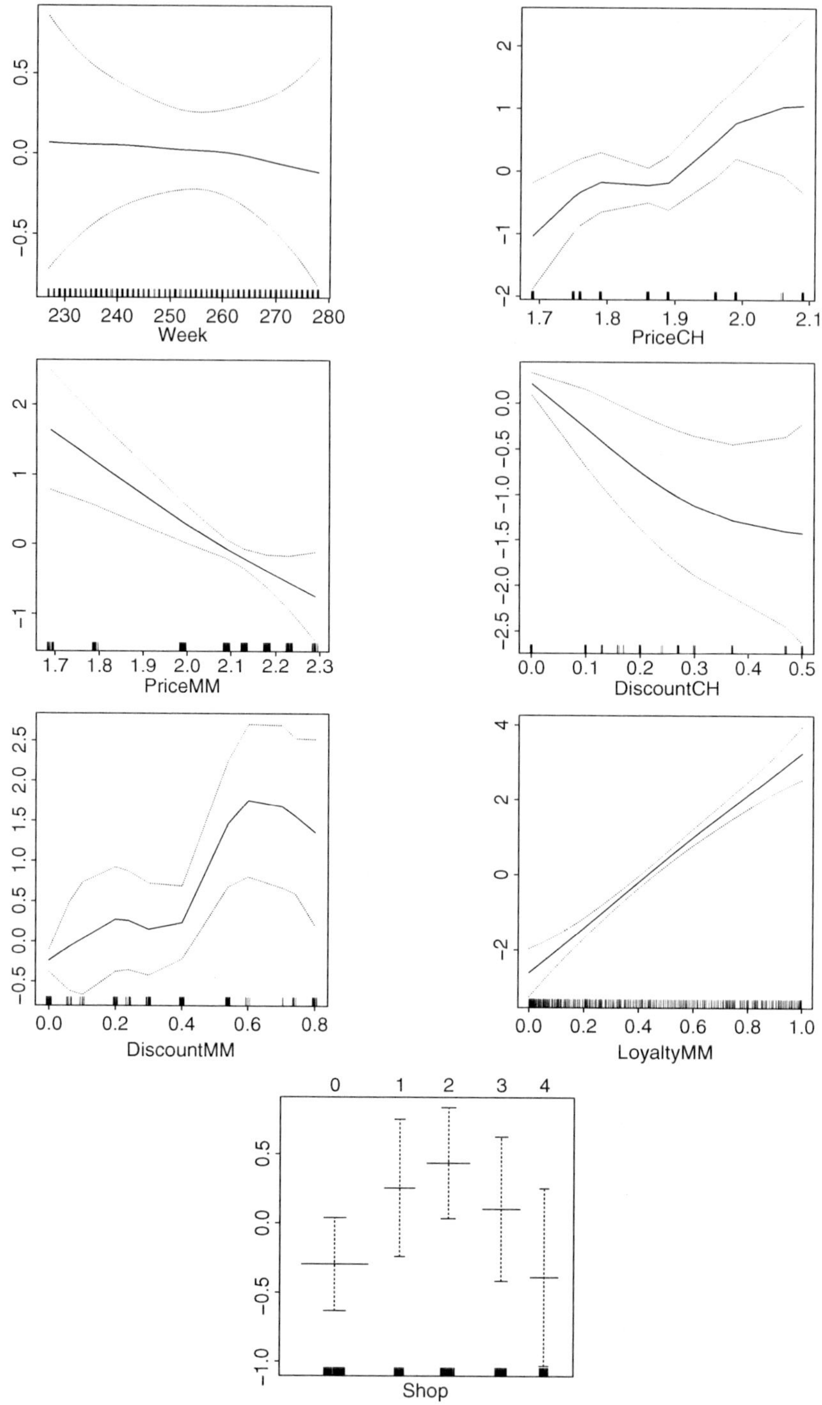

Figure 5.16 Fruit juice data: Effect of variables on classification with GAM model. For continuous variables, functions f_j are estimated by smoothing splines and yield partial effect of each covariate on the response; partial effect of qualitative variable is represented by estimated value for each level. Approximate 95% confidence bands for each function are also shown in different ways for continuous and discrete explanatory variables.

Table 5.8. FRUIT JUICE DATA: TABLE OF ANALYSIS OF VARIANCE FOR GAM MODEL

Component	Deviance	d.f.	*p*-value
s(week)	0.20	1.0	0.65
s(priceCH)	4.50	3.0	0.21
s(priceMM)	0.29	1.0	0.59
s(discountCH)	0.85	1.0	0.36
s(discountMM)	8.19	3.0	0.04
s(loyaltyMM)	0.52	1.1	0.52
store	7.44	4.0	0.11

Table 5.9. FRUIT JUICE DATA: CONFUSION MATRIX OF VERIFICATION SAMPLE WITH GAM MODEL

Prediction	Actual response		
	CH	MM	Total
CH	147	24	171
MM	22	75	97
Total	169	99	268

spline model on the model components. Table 5.8 lists the essential elements of the analysis of variance of the GAM model. Table 5.9 lists the confusion matrix for the resulting classifier, to predict classification on the test set, which gives a global misclassification error of 17.2%.

Also for MARS (see section 4.4.5), generalizations have been proposed to tackle the classification problem. In the case of $K = 2$, the simplest route consists of considering the classification variable as a quantitative variable that takes values 0 and 1 and uses the MARS algorithm for the regression. If $K > 2$, we can recode the response variable into K binary variables and apply the multivariate adaptive regression spline algorithm to each of them, as already seen in the linear model. We then assign each unit to the class that has the highest predicted value for the response variable associated with it.

Another way of generalizing MARS to the classification problem is PolyMARS, based on the multilogit model. As in the case of regression, the model grows when new basis functions are included, but in this case, a quadratic approximation of the multinomial log-likelihood is used to decide which basis function is to be included at each step. The expanded model is fitted to the data by maximum likelihood.

The confusion matrix for a PolyMARS model estimated on the fruit juice data is shown in table 5.10. The global misclassification error of this prediction method is 16.4%. Figure 5.17 shows the lift and ROC curves of the same model.

Table 5.10. Fruit Juice Data: Confusion Matrix of Test Set with PolyMARS Model

Prediction	**Actual response**		
	CH	**MM**	**Total**
CH	149	24	173
MM	20	75	95
Total	169	99	268

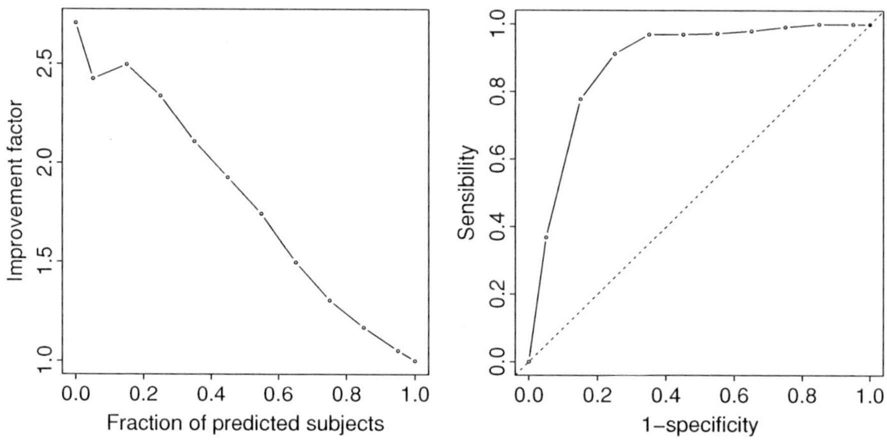

Figure 5.17 Fruit juice data: Lift and ROC curves for PolyMARS model.

Bibliographical notes
The reference base for GAM is the work by Hastie & Tibshirani (1990) in which the additive version of the proportional odds model is also discussed. PolyMARS was introduced by Stone et al. (1997).

5.7 Classification Trees

Let us adapt the idea of regression trees, presented in section 4.8, to the case in which the response variable is *qualitative* (categorical), with K levels. Figure 5.18 shows a simple case with $p = 1$ explanatory variables and $K = 2$. In real operations, we use this approach with larger p (and sometimes larger K).

Indicating the two classes by 0 and 1 and the probability that an individual with characteristics x belongs to class 1 by $p(x) = \mathbb{P}\{Y = 1|x\}$, we approximate $p(x)$ by means of a step function of the type

$$\hat{p}(x) = \sum_{j=1}^{J} P_j \, I(x \in R_j) \tag{5.13}$$

as in (4.15), where P_j now represents the probability that $Y = 1$ in region R_j.

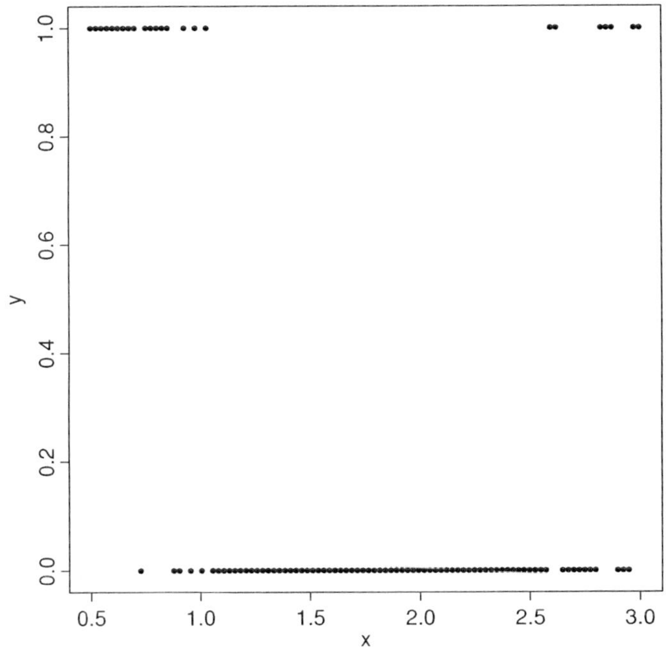

Figure 5.18 Simulated data of a categorical response with two levels and one explanatory variable.

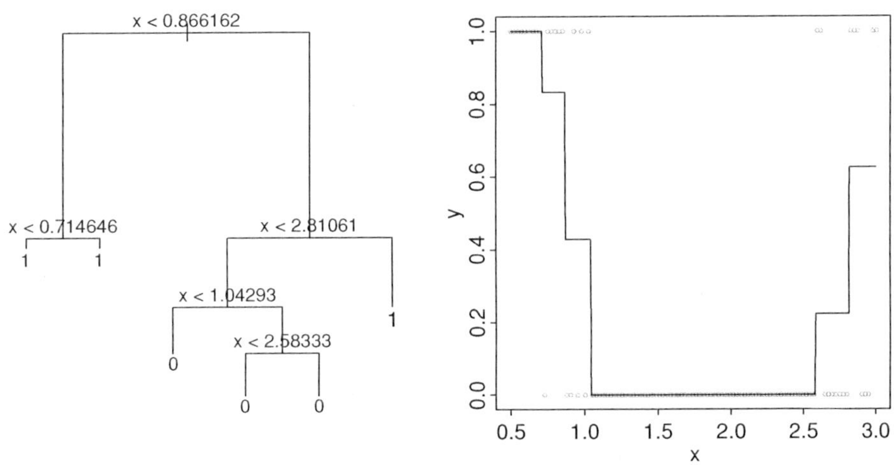

Figure 5.19 Simulated data of a categorical response variable with two levels and one covariate: Tree and estimate of $p(x)$.

The resulting tree is of the type shown in figure 5.19 (left) and the estimate of $p(x)$ (right). The only difference with respect to figure 4.21 is that a class indicator, which is 0 or 1, is associated with the leaves, instead of the values of function $p(x)$. In other words, when we drop a new observation x from the root of the tree to reach a leaf with associated probability $\hat{p}(x)$, this observation is allocated to class $C(\hat{p}(x))$, in which $C(p) = 0$ if $p \leq \frac{1}{2}$, and $C(p) = 1$ if $p > \frac{1}{2}$.

To estimate the P_j of (5.13), we use the arithmetic mean

$$\hat{P}_j = M(y_i : x_i \in R_j) = \frac{1}{n_j} \sum_{i \in R_j} I(y_i = 1),$$

which is the relative frequency of elements 1 in region R_j.

Given the binary nature of y, the deviance function as used for the linear model is not the most suitable. A more appropriate choice is the deviance of the binomial distribution

$$D = -2 \sum_{i=1}^{n} \{y_i \log \hat{p}_i + (1 - y_i) \log(1 - \hat{p}_i)\}$$

as given in (2.38). The deviance may be rewritten by pooling all units i belonging to region R_j, where the probability is constantly P_j, so that

$$D = -2 \sum_{j=1}^{J} n_j [\hat{P}_j \log \hat{P}_j + (1 - \hat{P}_j) \log(1 - \hat{P}_j)] = \sum_j D_j.$$

We reach an interesting interpretation by rewriting the deviance as

$$D = 2n \sum_j \frac{n_j}{n} Q(\hat{P}_j) \tag{5.14}$$

which, without constant $2n$, is an average of *entropies*:

$$Q(P_j) = - \sum_{k=0,1} P_{jk} \log P_{jk} = -\{P_{j1} \log P_{j1} + P_{j0} \log P_{j0}\}$$

$$= -\{P_{j1} \log P_{j1} + (1 - P_{j1}) \log(1 - P_{j1})\}$$

weighted with the relative size of leaves; here, P_{jk} is the probability of outcome k, which is $P_{j1} = P_j$ and $P_{j0} = 1 - P_j$. Terms $Q(\cdot)$ are called *impurity* measures because they indicate that the elements of a certain leaf are nonhomogenous with respect to the response variable. Clearly, $Q(p) = 0$ if $p = 0$ or $p = 1$, and increases gradually from the extremes of interval $(0, 1)$ toward $\frac{1}{2}$, which corresponds to maximum heterogeneity.

Expression (5.14) suggests that we can substitute the entropy with other impurity measures. Of the possible alternatives, one common variant is the *Gini index*

$$Q(P_j) = \sum_{k=0,1} P_{jk} (1 - P_{jk}). \tag{5.15}$$

Another simple index of misclassification error, often used as an alternative to the sum of impurities, is

$$\sum_{j=1}^{J} \frac{1}{n_j} \sum_{i \in R_j} I\left(y_i \neq C(\hat{p}(x_i))\right),$$

that is, the sum of the relative frequencies of errors.

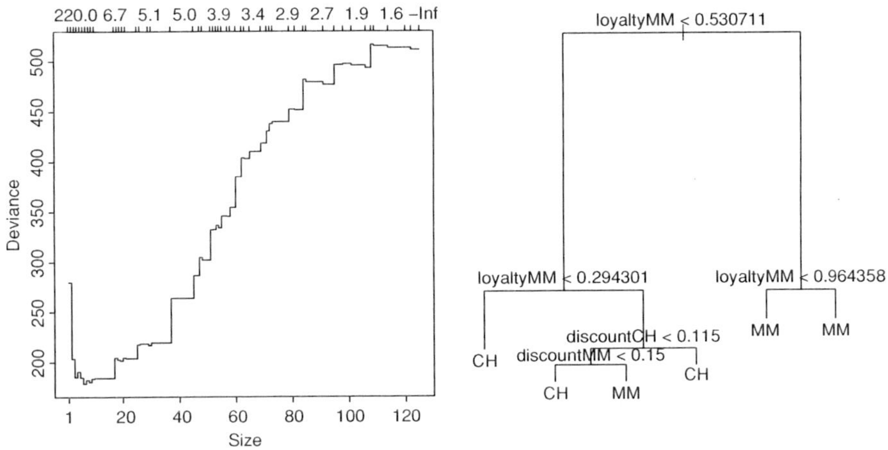

Figure 5.20 Fruit juice data: Classification tree.

Table 5.11. FRUIT JUICE DATA: CONFUSION MATRIX OF TEST SET WITH A CLASSIFICATION TREE MATRIX

Prediction	Actual response		
	CH	MM	Total
CH	135	18	153
MM	34	81	115
Total	169	99	268

We now use these tools for the fruit juice data. In the growth phase of the tree, we adopt entropy as impurity index and base the fit on a sample of 600 elements, taken from 802 observations of the training set. For pruning, we use the remaining 202 observations, for which we again use entropy as the adequacy measure. The corresponding deviance is shown in the left panel of figure 5.20, from which we select dimension $J = 6$ for the tree, shown in the right panel.

The resulting tree demonstrates the importance of `loyaltyMM` and the discount variables. We also note that the two leaves on the extreme right could be pruned, if we are merely focusing on classification, because their distinction deals with the associated probability value, which is 0.72 for the left leaf and 1 for the other leaf.

The confusion matrix is shown in table 5.11 and indicates a global error of 19.4%. Figure 5.21 shows the lift and ROC curves.

The adaption for the case of $K > 2$ is as follows. Function $p(x)$ takes values in the K-dimensional simplex—that is, its values are probabilities $p_0(x), \ldots, p_{K-1}(x)$, which total 1. The impurity indices used earlier take the form

$$\text{entropy} = -\sum_{k=0}^{K-1} P_{jk} \log P_{jk}, \qquad \text{Gini} = \sum_{k=0}^{K-1} P_{jk}\,(1 - P_{jk})$$

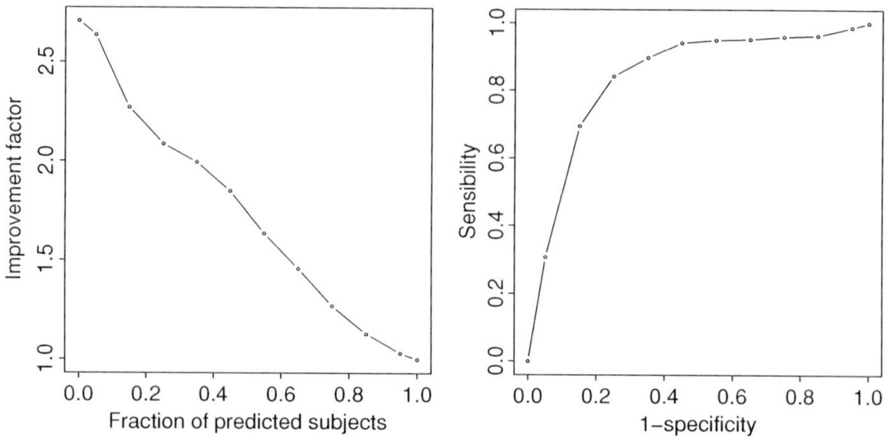

Figure 5.21 Fruit juice data. Left: lift curve; right: ROC curve for classification tree.

each one of which can be inserted into objective function (5.14) where $\hat{P}_j$ is now a K-dimensional vector of estimates of $p_0(x), \ldots, p_{K-1}(x)$.

For a numerical illustration, consider the data of figure 5.10, where $K = 3$. The fitting process gives the plots in figure 5.22, where the top-left plot shows the completely developed tree with entropy as an impurity measure, and the top-right plot shows the deviance obtained by cross-validation, dividing the set into 10 portions, leading to $J = 9$. The bottom plots refer to the pruned tree and the corresponding graphical representation in $\mathbb{R}^2$.

For a discussion of the pros and cons of classification trees, the remarks already made in section 4.8.4 for regression trees apply.

Bibliographical notes

The references provided at the end of section 4.8 are pertinent also here. In addition, the basic methodology described in CART is also used in C4.5 developed by Quinlan (1993) and the commercial version C5.0. Small differences with respect to CART are in tree structure (C4.5 may have multiway splits), splitting criteria (only entropy is allowed by C4.5), pruning method (C4.5 uses an error-based pruning, see, for example Ripley 1996, p. 227) and the way missing values are handled.

5.8 Some Other Topics

The set of classification techniques is vast. We have only presented some here; there are many others. In this section, we give a brief description of a few of them without attempting to cover the complete list.

5.8.1 Neural Networks

The extension of neural networks (section 4.9) to this context is immediate. Starting as usual from the case of $K = 2$, where class indicator y takes the value 0 or 1, the only important adaptation to be introduced is that in (4.18): activation

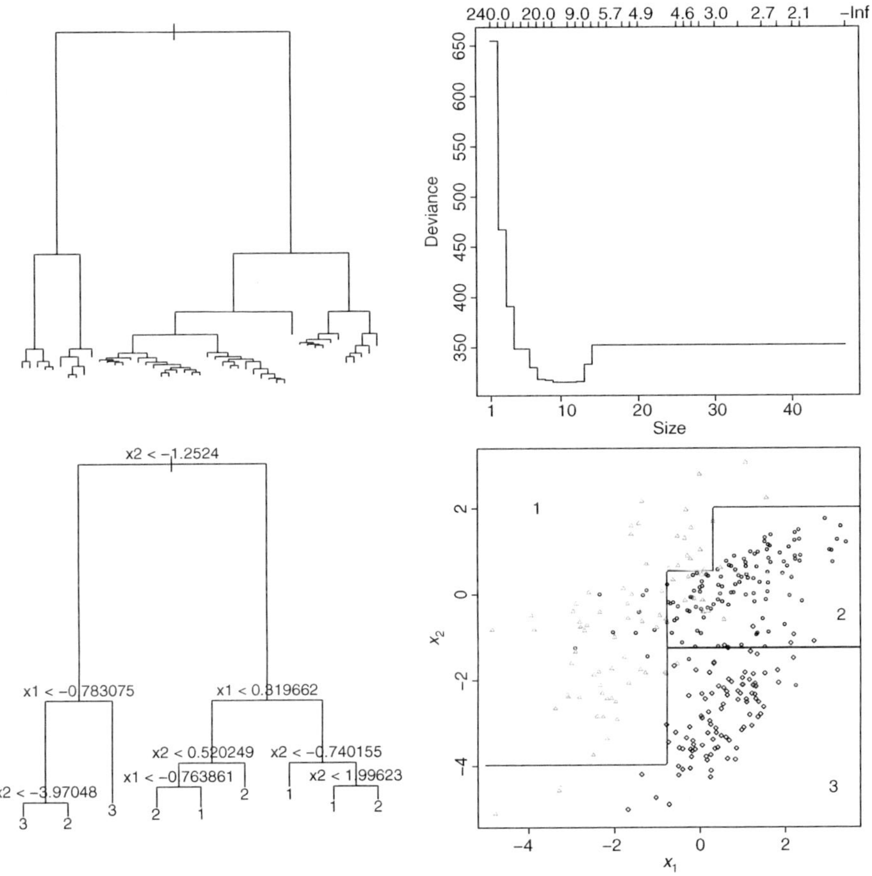

Figure 5.22 Simulated data with three groups: Classification by tree.

function f_1 must have interval $(0, \ 1)$ as a codomain; the most commonly used function is logistic function $\ell(x)$, defined in (2.40). When the two classes are encoded -1 and 1, we use the function

$$2\,\ell(x) - 1 = \frac{e^x - 1}{e^x + 1} = \tanh(x/2).$$

If $K > 2$, we proceed as in section 5.4.2, in the sense that we create K response variables with values 0 or 1. The new ingredient is the choice of the activation function. Putting

$$T_r = \sum_{j \to r} \beta_{jr} z_j,$$

the activation functions between the hidden layer and the output layer in (4.18) are

$$f_{1k} = \frac{\exp(T_k)}{\sum_{r=0}^{K-1} \exp(T_r)}, \qquad k = 0, \ldots, K-1.$$

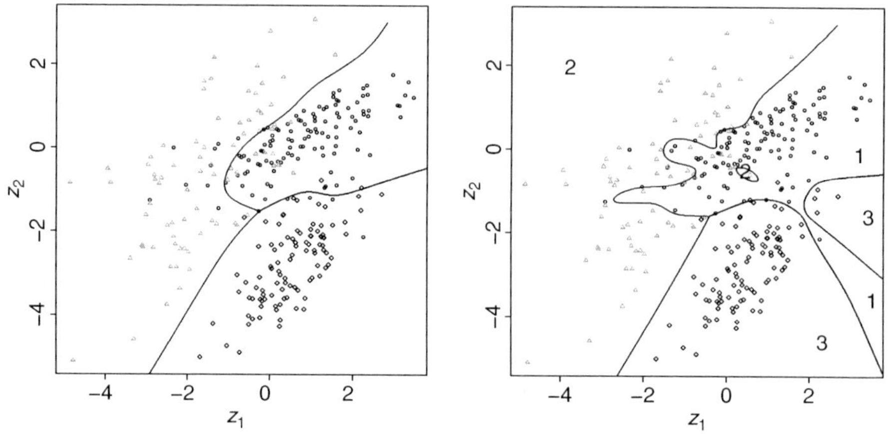

Figure 5.23 Simulated data with three classes: Classification with neural networks.

In this context, this type of function is called *softmax*, but it is essentially the same as (5.3). Term D in objective function (4.20) is no longer the Euclidean distance but rather entropy, as in many other classification methods. Correspondingly, the suggested choice of Ripley (1996, p. 163) for regulation parameter λ also changes: it must now be between 10^{-3} and 10^{-1}, referring to the second form of (4.21) for J.

Figure 5.23 shows the results of classifying the simulated data of three classes, varying the number of hidden nodes r: in the first panel $r = 4$ and in the second $r = 12$; in both cases regulation parameter λ in (4.20) is 10^{-2}. The second panel shows an overfit effect, indicated by the zones without or almost without any point within a zone that is well characterized by points in another class and also by the irregular form of the borders between the classes in some cases. This clear-cut overfit effect stresses the need for care in choosing regulation parameter λ and the number of hidden nodes.

5.8.2 Support Vector Machines

Figure 5.24 shows two sets of points in $\mathbb{R}^2$, whose elements are distinguished by different symbols, and many straight lines cut the plane, perfectly separating the two classes. As one line must be chosen, it is obvious that the line giving the cleanest separation is the best, in the sense of maximizing its distance from the closest point. Intuitively, this line will have the same distance m from the closest representative of each of the two classes. There are two other lines associated with it and parallel to it, which pass through the closest point of each of the classes.

This example is a simple illustration of the more general case of two sets of points in $\mathbb{R}^p$ that are *linearly separable*—that is, perfectly separable by a hyperplane. For these situations, there is an algorithm to determine the optimal separation hyperplane, that is, with maximum value m, in a finite number of operations. This algorithm and its connected aspects go back to the work of Frank Rosenblatt in the late 1950s on the *perceptron*, on which the development of neural networks was based.

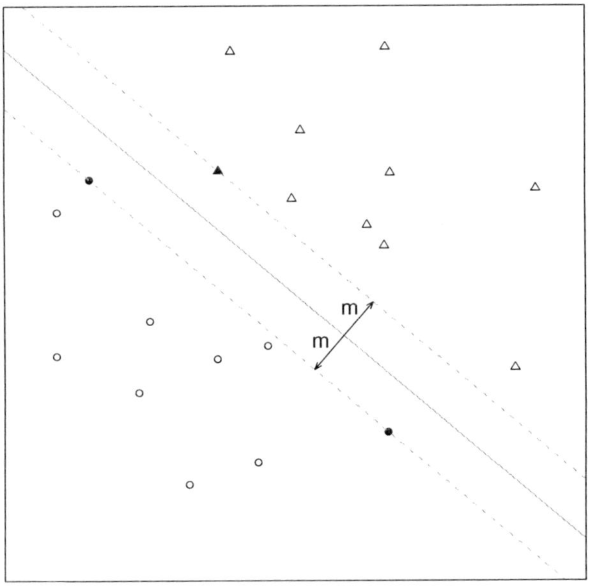

Figure 5.24 Maximum separation margin between two classes: Points belonging to different classes are marked by different symbols.

It is convenient to recall some geometric concepts. For a hyperplane in $\mathbb{R}^p$ with equation

$$a + b^\top x = 0, \qquad (x \in \mathbb{R}^p)$$

identified by coefficients a ($a \in \mathbb{R}$) and b ($b \in \mathbb{R}^p$), each of the following hold:

- for every point x' on the hyperplane, it follows that $b^\top x' = -a$;
- if x' and x'' are any two points on the hyperplane, $b^\top(x' - x'') = 0$;
- it follows that vector b is orthogonal to the hyperplane, and $\hat{b} = b/\|b\|$ is the corresponding unit-norm vector;
- the signed distance from a point $x \in \mathbb{R}^p$ to the hyperplane, that is, to projection x_0 of x on the hyperplane, is given by

$$\hat{b}^\top(x - x_0) = \frac{1}{\|b\|}(b^\top x + a)$$

We now examine the optimization problem more closely. We consider the case of $K = 2$ classes to which this time we assign the conventional values $y = -1$ and $y = 1$, and denote by

$$\beta_0 + x^\top \beta = 0 \qquad (x \in \mathbb{R}^p) \tag{5.16}$$

the equation that identifies a general hyperplane candidate to separate the two classes. Note that without loss of generality, we can let $\|\beta\| = 1$.

For a fixed choice of (5.16), unit $(\tilde{x}, y)$ is classified either correctly or incorrectly, depending on

$$\tilde{y}\,(\beta_0 + \tilde{x}^\top \beta) > 0 \quad \text{or} \quad \tilde{y}\,(\beta_0 + \tilde{x}^\top \beta) < 0.$$

Therefore, the optimization problem may be formulated as

$$\max_{\beta_0,\,\beta} m \qquad \text{subject to} \begin{cases} \|\beta\| = 1, \\ y_i\,(\beta_0 + \tilde{x}_i^\top \beta) \geq m, & i = 1, \ldots, n, \end{cases} \tag{5.17}$$

where $2m$ is called the *margin*, and it represents the width of the free band of points in figure 5.24. Problem (5.17) can then be conveniently rewritten as follows. To free ourselves from condition $\|\beta\| = 1$, we rewrite the constraints in the form

$$\frac{1}{\|\beta\|} y_i\,(\beta_0 + \tilde{x}_i^\top \beta) \geq m,$$

which implies a redefinition of β_0 or, equivalently,

$$y_i\,(\beta_0 + \tilde{x}_i^\top \beta) \geq m\,\|\beta\|.$$

Because multiplication of β and β_0 by a arbitrary positive constant does not change the constraints, we also presume condition $\|\beta\| = 1/m$, and rewrite (5.17) in the equivalent form

$$\min_{\beta_0,\,\beta} \tfrac{1}{2}\|\beta\|^2 \qquad \text{subject to } y_i\,(\beta_0 + \tilde{x}_i^\top \beta) \geq 1 \quad i = 1, \ldots, n. \tag{5.18}$$

Now the half-width m of the free band of points in figure 5.24. is given by $1/\|\beta\|$. Optimization problem (5.18) becomes a minimization problem of a quadratic function with linear constraints, which can be solved by known techniques.

A situation in which a hyperplane achieves perfect separation between the two classes is of course rare in practice. However, we can take the previous example to extend the criterion to more realistic cases. We do not treat it in detail but only outline the basic idea. In a case like that of figure 5.25, there is no straight line that perfectly separates the two classes, and we must therefore select the line using a less stringent requirement.

Because in this new case we have to accept the fact that some points will be wrongly classified, we introduce auxiliary nonnegative variables $\xi_1, \ldots, \xi_n$, which express how far the points are on the wrong side of the margin of their class; when a point is within its margin, $\xi_i = 0$. In figure 5.25, the ξ_i are represented by the length of the line segments connecting the margin of each class with those points

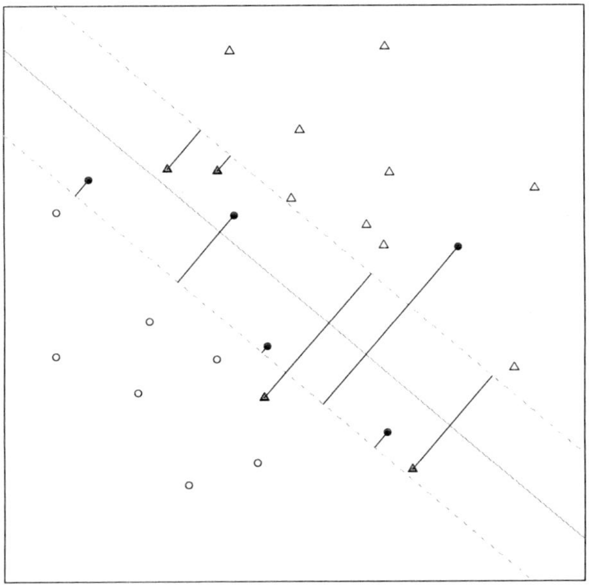

Figure 5.25 An example of two classes of points that cannot be separated by a straight line. Membership of points is distinguished by triangles and circles. Line segments between some points and dotted lines show auxiliary variables ξ_i.

that violate the margin of their membership class. So optimization problem (5.17) can be adapted, replacing constraints $y_i\,(\beta_0 + \tilde{x}_i^\top \beta) \geq m$ with the form

$$y_i\,(\beta_0 + \tilde{x}_i^\top \beta) \geq m(1 - \xi_i), \qquad i = 1, \ldots, n.$$

Reformulating the problem in a way similar to the linearly separable case, we reach the form

$$\min_{\beta_0,\,\beta} \tfrac{1}{2}\|\beta\|^2 + \gamma \sum_{i=1}^{n} \xi_i \qquad \text{subject to} \begin{cases} y_i\,(\beta_0 + \tilde{x}_i^\top \beta) \geq 1 - \xi_i \\ \xi_i \geq 0 \quad i = 1, \ldots, n \end{cases} \tag{5.19}$$

where γ represents a positive constant that plays the role of the regulation parameter and represents the cost of violating the barriers. It can be shown that the solution for β to the optimality problem is of the form

$$\hat{\beta} = \sum_{i=1}^{n} a_i\, y_i\, \tilde{x}_i \tag{5.20}$$

where only some of the a_i are nonzero. Therefore, the solution can be expressed through only some of the observations, which are called *support vectors*.

As in many other techniques, it is convenient to consider transforming the explanatory variables, as in

$$h(x) = (h_1(x), \ldots, h_q(x))^\top, \qquad (x \in \mathbb{R}^p)$$

where the number of components q may be less than, equal to, or greater than p. Correspondingly, (5.16) is substituted by the separation curve

$$f(x) = \beta_0 + h(x)^\top \beta = 0$$

which, in light of (5.20), becomes

$$\hat{f}(x) = \hat{\beta}_0 + \sum_{i=1}^{n} a_i y_i h(x)^\top h(\tilde{x}_i) = \hat{\beta}_0 + \sum_{i=1}^{n} a_i y_i \langle h(x), h(\tilde{x}_i) \rangle$$

In the second expression, we used the most commonly adopted notation in the machine learning literature for the inner product. The resulting method takes the name *support vector machines* (SVM).

Note that the observations enter these formulas only through the inner products of the form $\langle h(x), h(\tilde{x}_i) \rangle$ and the products between these and the y_i. Specification of the functions that form $h(x)$ can therefore occur through the *kernel function*

$$K(x, x') = \langle h(x), h(x') \rangle$$

which calculates the inner products in the space of the transformed variables. The most commonly used kernel functions are the following:

Kernel	$K(x, x')$
polynomial	$(1 + \langle x, x' \rangle)^d$
radial basis	$\exp(-d \, \|x - x'\|^2)$
sigmoidal	$\tanh(d_1 \langle x, x' \rangle + d_2)$

where d, d_1, d_2 are quantities that must be specified a priori.

For example, if $p = 2$ with $x = (x_1, x_2)$ and we adopt the polynomial kernel of order $d = 2$, we have

$$\begin{aligned} K(x, x') &= (1 + \langle x, x' \rangle)^2 \\ &= (1 + x_1 x_1' + x_2 x_2')^2 \\ &= 1 + (x_1 x_1')^2 + (x_2 x_2')^2 + 2 x_1 x_1' + 2 x_2 x_2' + 2 x_1 x_1' x_2 x_2' \end{aligned}$$

for which $q = 6$. The corresponding functions $h_j(x)$ are

$$h_1(x) = 1, \quad h_2(x) = \sqrt{2}\, x_1, \quad h_3(x) = \sqrt{2}\, x_2,$$
$$h_4(x) = x_1^2, \quad h_5(x) = x_2^2, \quad h_6(x) = \sqrt{2}\, x_1 x_2.$$

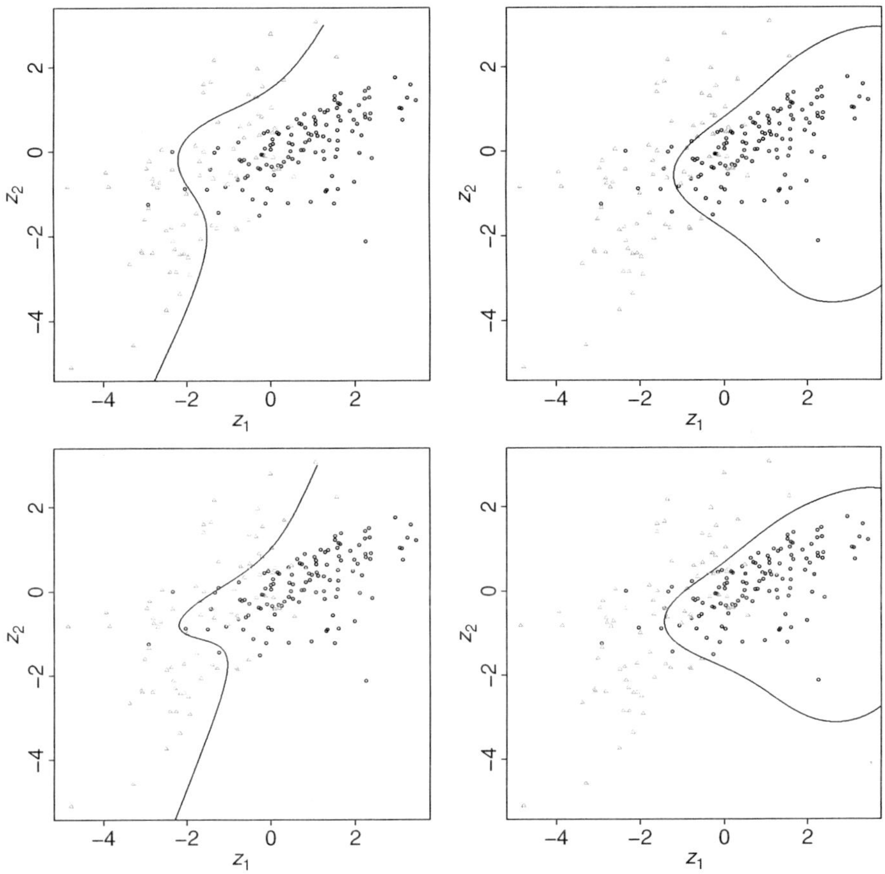

Figure 5.26 Simulated data with two groups. Left: classification by SVM with a polynomial kernel; right: a radial basis kernel. Top plots: $\gamma = 1$; bottom plots: values chosen by cross-validation.

To illustrate the results of the method, examine figure 5.26, in which the data points are the same as those in figure 5.8. A polynomial of order 3 is used in the two left panels, and the radial basis kernel with $d = \frac{1}{2}$ in those on the right. In the two top panels, value $\gamma = 1$ is fixed, whereas the two bottom panels show a scan of 25 values of γ, logarithmically equally spaced between 10^{-2} and 10^{4}. For each value, we proceed to evaluate the total misclassification error by cross-validation by rotating 10 data subgroups; the resulting optimal values are $\gamma = 70$ for the polynomial and $\gamma = 7.39$ for the radial basis.

Bibliographical notes

Hastie et al. (2009, ch. 12) provide more details of the foregoing discussion. Cristianini & Shawe-Taylor (2000) offer a systematic description of SVM, although they define it as an "introduction." Another authoritative text, by one of the main craftsmen of the approach, is that of Vapnik (1998).

5.9 Combination of Classifiers

In many real-life cases, several models fit the data equally well and none appear to be preferable to another. For example, in a problem with 50 explanatory variables, if we construct a logistic regression model with, say, five covariates, there are more than 2 million possible groups of five variables from which we can choose. If we calculate the prediction error on a test set to measure the adequacy of the model, we generally find several sets of five variables with very similar error rates. These models are essentially equivalent from the viewpoint of their prediction error, but they may be quite different when we consider the actual classification of the units in the two groups.

In a similar and perhaps even more obvious way, classification by more unstable methods—for example, trees or neural networks—is greatly influenced by the specific choice of the data set used by the estimate. If this set is modified slightly—for instance, by eliminating a small percentage (2–3%) of data—we can obtain a model that is markedly different from the original one with about the same prediction error. That is, many different models can give similar results for the prediction error.

To improve the predictive ability of each model, one possibility is to combine predictions obtained from various methods, and various paths have been proposed. Each of them produces a model that in some way gathers all the qualities of the single components and thus often gives more accurate predictions. This section presents the main features of the most popular methods.

5.9.1 Bagging

Let $Z = \{(\tilde{x}_1, y_1), (\tilde{x}_2, y_2), \ldots, (\tilde{x}_n, y_n)\}$ be the training set and $C(x)$ a classifier obtained with one of the methods presented earlier. In the following, the model associated with $C(x)$ is called the base model. For the sake of simplicity, we consider the case with $K = 2$.

Adopting a bootstrap procedure, examine sample Z_1^* obtained by extracting n elements from training set Z with replacement. We obtain a new classifier $C_1^*(x)$, by fitting to Z_1^* one of the models presented earlier in this chapter, for example, a classification tree. In general, for a fixed x, the new fitted model is different from the original one. Repeated application of this step, say, B times, produces a set of samples Z_b^* $(b = 1, \ldots, B)$, each of size n, and they in turn produce B new classifiers $C_b^*(x)$, $b = 1, \ldots, B$.

A new classifier that is an average of the results from each of the $C_b^*(x)$ on the given x can be introduced. The most natural form of averaging is the arithmetic mean

$$C_{bag}(x) = \frac{1}{B}\sum_{b=1}^{B} C_b^*(x)$$

which allocates the unit with explanatory variables x to $y = 1$ if $C_{bag}(x) > \frac{1}{2}$ and to $y = 0$ otherwise. As discussed for logistic regression, the choice of $\frac{1}{2}$ is not mandatory, and the method seen in section 5.2.2 still applies. If we think of every single classifier $C_b^*(x)$ as a voter who assigns a vote to one class or the other, we choose the class corresponding to the largest number of votes, so that this criterion is indicated as a *majority vote*. This classification procedure is called bootstrap aggregating, from which the abbreviated the term *bagging* is derived. The classification error of the new procedure is often lower than that of the base models.

Many classification procedures also yield a function $\hat{p}(x)$, which gives the probability that a unit with explanatory variables x belongs to each class. A variation of bagging works by averaging the $\hat{p}_b^*(x)$, which estimate the class probabilities for the model fitted to each of the B bootstrap samples Z_b^* and using this new $\hat{p}_{bag}(x) = \sum_b \hat{p}_b^*(x)/B$ as a probability indicator of class membership.

The bagging strategy can easily be adapted to the regression context, where in place of classifiers $C(x)$ we use the predictions derived from the models discussed in chapter 4. In this case, it is not necessary to return to the majority vote criterion, because we can directly use the average of the predictors obtained by bootstrap resampling as a new predictor. The new prediction may have variance smaller than that of the original model.

Bagging procedures often greatly improve predictive ability, particularly when the classifiers used are very unstable, for example, trees or neural networks. However, with more stable procedures, bagging can somewhat worsen prediction quality. It is also obvious that the operation of combining the results of the single models by way of the arithmetic mean involves losing whatever simple structure existed in the base model, leading to greater difficulty in interpreting the results.

A variant, called *bumping*, or *stochastic search of the model*, picks out, as a new classifier, the model with the smallest prediction error among all the models obtained in bootstrap resampling.

To illustrate how the method works, a bagging procedure with majority vote was applied to the classification tree fitted to the fruit juice data (see table 5.11). Table 5.12 shows the confusion matrix obtained from the test set after bagging based on 300 bootstrap samples taken from the training set and, for every sample, fitting a tree with growth and pruning carried out on two random subsets of each bootstrap sample.

The model obtained with this procedure is better than the original one, as shown by figure 5.27, which compares the error rates of the base model and the bagging procedure when the number of bootstrap samples used grows. Figure 5.28 shows

Table 5.12. Fruit Juice Data: Confusion Matrix of Bagging Procedure Based on Classification Tree on Test Set

Prediction with bagging	Actual response CH	MM	Total
CH	143	24	167
MM	26	75	101
Total	169	99	268

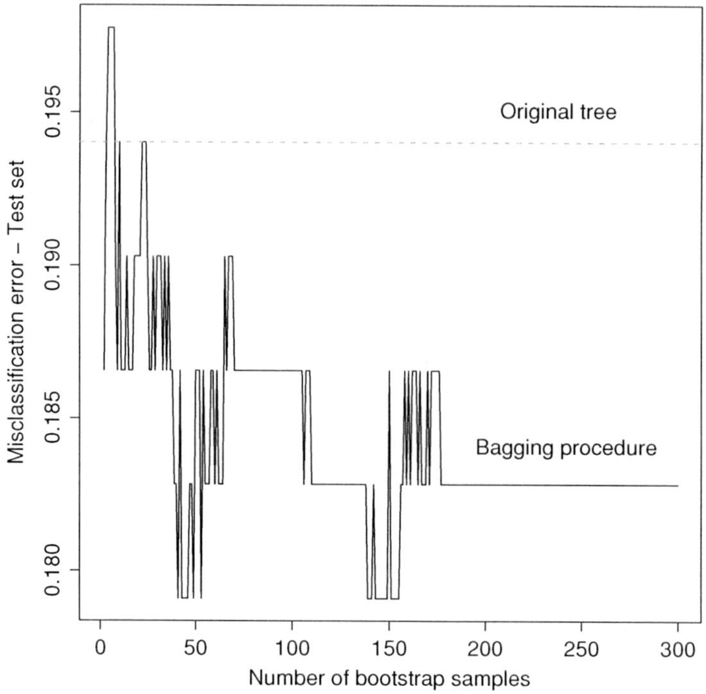

Figure 5.27 Fruit juice data: Estimation errors for bagging on a classification tree.

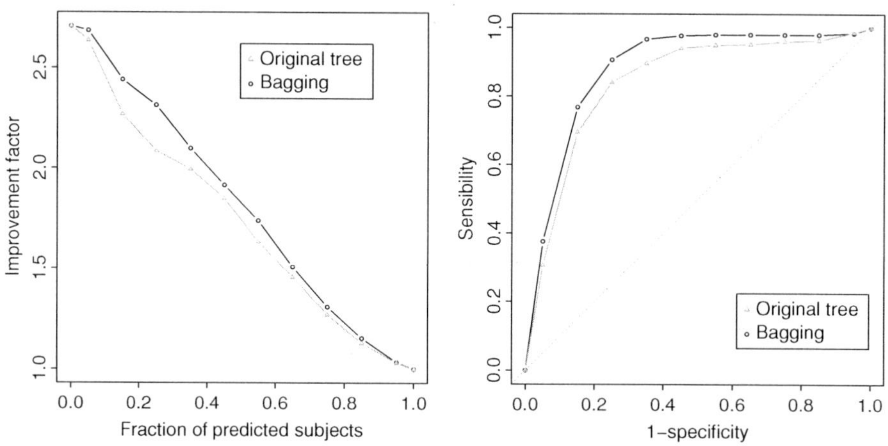

Figure 5.28 Fruit juice data: Lift (left) and ROC (right) curves of classification tree and classifier obtained by bagging a tree.

the lift and ROC curves for the bagging model, compared with similar curves for the initial classification tree, and $B = 300$.

Using random samples of observations allows the use of a technique called *out-of-bag* for easy estimation of prediction errors. In fact, in each bootstrap sample, some of the data of the original training set are excluded. Consequently, for each classifier $C_b^*(x)$, the data of training set Z that are not in sample Z_b^* can be used as a test set. We can therefore estimate, for instance, the misclassification error on these data outside the sample used for the fit (out-of-bag), without requiring a test set or having to choose computing-intensive solutions, such as cross-validation.

5.9.2 Boosting

The idea underlying bagging is to combine results from different data sets, extracted through equal-weight random sampling of available units, and fit them with the same type of model.

Analogously, *boosting* consists of combining the results of a model fitted from several data sets, but we assign a different probability of entering the sample to each unit. Specifically, we assign greater weight to observations classified poorly in the early stage. We thus aim at improving model performance, acting mostly on those subsets in which the original classifier had more problems.

The procedure is iterative. We start by choosing a base model among the classifiers discussed earlier. In the first step, the base classifier is fitted to the data by assigning the same weight to each observation. In the following iterations, the weight assigned to each observation is modified, depending on the classification error. A new classifier is then fitted at each iteration from the modified set of weights. At the end of the process, a new classifier is identified through a weighted majority vote among the classifiers fitted in all the iterations.

This logic has been implemented in many different ways. The most frequent procedure, also the original one, is called *AdaBoost*, presented in algorithm 5.2.

As the number of iterations increases, the importance of the choice of base classifier tends to fall, as the classification choice is more and more closely linked to iteration, that is, it concentrates on badly classified units. This explains the common choice of a tree grown with one or at most two levels, without pruning, as a base classifier. Because their error rate is only slightly better than random guessing, in this context they are usually called *weak classifiers*. When the weak classifier is a tree, the number of levels is connected to the order of interactions allowed by the final model. For example, if only one level of trees is included, only main effects are allowed.

Note that when a tree is fully grown, all its leaves are pure, the classifier makes no errors on the training data, and its error rate is therefore 0. This means that boosting will stop because there are no wrongly classified training units to be boosted. Clearly, the same thing occurs if the tree is just very large, without being fully grown, so that it probably overfits the data. For this reason, it is usually better not to use very large trees for boosting.

Boosting has shown a remarkable ability to produce accurate classifiers in a wide range of situations. They have also been studied theoretically, proving statistical properties that justify their excellent empirical performance; see Friedman et al. (2000).

To illustrate the method, we again use the fruit juice data and choose a tree with two levels as a weak classifier, corresponding to four final leaves. Boosting is stopped after 200 iterations. Figure 5.29 shows the error rate obtained in the test set as iterations increase, clearly demonstrating the improvement up a certain number of iterations, at which point the error stabilizes.

Algorithm 5.2 Boosting (AdaBoost)

1. Initialize weights $w_i = 1/n, \qquad i = 1, 2, \ldots, n.$
2. Cycle for $b = 1, \ldots, B$:

   a. Fit a classification model $C_b(x)$ to the training set, with target values 0 or 1, by weighting the observations by w_i.
   b. Obtain:

$$\text{err}_b \leftarrow \frac{\sum_{i=1}^{N} w_i \, I(y_i \neq C_b(x_i))}{\sum_{i=1}^{N} w_i},$$

$$\alpha_b \leftarrow \log \frac{1 - \text{err}_b}{\text{err}_b}.$$

   c. Assign the new weights:

$$w_i \leftarrow w_i \, \exp\{\alpha_b \, I(y_i \neq C_b(x_i))\}, \qquad i = 1, 2, \ldots, n.$$

3. The new classifier is:

$$C(x) = \begin{cases} 1 & \text{if } \dfrac{\sum_{b=1}^{B} \alpha_b \, C_b(x)}{\sum_{b=1}^{B} \alpha_b} > \frac{1}{2}, \\ 0 & \text{otherwise} \end{cases}$$

Table 5.13 shows the confusion matrix on the test set for the classifier obtained by boosting; the misclassification error is 16%. Figure 5.30 plots lift and ROC curves for the same classifier.

5.9.3 Random Forests

Both bagging and boosting construct different models they then combine by changing at each iteration the set of units or the weight assigned to each unit on which to fit the model, using all available p explanatory variables at each iteration. Another way of obtaining combinations of models consists of considering several subsets of the explanatory variables, instead of considering subsets of the units.

One strategy of this type has been proposed with trees as base classifiers, choosing the variables to put into each model by random selection: this procedure is called *random forest*. Note that this term is sometimes used with a more general meaning, referring to any classifier obtained as a combination of a set of classification trees. For example, in this interpretation of the term, bagging and boosting also belong to random forests when applied to trees.

The procedure consists of selecting at random, at every tree node, a small group of covariates, which are examined to find their best point of subdivision, according

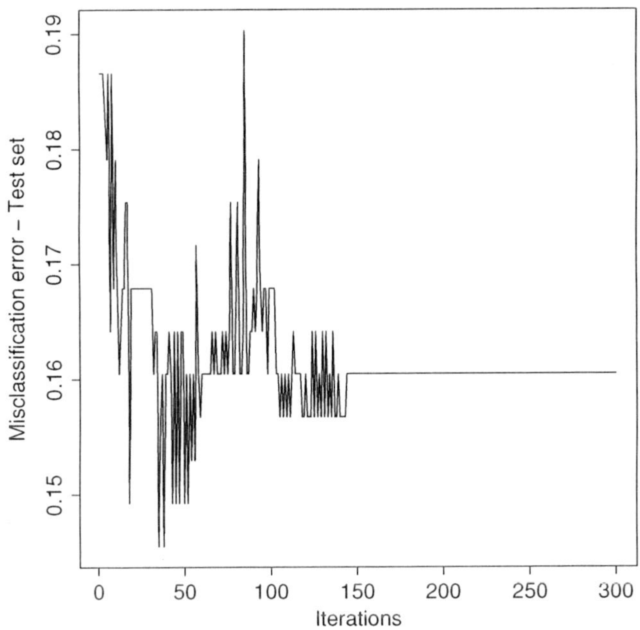

Figure 5.29 Fruit juice data: Estimated error for boosting on a classification tree.

Table 5.13. FRUIT JUICE DATA: CONFUSION MATRIX ON TEST SET OF BOOSTING CLASSIFIER, BASED ON A CLASSIFICATION TREE WITH FOUR LEAVES

Prediction with boosting	Actual response CH	MM	Total
CH	148	22	170
MM	21	77	98
Total	169	99	268

to the splitting criterion described in section 5.7. Therefore, rather than exploring all the possible variable in each node, only q $(q \ll p)$ randomly chosen variables are examined. The tree grows to maximum size but is not pruned. In fact, the resulting combination of various trees avoids overfitting.

The number q of variables to be selected in each node is a tuning parameter to be determined and is generally kept constant on all nodes. The number is often chosen considering forests constructed with different values of q and determining the value that minimizes the error on a test set.

The other tuning parameter is the number of trees, let's say B, that make up the forest. It can be shown that the global error converges to a lower bound when B increases and that it does not cause overfitting problems when additional trees are added. If, therefore, a sufficiently large value is chosen for B, we can be confident that the prediction error obtained will not be very far from its minimum.

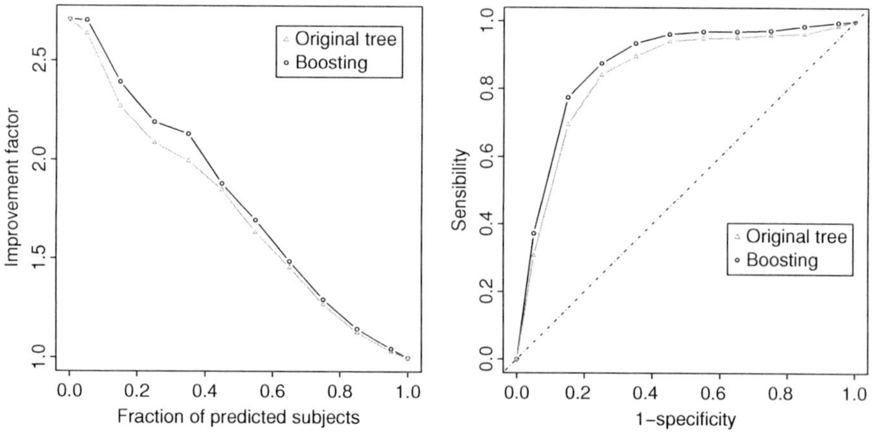

Figure 5.30 Fruit juice data. Left: lift curve; right: ROC curve of classification tree and classifier obtained by boosting on a tree with four leaves.

When constructing a forest, a bagging procedure is usually also associated with random selection of variables. Each tree is made to grow on a different bootstrap sample with a number q of randomly selected variables for each node. Here, bagging, for which the main aim is to improve prediction accuracy, also allows us to use the out-of-bag technique to choose regulation parameter q and obtain the importance measures of the covariates. We can use the prediction error obtained from the out-of-bag data when determining q, instead of the error on a test set.

To obtain a measure of the importance of each explanatory variable in predicting the response, we can proceed using out-of-bag data in the following way. For each tree, the misclassification error on the out-of-bag portion of the data is obtained. The same is done after randomly permuting the values of each explanatory variable. The average of the difference between the two misclassification errors is computed and divided by the standard deviation of the differences, providing an indicator of how that variable influences predictions.

Another indicator of the relevance of variables is based on the importance measure for a single tree, $\sum_h g_h^2$, introduced at the end of section 4.8.3. This is obtained as the average over all the trees in the forest of that importance measure, calculated separately for each variable.

With respect to other methods of model combination, random forests have some interesting advantages. The accuracy of their predictions is comparable to that of boosting and in some cases is better, but they are much faster because every single tree is based on fewer variables and the computational burden is therefore lower. It is also relatively simple to build an algorithm that, taking advantage of parallel computing, can further accelerate the random forest procedure.

For illustration, we present the result of a random forest obtained for the fruit juice data. Note, however, that the presence of only eight covariates would not justify using this strategy, which really only produces interesting results when some hundreds of variables are involved.

Table 5.14. FRUIT JUICE DATA: CONFUSION MATRIX OF RANDOM FOREST USING TEST SET AND OUT-OF-BAG SAMPLES

Test set				Out-of-bag samples			
Prediction	Actual response			Prediction	Actual response		
random forest	CH	MM	Total	random forest	CH	MM	Total
CH	145	21	166	CH	414	77	491
MM	24	78	102	MM	70	241	311
Total	169	99	268	Total	484	318	802

In this example, we constructed a forest of 500 trees, and every tree was made to grow through the Gini index as an impurity measure, with $q = 2$ variables randomly chosen for each node of every tree. The left part of table 5.14 shows the confusion matrix for the forest with the test set; the total classification error is 16.79%. The right side shows the confusion matrix resulting from out-of-bag samples, with a classification error of 18.33%.

The top panel of figure 5.31 plots the error rates obtained on the test set and the out-of-bag samples when the number of iterations increases. The bottom panel plots the importance measures based on the out-of-bag data of the variables in predicting purchases of MM. In this case, `loyaltyMM` is by far the most important variable for predicting MM purchases; `discountMM` and `store` are less important.

Bibliographical notes

Combination methods of classifiers have been proposed by many authors in both statistical and machine learning literature. Bagging was introduced by Breiman (1996), taking advantage of the statistical results on bootstrap. For a presentation of the bootstrap method, see, for example, Efron & Tibshirani (1993) and Davison & Hinkley (1997). The out-of-bag technique was introduced by Wolpert & MacReady (1999) and later exploited by Breiman (2001a). Boosting was initially introduced in the machine learning environment as AdaBoost by Freund & Schapire (1996), and its statistical properties were examined by Hastie et al. (2009). Random forests have been introduced and discussed by Breiman (2001b).

5.10 CASE STUDIES

We now present some real-life cases in which some of the tools described in this chapter are applied to resolve business problems. Because the method follows the lines expressed in section 4.10, we simply outline the problems, list the models used, and present the results and the choice of the final models.

5.10.1 The Traffic of a Telephone Company

We return to the real-life case analyzed in section 4.10.1 and concentrate on identifying the customers who used services offered by the telephone company. If we refer to the variable `total duration of outgoing calls` in a certain

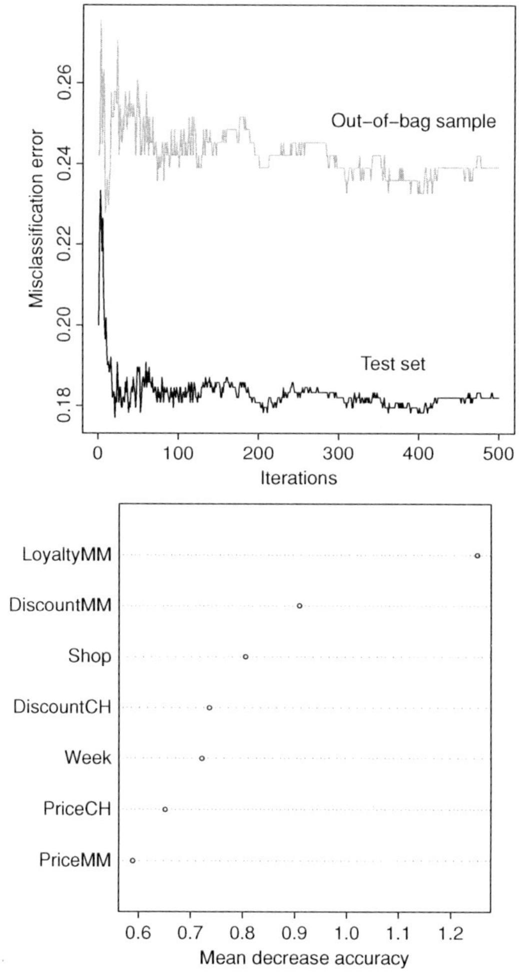

Figure 5.31 Fruit juice data: Estimation errors and importance measures of variables for random forests.

month for a certain population of interest (see section 4.10.1), we must subdivide customers into two classes: those with 0 or positive call durations.

We use as the response variable a new binary variable with value 1 for customers who made calls lasting at least 1 second in the month of interest, and value 0 for customers with no traffic. The following models were fitted to the data:

- linear regression model (see section 5.4), with threshold $\frac{1}{2}$, in two variant forms: (i) with all 98 available explanatory variables; (ii) with only the 55 most significant variables (p-value lower than 0.1);
- logistic regression model (see section 2.4), again using all 98 available explanatory variables and only the 55 most significant ones (p-value lower than 0.1);
- linear discriminant analysis (see section 5.5.2);

- logistic additive model (see section 5.6), with smoothing splines with 4 effective degrees of freedom as smoothers for each variable; this model was also fitted to the data with all 98 variables and only the 29 most significant ones (p-value lower than 0.1);
- MARS (see section 5.6), with linear regression splines with single nodes as elements;
- a classification tree (see section 5.7), with entropy as an impurity index and the number of leaves for the final tree selected by growing and pruning that tree on two separate sets of equal size, randomly chosen from the training set;
- neural networks, with five nodes in the hidden layer and with weight decay parameter $\lambda = 10^{-2}$;
- support vector machine, with radial kernel and tuning parameter ($\gamma = 4$) selected on the test set;
- random forest with 500 trees; the number of variables sampled as candidate at each split (60) was selected on the test set;
- bagging with 500 trees;
- boosting with 50 trees; to include higher order interactions, each tree was grown up to 8 leaves (a test set was used to select it).

We compared the various models according to the percentages of misclassification error, false positives and false negatives, listed in table 5.15. The models were also compared with the lift and ROC curves of figure 5.32.

Comparison of error rates and curves shows that the classifier that predicts best is bagging, because it has the lowest total error rate and the lowest percentage

Table 5.15. Telecommunications Customer Data: Prediction Errors (%) for Models Described in section 5.10.1. Where Necessary Thresholds were Set at $\frac{1}{2}$

Model	Total error	False negatives	False positives
Linear model	22.56	29.69	19.92
Linear model – selected variables	22.61	29.64	20.04
Logistic regression model	17.58	27.10	12.50
Logistic regression model – selected variables	20.20	34.87	8.69
Discriminant analysis	22.30	30.25	19.16
GAM	16.06	23.37	12.49
GAM – selected variables	16.13	23.75	12.36
MARS	15.61	18.75	14.34
Classification tree	15.79	21.95	12.95
Neural network	21.39	21.59	21.33
SVM	16.72	24.41	12.96
Random forest	15.40	18.69	14.07
Bagging	15.04	18.51	13.60
Boosting	15.59	19.48	13.98

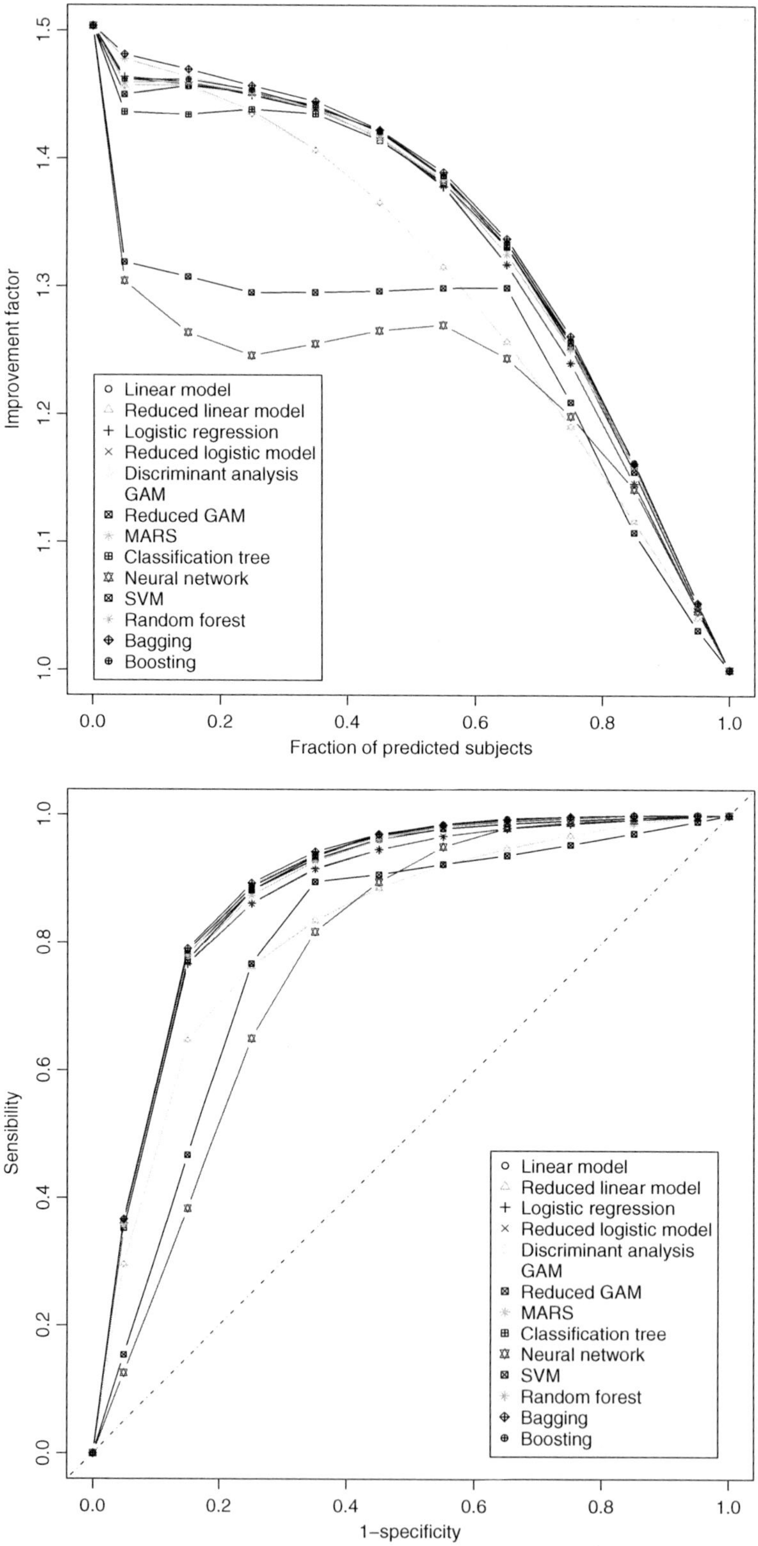

Figure 5.32 Telecommunications data: Comparison of lift (top) and ROC (bottom) curves for various models.

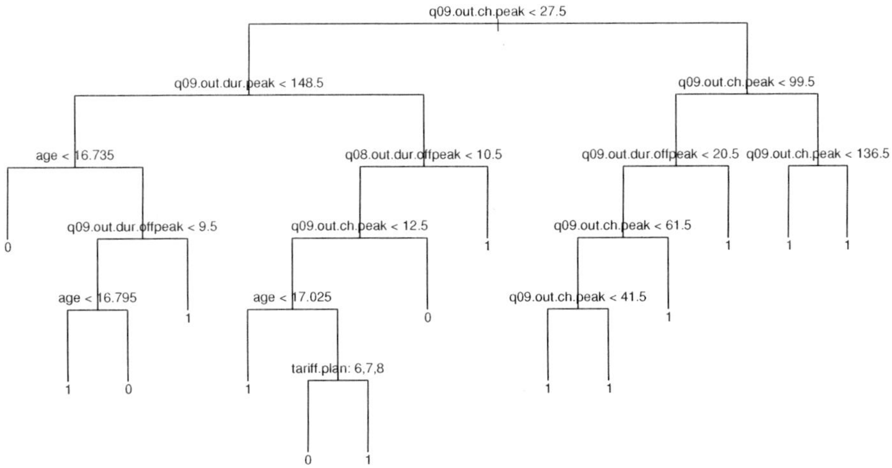

Figure 5.33 Telecommunications customer data: Final classification tree.

of false negatives. Although its percentage of false positives is not the lowest, it has an acceptable value. However, the classification tree has not only a low misclassification error rate and a percentage of false positives lower than that obtained by bagging, but also a ROC curve that is essentially equal to that of bagging and is easier to interpret. We therefore chose the tree to predict customers who do not generate telephone traffic.

Figure 5.33 shows the final version of the tree. To predict which customers will have traffic in the next month, predictive variables are customer's age, phone tariff plan, and several variables linked to traffic in the current and previous months. One interpretation of this evidence is that a customer does not suddenly stop using a telephone but generally reduces traffic slowly until it stops completely.

5.10.2 Churn Analysis

A typical CRM problem for many companies with a large customer base is how to evaluate customer loyalty and, in particular, how to predict which customers are most likely to abandon the company and transfer to another supplier. These customers are often described as being *churners*. This problem is prominent in sectors where customers have ongoing relationships with companies, such as banks, insurance companies, telecommunications services, and services companies in general. Companies of this type must have good models for predicting deactivation by their customers to be able to carry out appropriate retention actions later on.

It is also very useful to understand what reasons customers have for leaving the company. Constructing a model is therefore inspired not only by the need to fit the data but also by the need for that model to indicate *marketing actions*, for example, customer retention strategies, because this obviously translates into profit.

To handle a real-life case in which this problem was tackled, the same data analyzed in sections 4.10.1 and 5.10.1 were used for the customers of a telecommunications operator. The aim of the analysis was to predict deactivation by a customer in a given month with at least 2 months' notice. This requirement is used to plan and implement loyalty actions toward this customer in those 2 months. In other words, we search for the indicator preluding the decision to abandon the company, using information on the structural characteristics of customers, their behavior in terms of use of services offered by the company (if any), their change in usage style, and any other information available in the data mart.

Our data mart contains a `status` variable (not used in the previous analyses) that indicates customer status in terms of deactivation 2 months after the last month of available traffic (indicated in the data by the number 10). This variable takes value 1 if a customer has deactivated and 0 if the customer remains active. Our objective is therefore to predict this indicator variable by using the other 108 available variables.

A simple inspection reveals that the percentage of customers who deactivate is about 13.8% in the training set. In this case, the percentage of events in the population is fairly small—lower than the total prediction rate reasonably envisioned for this problem. This fact causes problems: if we classify all cases as nonevents, irrespective of their individual features, this strategy would appear to be acceptable or possibly even superior to methods that use customer information. This sort of problem is exacerbated in cases in which the percentage of events is even smaller and becomes extreme in cases of rare events: if the percentage of nonevents is 1%, a flat classification scheme of all customers as nonevents has a total error rate of 1%. All this requires us to change our strategy with respect to the previous problems.

Clearly, a prediction of this type, although it minimizes misclassification errors, is not useful for those who want to identify customers who intend to abandon the operator, together with their characteristics. Instead, we need a strategy allowing us to fit a model that can identify customers as accurately as possible, even at the expense of a relatively bad classification of loyal customers, which therefore translates into an increase in the global misclassification error.

The strategy applied here, commonly used in data mining, consists of using a sample stratified by the values of the response variable in the training stage. We thus select all "rare event" customers, that is, all the deactivators or "churners," and a random sample with a similar number of customers with the more common event, that is, customers who are still active.

In this strategy, most of the data are discarded and not used for the estimate. There are alternative proposals for using all the available data, based on evaluation of various costs involved in various types of misclassification. In the present case, for example, we could decide to assign a higher cost to misclassifying a deactivated customer as active, compared with that of an active customer classified as nonactive. These costs may be included as weights of the terms composing the objective function of most of the models discussed in this chapter.

In this analysis, considering the abundance of available data, we preferred to carry out balanced sampling from the original data mart, to obtain the set to be used for the estimate. Obviously, the test set must retain its original proportion of units, so we can evaluate and compare the results of the various models correctly. The following models were fitted to the new balanced training set:

- linear regression model, with threshold $\frac{1}{2}$;
- logistic regression model;
- linear discriminant analysis;
- logistic additive model, with smoothing splines with 4 effective degrees of freedom as smoother for each variable; this model was also fitted to the data with all 108 variables and also the 36 variables that turned out to be the most significant in the previous model;
- MARS, with linear regression splines with singles node as elements;
- classification tree, with entropy as an impurity index and number of leaves for the final tree selected by growing and pruning two separate sets of equal size, randomly selected from the training set;
- neural networks, with five nodes in the hidden layer and with weight decay parameter $\lambda = 2 \times 10^{-2}$;
- support vector machine, with radial kernel and tuning parameter ($\gamma = 4.5$) selected on the test set;
- random forest with 500 trees; the number of variables sampled as candidate at each split (50) was selected on the test set;
- bagging with 500 trees;
- boosting with 50 trees; to include higher order interactions, each tree was grown up to 16 leaves (a test set was used to select it).

Table 5.16 lists the percentages of total misclassification errors, false positives and false negatives obtained on the test set. To further appreciate the usefulness of balanced samples, the error rates for a linear model, a logistic regression model and a classification tree fitted to the original nonbalanced sample are also listed.

As expected, the models fitted to the nonbalanced training set all have a lower total error than the other models, but they are all around the percentage obtainable by classifying all customers as active, which is 13.9% in the test set. However, these predictions have a higher percentage of false negatives with respect to other models fitted on balanced samples.

Comparing the percentages of false positives and false negatives shows that bagging classification is preferable. The percentages in table 5.16 show that the logistic additive model, which is easier to interpret than bagging, gives slightly worse predictions for all three three indicators, so it seems reasonable to consider this simpler model.

Figure 5.34 shows the lift curves of some models. The bottom panel enlarges the low fractions of predicted customers, which is the important part of the curve, because it is the portion with the greatest differences among models.

As the fraction of predicted subjects varies, the classifier with the highest lift is almost always bagging. Only for the first percentile does boosting have a higher lift.

Table 5.16. CHURN PREDICTION: ERRORS (%) FOR MODELS DESCRIBED IN SECTION 5.10.2. WHERE NECESSARY THRESHOLDS WERE SET AT $\frac{1}{2}$

Model	Balance	Total error	False negatives	False positives
Linear model	yes	33.97	9.02	77.59
Logistic regression	yes	34.46	9.08	77.87
Discriminant analysis	yes	33.97	9.02	77.59
GAM	yes	31.75	8.49	75.86
Restricted GAM	yes	32.16	8.07	75.60
MARS	yes	31.84	8.45	75.86
Classification tree	yes	24.94	9.55	72.64
Neural network	yes	40.32	10.90	81.88
Support vector machine	yes	33.69	8.28	76.62
Random forest	yes	31.78	8.02	75.34
Bagging	yes	30.24	7.81	74.19
Boosting	yes	31.43	8.46	75.64
Linear model	no	13.94	13.75	54.93
Logistic regression	no	13.86	13.41	48.73
Classification tree	no	14.13	12.32	52.62

With this model, at its first percentile, we can choose customers who have almost five times the probability of deactivating with respect to average customers. A fraction of 1% of customers may not seem much, but if—for example—we have a customer base of 1,000,000, this 1% corresponds to 10,000 customers, which already means a nontrivial cost for retention actions on all selected customers. The lift curve can be used to select these 10,000 customers to whom retention action (for example, sending a letter or a gift) will be more profitable, because it indicates those customers most likely to churn.

Among the easily interpretable models, the classification tree has the highest lift curve (about 4 at the first percentile) and falls more slowly than the other models until the first decile of predicted units. The corresponding tree of figure 5.35 shows that the traffic of the last and previous months and customer age are the variables most closely linked to churning. Less important but still relevant are the sales activation channel and the chosen method of payment. These are *actionable variables*, in the sense that the telecommunications operator can act directly on them. For example, the company can try to dissuade customers from paying bills by mail (the tree in figure 5.35 shows that the probability of churning is higher for such customers) or have greater control over those sales channels that provide potential churner customers.

This last remark empirically highlights the important fact that the final model should also suggest commercial actions. In our case, for example, models with actionable variables are preferable to ones in which prediction variables are not easily translatable into actions, for example, gender. If we were to discover that

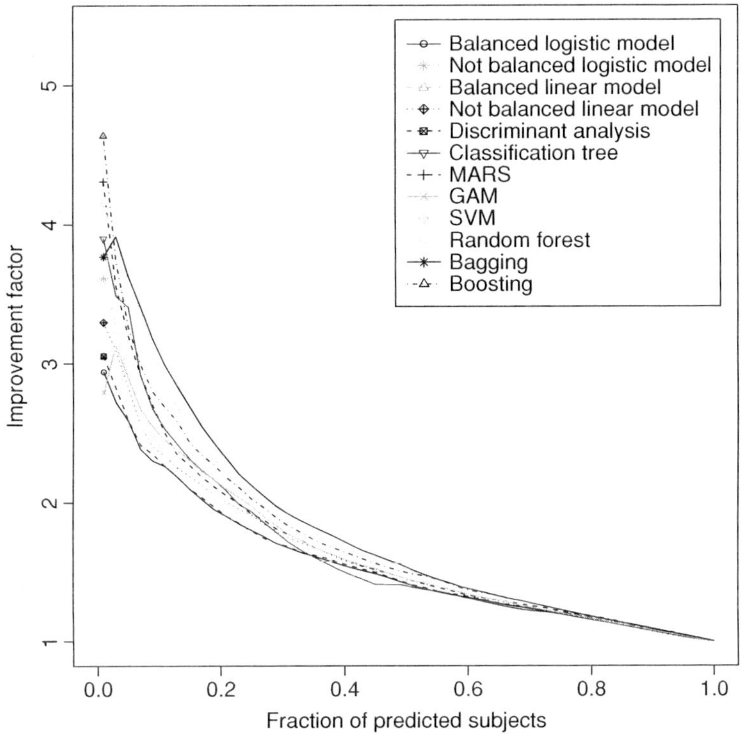

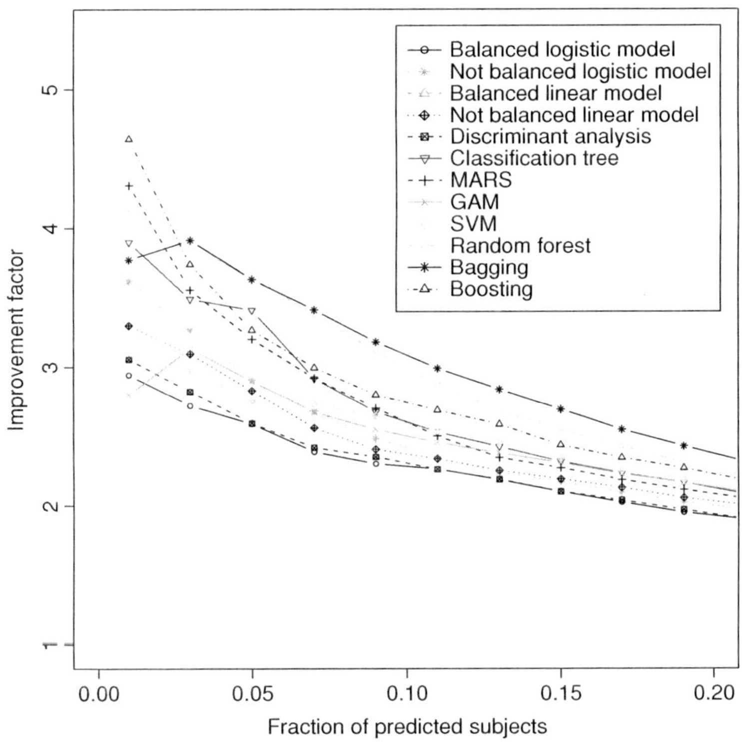

Figure 5.34 Churn prediction: Lift curves.

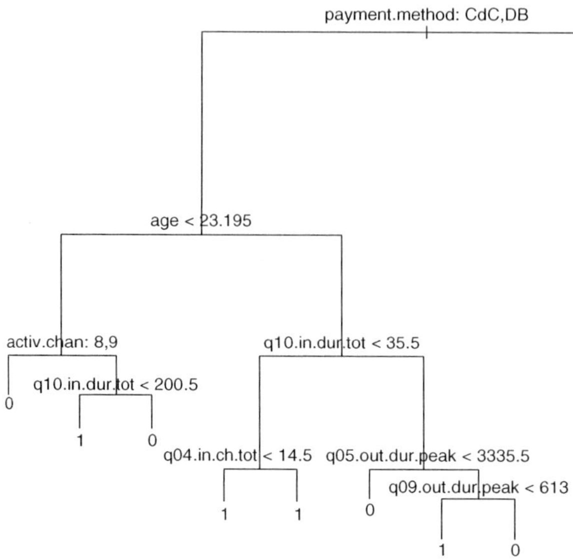

Figure 5.35 Churn prediction: Classification tree.

men churn more easily, we certainly cannot decide to stipulate fewer contracts with men, as they make up about half the population of interest.

In cases like these, the importance of choosing classifiers that are easily translatable into actions emphasizes the essential fact that a human being carries out the analysis and selects models that are easy to interpret and not based on *black-box* procedures or ones of the type "press a button and the computer will do it for you."

5.10.3 Customer Satisfaction

Quantitative measurement of customer satisfaction is one of the most important key performance indicators for companies in many business sectors.

Customer satisfaction surveys are typically implemented by questionnaires containing many items detailing various aspects of customers' feelings toward the company and of their expectations regarding services offered by that company.

Data and background problem

The data analyzed here are described in detail in section B.7 and represent a random sample of 4,000 questionnaires submitted to the customers of an IT (information technology) company producing software and offering consulting services. Opinions about a large number of items are collected by asking customers to score the importance attributed to every aspect characterizing the relationship between customers and company and their actual degree of satisfaction.

Overall satisfaction was investigated by a single question at the end of the questionnaire: "Recalling all the aspects analyzed in this questionnaire, how satisfied are you with the company, overall?" The answer was coded in six levels:

Level	Description
1	Extremely satisfied
2	Very satisfied
3	Quite satisfied
4	Quite dissatisfied
5	Very dissatisfied
6	Extremely dissatisfied

The answers clearly show that overall satisfaction is a ordinal categorical variable. Marketing managers are interested in identifying the specific aspects most closely connected with answers to this question. We describe and predict such a variable by fitting models according to three strategies:

a. the response variable considered as ordinal categorical, as seen in section 5.3.2;
b. the response variable considered as categorical by ignoring level order and setting a classification problem with six classes;
c. the response variable considered as quantitative discrete by assigning to each level of overall satisfaction a numeric mark and applying the methods introduced in chapter 4.

As usual, when data are collected by questionnaire, the number of units is not as enormous as may occur in other contexts of data mining. In our case, we decided to set aside one-quarter of the observations (1,000 customers) in the validation set for the final operation of comparing different models. To tune and test models, we preferred not to divide the training set into two parts but to apply fivefold cross-validation to the entire training set of 3,000 customers (see section 3.5.2).

Figure 5.36 shows the percentage of customers by satisfaction level in the training set. About 69% of them were "quite satisfied," and only 0.47% were "extremely satisfied." The overall percentage of dissatisfied customers was 14.93%, of which only 2.93% are very or extremely dissatisfied.

Some prediction models

As discussed in section 5.3.2, the simplest model allowing for the ordinal nature of the response variable is the proportional odds version of the cumulative logit model. We fitted such a model to the data by selecting important variables with a stepwise procedure (see section 3.6.1) based on AIC (see section 3.5.3). Table 5.17 shows the final fitted proportional odds model.

A useful feature of this model is its interpretability. The categories of response variables are in inverse order with respect to common sense—that is, 1 for the most satisfied and 6 for the least satisfied, so the parameter signs must be interpreted inversely. For example, table 5.17 shows that given all other

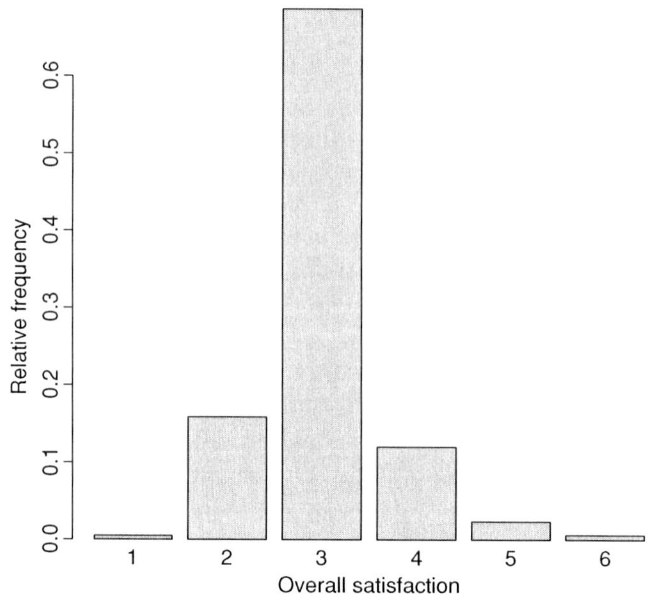

Figure 5.36 Customer satisfaction: Bar graph of overall satisfaction on training set.

variables, customers who often had direct contacts with company personnel (question V11) were less satisfied, and older customers were more likely to be less satisfied than younger ones. The importance attributed to product quality and flexibility (questions V33 and V35), efficiency, and the capacity to understand customers' needs (questions V37 and V39) were negatively related with satisfaction; satisfaction about single aspects such as efficiency, speed of problem solving (questions V45, V47, and V48), and the capacity to understand and respond to customers' needs (questions V44, V52, and V54) were positively related with overall satisfaction. Using some specific products/services (the numbers 5 and "others"—variables V6 and V9) were positively related with satisfaction, whereas product 6 (variable V7) was sometimes a cause of dissatisfaction.

The first part of table 5.18 lists the confusion matrix of the linear proportional odds model in the validation set with classification errors for each predicted level. The overall classification error was 26.3%. This is a weighted average of the specific classification errors of each predicted level. Marketing managers, in addition to receiving good predictions for each category of satisfaction, are particularly interested in reducing the classification error in one of the "satisfaction" categories (1, 2, and 3) of customers who express some level of dissatisfaction (3, 4, and 6). Using the proportional odds model, among the "very dissatisfied" we can predict 1 customer as "very satisfied" and 5 as "quite satisfied," and we can classify 62 customers as "quite satisfied" among the "quite dissatisfied." The total of all these particular misclassification errors was 6.8% in the validation set.

Table 5.17. Customer Satisfaction: Summary of Proportional Odds version of Cumulative Logit Model. Variables are Described in section B.7

	Estimate	SE	Wald 95% conf. limits	
`(Intercept 1|2)`	−15.83	0.65	−17.10	−14.56
`(Intercept 2|3)`	−11.26	0.56	−12.36	−10.17
`(Intercept 3|4)`	−4.85	0.50	−5.83	−3.86
`(Intercept 4|5)`	−1.19	0.51	−2.18	−0.20
`(Intercept 5|6)`	1.36	0.56	0.25	2.46
`V6`	−0.24	0.11	−0.44	−0.03
`V7`	0.23	0.10	0.03	0.43
`V9`	−0.16	0.10	−0.36	0.05
`V11-2`	0.33	0.20	−0.06	0.72
`V11-3`	0.27	0.11	0.06	0.48
`V11-4`	0.50	0.19	0.13	0.88
`V25`	−0.20	0.05	−0.30	−0.10
`V26`	−0.41	0.05	−0.52	−0.31
`V27`	−0.14	0.04	−0.23	−0.05
`V28`	−0.08	0.04	−0.15	−0.01
`V33`	0.10	0.05	0.00	0.20
`V35`	0.12	0.06	0.00	0.24
`V37`	0.09	0.06	−0.02	0.21
`V38`	−0.08	0.05	−0.18	0.02
`V39`	0.13	0.06	0.01	0.25
`V41`	−0.08	0.05	−0.18	0.02
`V44`	−0.09	0.04	−0.17	−0.00
`V45`	−0.12	0.06	−0.24	−0.00
`V47`	−0.25	0.07	−0.38	−0.12
`V48`	−0.10	0.05	−0.20	−0.00
`V52`	−0.24	0.05	−0.34	−0.14
`V54`	−0.18	0.05	−0.27	−0.08
`V60`	0.03	0.01	0.00	0.05
`V61`	−0.02	0.01	−0.04	0.00

$D = 3406.630$ on 29 d.f.

A nonparametric generalization of the proportional odds model follows directly from generalized additive models with logit link function. We replace the linear predictor in (5.5) with the additive predictor $\sum_{j=1}^{p} f_j(x_j)$, as we did for the logit model in (5.12)

$$\log \frac{\mathbb{P}\{Y \leq k\}}{1 - \mathbb{P}\{Y \leq k\}} = \beta_{0k} - \sum_{j=1}^{p} f_j(x_j), \qquad k = 1, \ldots, K-1.$$

Table 5.18. CUSTOMER SATISFACTION: CONFUSION MATRIX AND CLASSIFICATION ERRORS FOR EACH PREDICTION LEVEL FOR LINEAR AND ADDITIVE PROPORTIONAL ODDS MODELS

Linear proportional odds model

	Actual response						**Classif.**
$\hat{y}$	**1**	2	3	4	5	6	**error**
1	0	0	0	0	0	0	—
2	9	62	28	0	1	0	0.380
3	3	107	619	62	5	0	0.222
4	0	0	23	51	16	1	0.440
5	1	0	0	6	3	1	0.727
6	0	0	0	0	0	2	0.000

Additive proportional odds model

	Actual response						**Classif.**
$\hat{y}$	1	2	3	4	5	6	**error**
1	0	0	0	0	0	0	—
2	9	69	35	0	1	0	0.395
3	3	100	613	64	5	0	0.219
4	0	0	22	51	15	1	0.427
5	0	0	0	4	4	2	0.600
6	1	0	0	0	0	1	0.500

A version of the local scoring algorithm (see algorithm 5.1) has been developed to fit this model. The estimated intercepts for the customer satisfaction problem are:

1|2	2|3	3|4	4|5	5|6
16.027	11.497	4.882	1.266	−1.253

and figure 5.37 plots the estimated effects of the covariates. The second part of table 5.18 lists the misclassification errors for the additive proportional odds model in the validation set. The overall misclassification error was 26.2% and the percentage of dissatisfied customers classified as satisfied 7.0%.

We then fitted some typical classification models by ignoring the ordering of the response variable. A multinomial model (see section 5.3.1) was fitted by selecting important variables with a stepwise procedure based on AIC. Table 5.19 summarizes the estimation procedure, and the first part of table 5.20 shows the misclassification errors in the validation set. The overall error was 27.4% and the error among dissatisfied customers 7.2%.

We also fitted a multivariate multiple linear model, as discussed in section 5.4.2, considering each of the six variables as indicating a single level of satisfaction. The second part of table 5.20 shows the misclassification errors for this model. The overall error was 31.9%, and dissatisfied customers who are misclassified 12.6%.

Misclassification errors for linear and quadratic discriminant analysis (see section 5.5.2) are shown in table 5.21, with an overall misclassification error of 28.3% for the linear version and 33.2% for the quadratic one. The percentage of dissatisfied customers classified as satisfied was 6.4% for LDA and only 5.9% for QDA.

A k-nearest-neighbor estimator (section 5.6) was fitted to the customer satisfaction data by assigning to each customer the satisfaction level chosen by the majority of k closest customers (with respect to the covariates). The number k of customers to be considered in each neighborhood was selected by fivefold cross-validation (section 3.5.2) on the training set. The optimal choice for k was 20, and the first part of table 5.22 shows the misclassification errors for this procedure. The overall error was 29.3% and the error for dissatisfied customers 8.7%.

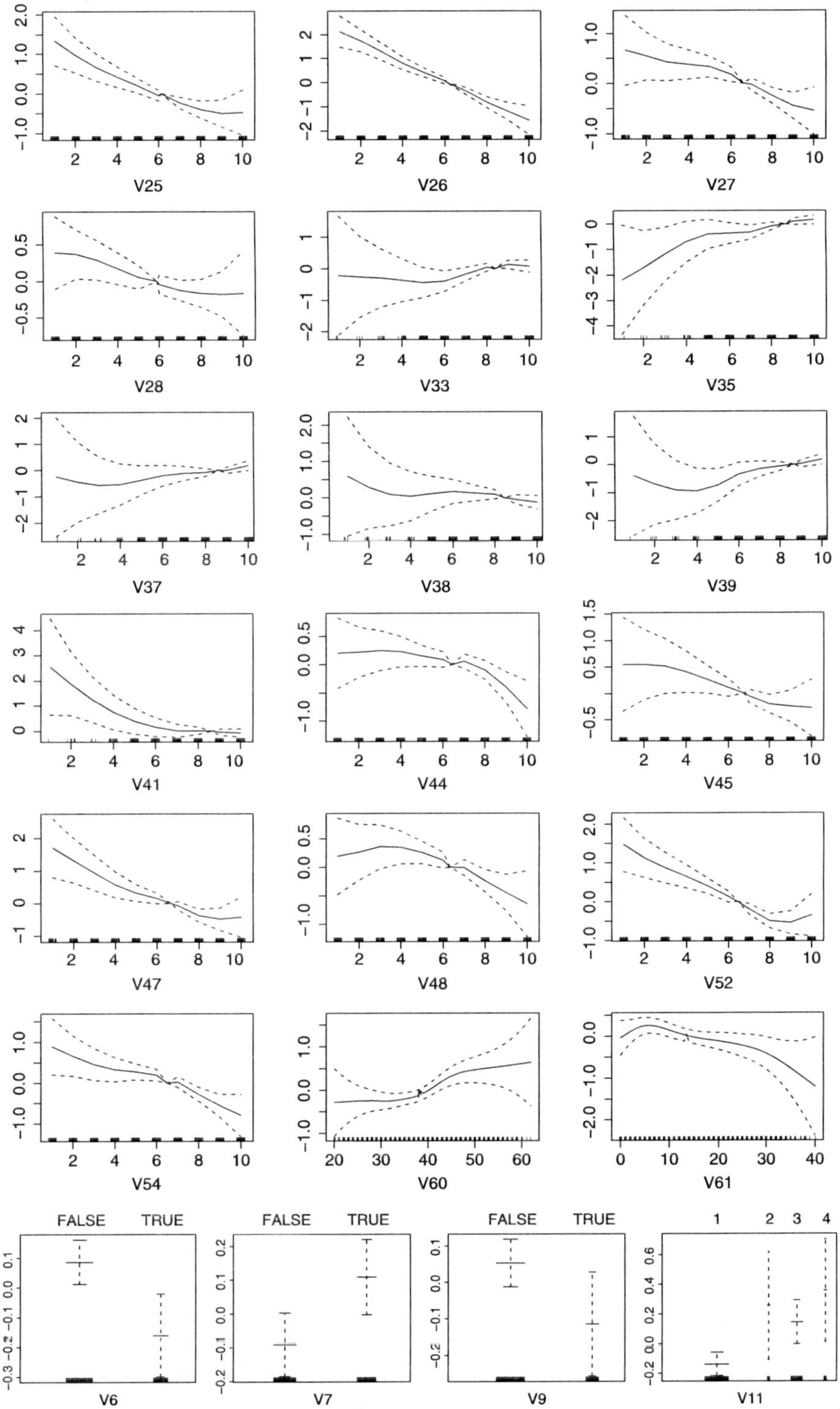

Figure 5.37 Customer satisfaction: Effect of variables on classification with proportional odds additive model.

Table 5.19. CUSTOMER SATISFACTION: ESTIMATED COEFFICIENTS OF MULTINOMIAL LOGIT MODEL (STANDARD ERRORS IN PARENTHESES)

	$\log(\pi_2/\pi_1)$	$\log(\pi_3/\pi_1)$	$\log(\pi_4/\pi_1)$	$\log(\pi_5/\pi_1)$	$\log(\pi_6/\pi_1)$
(intercept)	9.44 (3.073)	20.66 (3.101)	25.18 (3.180)	28.47 (3.421)	25.63 (3.871)
V5	−0.34 (0.703)	−0.13 (0.708)	0.16 (0.729)	−0.57 (0.793)	0.47 (0.946)
V6	−0.53 (0.634)	−0.73 (0.639)	−0.97 (0.660)	−1.41 (0.754)	−3.36 (1.269)
V25	−0.07 (0.329)	−0.18 (0.332)	−0.47 (0.339)	−0.73 (0.364)	−0.53 (0.421)
V26	−1.49 (0.455)	−1.92 (0.458)	−2.28 (0.464)	−2.54 (0.483)	−2.70 (0.543)
V27	0.79 (0.369)	0.58 (0.368)	0.37 (0.372)	0.27 (0.384)	0.17 (0.420)
V34	0.94 (0.381)	1.16 (0.382)	1.22 (0.390)	1.49 (0.412)	1.03 (0.470)
V39	0.39 (0.362)	0.55 (0.364)	0.75 (0.371)	0.55 (0.390)	0.98 (0.461)
V47	−0.03 (0.493)	−0.35 (0.495)	−0.69 (0.501)	−1.00 (0.515)	−0.88 (0.548)
V48	−0.19 (0.449)	−0.38 (0.451)	−0.55 (0.456)	−0.37 (0.472)	−0.56 (0.517)
V52	−0.62 (0.513)	−0.99 (0.514)	−1.17 (0.519)	−1.45 (0.530)	−1.49 (0.562)
V54	−0.42 (0.522)	−0.63 (0.523)	−0.75 (0.527)	−0.88 (0.538)	−1.04 (0.570)
V60	0.01 (0.032)	0.03 (0.033)	0.03 (0.034)	0.01 (0.039)	0.07 (0.048)
V63M	−0.86 (0.831)	−1.40 (0.834)	−0.81 (0.850)	−0.88 (0.901)	−0.82 (1.026)

$D = 3372.143$ on 70 d.f.

Table 5.20. CUSTOMER SATISFACTION: CONFUSION MATRIX AND CLASSIFICATION ERRORS FOR EACH PREDICTION LEVEL FOR MULTINOMIAL AND LINEAR MULTIVARIATE MODELS

Multinomial logit model

	Actual response						Classif. error
$\hat{y}$	1	2	3	4	5	6	
1	0	1	2	0	0	0	1.000
2	8	62	40	0	1	0	0.441
3	4	106	608	65	6	0	0.229
4	0	0	20	49	12	1	0.402
5	1	0	0	5	6	3	0.600
6	0	0	0	0	0	0	—

Linear multivariate model

	Actual response						Classif. error
$\hat{y}$	1	2	3	4	5	6	
1	0	0	0	0	0	0	—
2	0	1	1	0	0	0	0.500
3	13	168	666	105	21	0	0.316
4	0	0	3	14	4	4	0.440
5	0	0	0	0	0	0	—
6	0	0	0	0	0	0	—

Another approach to modeling overall customer satisfaction is to consider this response variable as a quantitative discrete variable and fit regression models. In this case, class prediction is obtained by rounding the predicted continuous values to the nearest integer. The second part of table 5.22 shows the classification errors for a linear model, in which the explanatory variables were selected by a stepwise procedure based on AIC. The overall error on the validation set was 27.3%, and the classification error, predicted as satisfied customers who are in fact dissatisfied, was a very low 4.8%.

Several nonparametric models were fitted by following both strategies, considering responses as either categorical or quantitative variables. To balance bias and variance in the choice of tuning parameters, cross-validation was applied to the training set.

Table 5.21. CUSTOMER SATISFACTION: CONFUSION MATRIX AND CLASSIFICATION ERRORS FOR EACH PREDICTION LEVEL FOR LINEAR DISCRIMINANT ANALYSIS AND QUADRATIC DISCRIMINANT ANALYSIS

Linear discriminant analysis

$\hat{y}$	Actual response 1	2	3	4	5	6	Classif. error
1	0	0	0	0	0	0	—
2	9	50	23	0	1	0	0.398
3	3	119	620	55	4	0	0.226
4	1	0	26	55	15	1	0.439
5	0	0	1	7	5	3	0.688
6	0	0	0	2	0	0	1.000

Quadratic discriminant analysis

$\hat{y}$	Actual response 1	2	3	4	5	6	Classif. error
1	0	1	0	0	0	0	1.000
2	9	95	95	0	1	0	0.525
3	3	73	518	55	4	0	0.207
4	1	0	54	50	15	1	0.587
5	0	0	3	13	5	3	0.792
6	0	0	0	1	0	0	1.000

Table 5.22. CUSTOMER SATISFACTION: CONFUSION MATRIX AND CLASSIFICATION ERRORS FOR EACH PREDICTION LEVEL FOR *k*-NEAREST-NEIGHBORS AND LINEAR MODEL, CONSIDERING RESPONSE AS QUANTITATIVE

***k*-nearest-neighbors**

$\hat{y}$	Actual response 1	2	3	4	5	6	Classif. error
1	0	0	0	0	0	0	—
2	10	51	44	0	1	0	0.519
3	3	118	615	78	8	0	0.252
4	0	0	11	40	15	3	0.420
5	0	0	0	1	1	1	0.667
6	0	0	0	0	0	0	—

Linear model, quantitative response

$\hat{y}$	Actual response 1	2	3	4	5	6	Classif. error
1	0	0	0	0	0	0	—
2	11	70	40	0	1	0	0.426
3	1	99	582	45	2	0	0.202
4	1	0	48	73	20	1	0.490
5	0	0	0	1	2	3	0.667
6	0	0	0	0	0	0	—

Two tree models were fitted, one classification tree (section 5.7) with entropy as splitting criterion, and a regression tree (section 4.8) with a quantitative response. In both cases, the pruned trees were selected by fivefold cross-validation, as shown in figure 5.38. Table 5.23 lists misclassification errors for both procedures. The overall misclassification error was 29.5% for the classification tree and 29.0% for the regression tree; the error for dissatisfied customers was 9.6% for the categorical response variable and 8.5% when it was quantitative.

Neural networks were also fitted, in two versions, with categorical and quantitative responses. The number of units in the hidden layer and weight decay were jointly chosen by fivefold cross-validation. The best classification network had three nodes with weight decay 0.05, a misclassification error of 27.9%, and an error for the dissatisfied customers of 8.2%. The best regression network had three nodes and a weight decay of 0.0005. The overall error on the validation set was 29.1%, and dissatisfied customers who were misclassified 7.3%. The entire set of misclassification errors is shown in table 5.24.

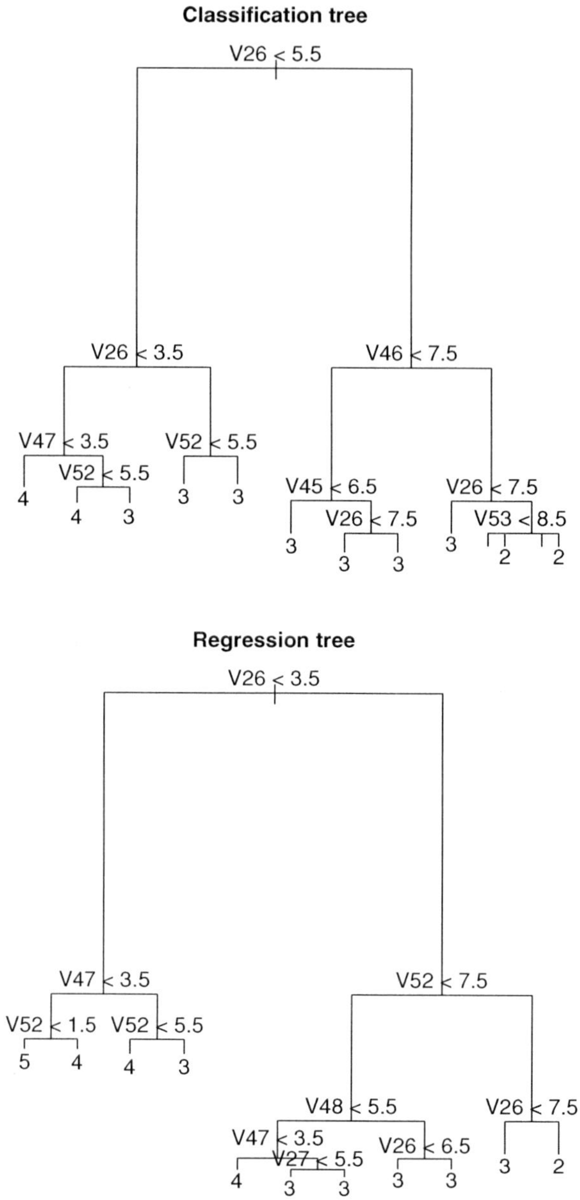

Figure 5.38 Customer satisfaction. Top: pruned classification tree; bottom: regression tree.

Projection pursuit (section 4.6) was also fitted to the data, the number of terms being selected by fivefold cross-validation. Table 5.25 shows the misclassification errors for this model. When the response variable was categorical, the best number of terms was 2, producing an overall error of 30.4% and an error for satisfied customers of 6.7%. When response was considered as quantitative, the best number of terms was 1, the overall error 28.1%, and the error for dissatisfied customers 6.2%.

Table 5.23. Customer Satisfaction: Confusion Matrix and Classification Errors for Each Prediction Level for Tree Models

Classification tree

	Actual response						Classif. error
$\hat{y}$	1	2	3	4	5	6	
1	0	0	0	0	0	0	—
2	11	73	61	0	1	0	0.500
3	1	96	597	84	11	0	0.243
4	1	0	12	35	13	4	0.462
5	0	0	0	0	0	0	—
6	0	0	0	0	0	0	—

Regression tree

	Actual response						Classif. error
$\hat{y}$	1	2	3	4	5	6	
1	0	0	0	0	0	0	—
2	9	80	67	0	1	0	0.490
3	3	88	590	76	8	0	0.229
4	0	1	13	40	16	3	0.452
5	1	0	0	3	0	1	1.000
6	0	0	0	0	0	0	—

Table 5.24. Customer Satisfaction: Confusion Matrix and Classification Errors for Each Prediction Level for Neural Networks

Classification neural network

	Actual response						Classif. error
$\hat{y}$	1	2	3	4	5	6	
1	0	0	0	0	0	0	—
2	11	71	48	0	1	0	0.458
3	1	98	605	73	8	0	0.229
4	0	0	16	42	13	1	0.417
5	1	0	1	4	3	3	0.750
6	0	0	0	0	0	0	—

Regression neural network

	Actual response						Classif. error
$\hat{y}$	1	2	3	4	5	6	
1	0	0	0	0	0	0	—
2	10	63	57	0	1	0	0.519
3	2	106	595	65	7	0	0.232
4	0	0	17	45	12	1	0.400
5	1	0	1	9	5	2	0.722
6	0	0	0	0	0	1	0.000

Table 5.25. Customer Satisfaction: Confusion Matrix and Classification Errors for Each Prediction Level for Projection Pursuit. Response Variables Categorical and Quantitative

Projection pursuit with categorical response

	Actual response						Classif. error
$\hat{y}$	1	2	3	4	5	6	
1	0	0	0	0	0	0	—
2	9	85	86	0	1	0	0.530
3	3	84	552	60	6	0	0.217
4	1	0	32	59	18	4	0.482
5	0	0	0	0	0	0	—
6	0	0	0	0	0	0	—

Projection pursuit with quantiative regression

	Actual response						Classif. error
$\hat{y}$	1	2	3	4	5	6	
1	0	0	0	0	0	0	—
2	11	69	47	0	1	0	0.461
3	1	100	590	57	4	0	0.215
4	1	0	33	56	16	1	0.477
5	0	0	0	6	4	3	0.692
6	0	0	0	0	0	1	0.000

Table 5.26. CUSTOMER SATISFACTION: CONFUSION MATRIX AND CLASSIFICATION ERRORS FOR EACH PREDICTION LEVEL FOR POLYMARS AND QUANTITATIVE MARS

PolyMARS

	Actual response						Classif. error
$\hat{y}$	1	2	3	4	5	6	
1	0	0	0	0	0	0	—
2	9	59	42	0	1	0	0.468
3	3	110	605	72	6	0	0.240
4	0	0	23	43	13	1	0.463
5	0	0	0	4	5	3	0.583
6	1	0	0	0	0	0	1.000

Quantitative MARS

	Actual response						Classif. error
$\hat{y}$	1	2	3	4	5	6	
1	0	0	0	0	0	0	—
2	9	57	35	0	1	0	0.441
3	3	112	593	50	1	0	0.219
4	0	0	42	67	19	1	0.481
5	1	0	0	2	4	3	0.600
6	0	0	0	0	0	0	—

Table 5.27. CUSTOMER SATISFACTION: CONFUSION MATRIX AND CLASSIFICATION ERRORS FOR EACH PREDICTION LEVEL FOR SVM, WITH RADIAL KERNEL AND LINEAR KERNEL

SVM, radial kernel

	Actual response						Classif. error
$\hat{y}$	1	2	3	4	5	6	
1	0	0	0	0	0	0	—
2	9	59	26	0	1	0	0.379
3	3	110	635	74	8	0	0.235
4	0	0	9	43	14	1	0.358
5	0	0	0	2	2	3	0.714
6	1	0	0	0	0	0	1.000

SVM, linear kernel

	Actual response						Classif. error
$\hat{y}$	1	2	3	4	5	6	
1	0	0	0	0	0	0	—
2	10	62	38	0	1	0	0.441
3	2	107	620	70	7	0	0.231
4	0	0	12	45	11	1	0.348
5	0	0	0	4	6	3	0.538
6	1	0	0	0	0	0	1.000

Next, PolyMARS (section 5.6) and quantitative MARS (section 4.4.5) were fitted to the data by selecting the number of basis functions by generalized cross-validation, as discussed in section 4.4.5. Table 5.26 lists misclassification errors, with an overall error of 28.8% for PolyMARS, 27.9% for quantitative MARS, and an error for dissatisfied customers of 7.9% for PolyMARS and 5.2% when the response variable was quantitative.

SVM classification errors are shown in table 5.27. One radial and one linear function were selected as kernels, with tuning parameters γ selected by fivefold cross-validation. Overall misclassification errors were 26.1% for the radial kernel and 26.7% for the linear one; the error for dissatisfied customers was 8.3% for the radial kernel and 7.8% for the linear one.

Methods based on combinations of trees were also fitted to the data by considering both classification and regression trees. We present here only misclassification errors for the combinations of classification trees, because they are all lower than those for combinations of regression trees. Bagging and random forests were fitted by selecting the tuning parameters by fivefold cross-validation (table 5.28). Overall errors were 27.8% for bagging and 27.4% for random

Table 5.28. CUSTOMER SATISFACTION: CONFUSION MATRIX AND CLASSIFICATION ERRORS FOR EACH PREDICTION LEVEL FOR BAGGING TREES AND RANDOM FORESTS

Bagging trees

$\hat{y}$	Actual response 1	2	3	4	5	6	Classif. error
1	2	0	0	0	0	0	0.000
2	8	68	41	0	1	0	0.424
3	2	101	605	70	8	0	0.230
4	0	0	24	45	14	1	0.464
5	0	0	0	4	2	3	0.778
6	1	0	0	0	0	0	1.000

Random forests

$\hat{y}$	Actual response 1	2	3	4	5	6	Classif. error
1	2	0	0	0	0	0	0.000
2	7	64	36	0	1	0	0.407
3	3	105	614	69	6	0	0.230
4	0	0	20	48	16	1	0.435
5	0	0	0	2	2	3	0.714
6	1	0	0	0	0	0	1.000

Table 5.29. CUSTOMER SATISFACTION: PREDICTION ERRORS (%) FOR MODELS DESCRIBED IN SECTION 5.10.3

Model	Type of response	Overall classification error	Misclassification error of dissatisfied customers
Linear proportional odds	ordered	26.3	6.8
Additive proportional odds	ordered	26.2	7.0
Multinomial	categorical	27.4	7.2
Multivariate linear	categorical	31.9	12.6
k-nearest neighbour	categorical	29.3	8.7
Linear discriminant analysis	categorical	28.3	6.4
Quadratic discriminant analysis	categorical	33.2	5.9
Linear regression	quantitative	27.3	4.8
Classification tree	categorical	29.5	9.6
Regression tree	quantitative	29.0	8.5
Neural network	categorical	27.9	8.2
Neural network	quantitative	29.1	7.3
Projection pursuit	categorical	30.4	6.7
Projection pursuit	quantitative	28.1	6.2
PolyMARS	categorical	28.8	7.9
MARS	quantitative	27.9	5.2
Pupport vector machine – radial kernel	categorical	26.1	8.3
Support vector machine – linear kernel	categorical	26.7	7.8
Bagging trees	categorical	27.8	7.9
Random forests	categorical	27.0	7.6

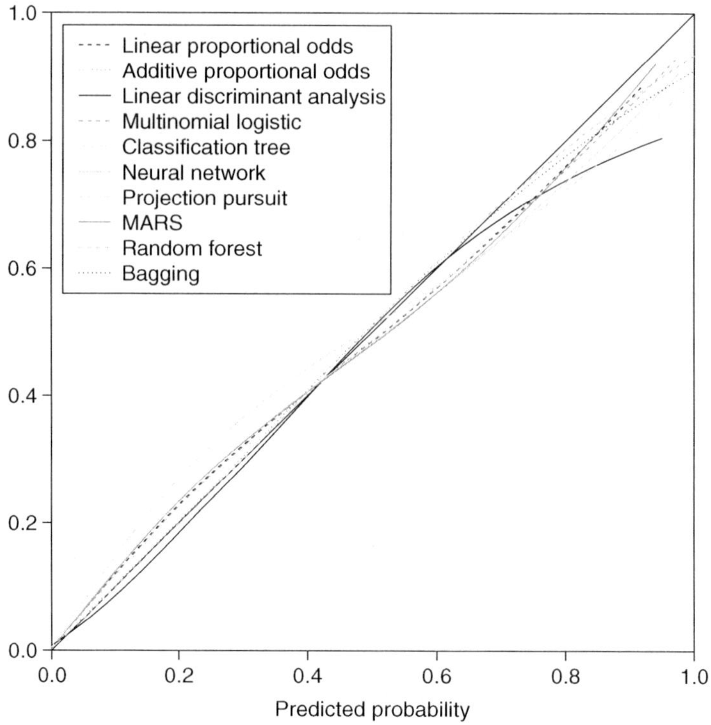

Figure 5.39 Customer satisfaction: Calibration plot.

forests, and the percentages of dissatisfied customers predicted as satisfied were 7.9% for bagging and 7.4% for random forests. Table 5.29 summarizes all these results.

Lift and ROC curves are not appropriate in this multiclass context. However, a tool to evaluate more finely the adequacy of a classification criterion is the *calibration plot* introduced by Dawid (2006) and presented by Venables & Ripley (2002, p. 349) for multicategory prediction. A model is suitable if its predicted probabilities are well calibrated—that is, if we predict an event with probability p, a fraction of about p of the events predicted actually occur. We can therefore plot the predicted probabilities against the actual proportion of events by comparing all the predicted p with the observed relative frequency of occurrence for each single event. If these two quantities are approximately equal, the set of predictions may be regarded as probably valid or well calibrated. Figure 5.39 shows calibration plots for each of the models fitted in this section. We applied a smoothing method with adaptive bandwidth (`loess`, see section 4.2.4) to estimate the relative frequency of occurrence of each event. Clearly, calibration plots can be obtained only for methods in which the response is categorical. Most of the curves are very close to the diagonal line, showing that almost all the models are substantially well calibrated. Only SVM (not shown in the figure) and MARS show slight over-confidence in predictions, especially at probabilities close to 1.

The lowest overall error, 26.1%, was obtained by the SVM with radial kernel. Unfortunately, this model has two drawbacks from the viewpoint of marketing managers: the percentage of dissatisfied customers misclassified as satisfied is quite high (8.3%, compared with the lowest 4.8% of linear regression), and it is not easy, in terms of questionnaire responses, to identify the characteristics of the customers allocated to each category of satisfaction.

As already mentioned, a much easier interpretation is available for linear and additive proportional odds models, which in our case had an overall misclassification error almost as low as that of SVM (the linear proportional odds model had an overall error of 26.2%, and 6.8% for dissatisfied customers).

If we consider the misclassification error for dissatisfied customers, the best prediction is obtained by the linear regression model, with an error of only 4.8%. The overall error of this model is 27.3%, which is not very high.

In our case, two models were recommended to marketing managers: proportional odds and linear regression with quantitative responses. Both models are easy to interpret and have specific optimality, one for the overall misclassification error and the other the error for dissatisfied customers, and neither has a very high value for the other percentage of error.

5.10.4 Web Usage Mining

We now examine the website of a consulting company interested in better understanding its visitors, to identify appropriate marketing actions. Analysis of raw data (information about each hit and visitor) is extremely useful for decision making. Web programming tools allow companies to personalize their relationship with customers by configuring every page to be shown to visitors differently. Profiling visitors by their hit pattern is a simple way of identifying differences in interests between potential customers.

In this context, applying data analysis and data mining techniques to discover patterns from the web is called *web mining* and is generally divided into three main families, according to the primary type of data used: (i) *web usage mining*, to discover user access patterns from information about hits and the number of clicks made by each user (often called *click-stream data*); (ii) *web content mining*, to extract useful knowledge from web page contents (text, image, audio, or video data); and (iii) *web structure mining*, for useful knowledge from hyperlinks and document structures.

In the following, we only face some typical problems of interest for decision makers analyzing raw web usage data and refer to specific works for examples of web content and web structure mining. The data set we analyze contains data on about 26,157 anonymous hits on the website of a consulting company. For each hit, the pages of the site visited over a fixed time interval are known. Visitors are identified by number and no personal information is given. There are 231 pages in the website, and the total number of page views for the entire site is 47,387, so that every page was visited on average 205 times and each visitor hit an average of 1.81 pages. Because some of the pages have similar content, they were grouped into eight categories (home, contacts, communications, events, company, white papers, business units, consulting). The data are presented in greater detail in section B.8.

Prediction of visits to "contacts" pages

In this section, we consider the problem of predicting customers visiting one area of the website: "contacts." Section 6.3 extends analysis of the same data set with some of the tools presented in chapter 6.

The company's marketing managers are interested in identifying the characteristics of visitors who finish their visit in the contacts area: they are likely to be interested in the consulting services offered by the company. Identifying potentially interested customers before they visit the contacts page may be useful to the company, because they can be intercepted and offered some services before they explicitly request them.

This marketing objective can be achieved by considering the statistical problem of predicting the indicator variable, that is, the last page hit by each visitor in the contacts area, by examining previously visited pages. Clearly, we need to identify visitors who saw at least two pages, that is, 4,572. Figure 5.40 shows the frequency distribution of the number of web pages seen by each visitor who visited at least two pages. Because the number of visitors who went through more than 10 pages is small (about 500), we decided to keep only nine visited pages before the last one as predictors of the final hits on the contacts area.

Among the 4, 572 last pages, only 229, or about 5%, fall in the contacts area. Such a low percentage and the small absolute value suggest using cross-validation to trade off bias and variance. We assessed the performance of different models by 10-fold cross-validation, using the same random partition for all methods. To compare the actually observed data, we predicted each of the 10 parts using the best model fitted by using the other parts, thus avoiding having to divide the data set into training and validation sets. We typically also used inner cross-validation to choose a model within each class.

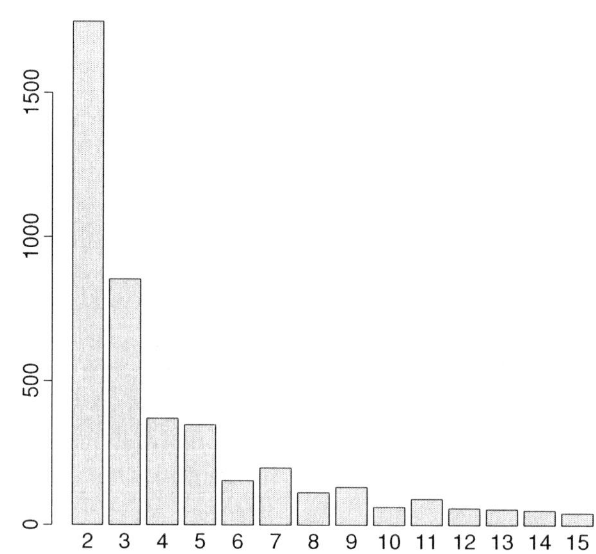

Figure 5.40 Web usage mining: Bar graph of number of web pages hit by each visitor who saw at least two pages.

Table 5.30. WEB USAGE MINING: CROSS-VALIDATION PREDICTION ERRORS (%) FOR MODELS DESCRIBED IN SECTION 5.10.4

Model	Overall error	False negatives	% of correct predictions over positive predictions
Linear discriminant analysis	9.14	47.16	28.07
Linear regression	9.49	46.29	27.27
Logistic regression	8.68	51.96	28.35
Classification tree	9.38	45.41	27.78
PolyMARS	9.53	44.98	27.45
Neural network	9.31	51.96	26.38
Support vector machine	9.38	45.41	27.78
Bagging trees	9.58	79.91	15.28
Random forests	42.78	43.67	6.50
Boosting trees	9.01	52.40	27.18

The response variable has another characteristic: the distribution of visits to the contacts area is very unbalanced. Given the relatively small number of total observations available, we cannot obtain a sample stratified by the values of the response variable if we want to keep the training set a reasonable size. We therefore modify the fitting procedure slightly, to consider this characteristic, described next.

Although all the explanatory variables are categorical, linear discriminant analysis can be fitted. These models do not need any particular specification to process unbalanced response variables. The overall error is 9.14%, false negatives (the error for actual visitors who finished their visit to the contacts area) is about 47.16%, and the percentage of visitors correctly predicted as positive is only 28.07%. Table 5.30 lists these errors for all fitted models.

Linear and logistic models can easily be modified to process unbalanced training sets. For linear models, it is sufficient to change the threshold separating the classes in (5.8): instead of $\frac{1}{2}$, we select a smaller number, closer to the observed proportion of visitors to the contacts area, for example, 0.2. Similarly, for logistic regression, we assign visitors to one category or the other according to whether the estimated probability is greater or less than a smaller number with respect to $\frac{1}{2}$—for example, again 0.2. All errors are listed in table 5.30.

To fit a classification tree, we need to cross-validate with respect to the choice of tree size. Ten-fold cross-validation is used for pruning. Given the unbalanced response, the consequences of misclassifying observations is more serious for visitors who finish in the contacts area. To take this into account, we define a 2×2 loss matrix L, where L_{jk} is the loss incurred in classifying a class j observation as class k. In our case, we consider $L_{kk} = 0$ and assign a higher loss for classifying a visitor in the contacts area who actually finishes in another area. The losses are incorporated in the model process, and the observation in class j is weighted by L_{jk}.

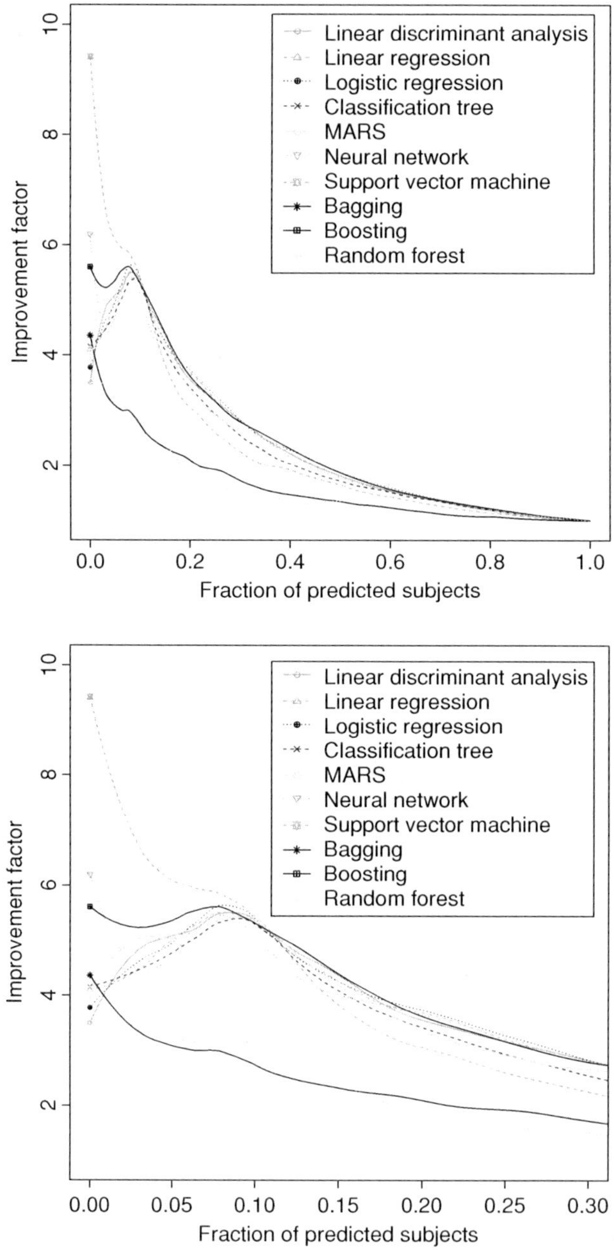

Figure 5.41 Web usage mining: Lift curves.

The pruning level for PolyMARS is also identified by an inner cycle of cross-validation. A loss matrix can be used to incorporate the various weights of the response classes in model selection.

We also fit a neural network by averaging across several fits (10) to overcome the problem of finding multiple local maxima of the likelihood. We choose

Table 5.31. WEB USAGE MINING: ESTIMATES FOR LOGISTIC MODEL

	Estimate	SE	z-value	p-value
(intercept)	−2.5313	0.1226	−20.64	0.0000
page −4: white papers	1.5103	0.6248	2.42	0.0156
page −3: business area	0.5396	0.3070	1.76	0.0788
page −2: business area	−1.2908	0.3323	−3.88	0.0001
page −2: consulting	−0.8367	0.2928	−2.86	0.0043
page −1: business area	−1.8203	0.2754	−6.61	0.0000
page −1: home	1.7043	0.1596	10.68	0.0000
page −1: white papers	−3.5884	0.7345	−4.89	0.0000

the number of hidden units and the amount of weight decay by inner cross-validation. As in the linear model, here, too, we tackle unbalanced classes in responses by changing the threshold separating predicted classes. SVM tuning parameters are selected by inner cross-validation, and weights are supplied. In our example, an SVM with a radial kernel was fitted. Random forests, bagging and boosting were fitted by including trees modified to take unbalanced responses into account.

Figure 5.41 compares smooth versions of lift curves for all fitted models. The top panel shows the entire lift curve, and the bottom one the same function enlarged for small fractions of predicted visitors.

Logistic regression has the lowest overall error and predicts that 388 visitors are interested in contacts, although only 110 of them actually finished their visits on a page of the contacts area, which is the highest percentage among the various models. Nevertheless, the percentage of false negatives suggests using a random forest (which, however, has an overall error that is too high) and PolyMARS or SVM. An SVM seems to be preferable, although we have a small percentage of predicted subjects in the lift curve.

Conversely, if we are interested in identifying a small number, say, less than 10%, of subjects who are most likely to visit the contacts area, an SVM is the most preferable predictor. Alternatively, logistic regression and boosting trees show the best lift curves for fractions of predicted subjects over 10%. The particular ease of interpretation of logistic regression results leads us to choose this model for the final prediction, if we are not interested in very small numbers of predicted subjects. To obtain a set of interpretable parameters, the simplest method is fitting logistic regression with all available data. Table 5.31 lists the estimates of parameters for such a model.

Given all the other variables, homepage visits increase to more than five times the probability of concluding the path in the company website with a page in the contacts area. Visits to business area pages or white papers is negatively related to the response, even in cases when these pages had been visited some clicks earlier, when they may have had a positive impact. Visits to pages in the consulting area have a decreasing impact on the probability of visiting the contacts area later.

Exercises

5.1 Consider a classification model with two categories and assume that we know the fitted probabilities of the vector of successes for the units of a test set, say, $p = (p_1, \ldots, p_n)$. Also assume that we know the correct membership category of all units and they are $y = (y_1, \ldots, y_n)$. Compute the lift curve from vectors p and y.

5.2 For every classification problem in two categories we have two lift curves, one for each of two classes (success and failure). What is the relationship between these curves? If we know vectors p and y of exercise 5.1 for the class identified by (successes), is it also possible to construct the lift curve for the other class? If so, what is explicit form of the lift curve for failures?

5.3 Is it possible to have a lift curve that is not monotone-decreasing? If so, how do we interpret this fact?

5.4 Explain why, in ROC curves, the diagonal corresponds to random classification.

5.5 Is it possible to have a ROC curve that is completely or partly under the diagonal? If so, how do we interpret this fact?

5.6 Write the equation of the line and the dashed curve in the two panels of figure 5.8.

5.7 Prove (5.3), where $\eta_r(x_i) = \beta_0 + \sum_{j=1}^{p} x_{ij}\beta_{jr}$, using the fact that (5.2) also holds with $r = 0$, by setting $\beta_{j0} = 0$, for $j = 0, \ldots p$.

5.8 In a classification problem with $K = 2$ classes, multiple linear regression can be used as a classification method in two ways, using either a column of indicator variables and selecting the class closest to the interpolated value, or two indicator variables (and two regression models) and selecting the one corresponding to the higher estimate. Why are the two procedures equivalent?

5.9 In the case of linear discriminant analysis with $K = 2$ and equal a priori probability for the two groups, show that $d_1(x) > d_2(x)$ takes the form

$$(\mu_1 - \mu_2)^\top \Sigma^{-1}(x - \mu) > 0$$

where $\mu = \frac{1}{2}(\mu_1 + \mu_2)$.

5.10 Show that the classification rule of discriminant analysis described in section 5.5.2, in which $K = 2$ and group sizes are equal, coincides with the rule obtained though the linear model presented in section 5.4; show also that this statement does not hold if classes have different numbers of observations (Fisher 1936).

5.11 In classification trees, if the impurity measure is entropy, how can we compute the gain achieved when passing from J to $J+1$ leaves, corresponding to $D_j^* - D_j$ in the case of regression trees?

5.12 Show that the Gini index is a first-order approximation of entropy. Show also that entropy is not smaller than the Gini index.

6
Methods of Internal Analysis

This chapter differs from the previous two in that we no longer presume the existence of a response variable related to explanatory variables. Here, all variables are on the same level.

In the terminology of machine learning literature, the following themes come under the heading of unsupervised learning, in the sense that learning is not driven by a set of observed cases; the themes of the previous two chapters cover supervised learning.

6.1 Cluster Analysis

6.1.1 General Remarks

We wish to group n available units into K groups, but—unlike the case of classification problems—we have no preassigned system of classification and therefore no response categorical variables. We are speaking of *cluster analysis.*

Typically, because we have no information about the number or nature of the groups, we look for a method to form them by starting from the available variables. Sometimes, a posteriori, we try to interpret the resulting groups. A typical example is that of the segmentation of a company's customer base. From data on how the chosen products are used, personal data, responses to questionnaires, and other sources of information, we arrive at customer groups, called "segments." To define *marketing actions,* we must be able to characterize each group, identifying some of the main explicit aspects, called "profiles."

In other cases, we have in mind certain customer profiles, but cannot directly observe whether an individual fits into a given profile, and we try to construct the closest fit of such profiles from the groups of individuals. In this case, the number K can be given as known.

For the ith individual, we have available p variables, $\tilde{x}_i = (x_{i1}, \ldots, x_{ip})^\top$, of which some are quantitative and some qualitative $(i = 1, \ldots, n)$. We allocate individuals who are more similar to each other in the same group, and dissimilar individuals in other groups. Note that these methods are essentially descriptive, at least regarding the more classical ones presented here.

6.1.2 Distances and Dissimilarities

A fundamental role is clearly played by the way we measure the "nearness" between individuals or, equivalently, their "distance." There are many possible ways of quantifying this, and they vary in nature of the variables in question and what our aims are. We use the general term *dissimilarity* to refer to these measures of distance.

In any case, the dissimilarity $d(i, i')$ between individuals i and i' is based on the composition evaluated for dissimilarities in each of the p observed variables, say, $d_j(x_{ij}, x_{i'j})$ for $j = 1, \ldots, p$. There are many options to define functions $d_j(x, x')$, but in all cases some conditions must be respected

$$d_j(x, x) = 0, \qquad d_j(x, x') \geq 0.$$

We often also need a condition of symmetry, $d_j(x, x') = d_j(x', x)$. A further condition that is often respected is triangle inequality

$$d_j(x, y) + d_j(y, z) \geq d_j(x, z)$$

and in this case dissimilarity qualifies as a *distance*.

For quantitative variables, the main choice for dissimilarity is given by the square of the Euclidean distance

$$d_j(x, x') = (x - x')^2$$

although it is by no means the only one. For qualitative variables we often use

$$d_j(x, x') = 1 - I(x = x')$$

where $I(x = x')$ is 1 if x and x' coincide, and 0 otherwise. For ordinal qualitative variables, that is, ones with levels ordered naturally, we assign a conventional score, such as $1, 2, \ldots, m$, and then treat them as if they were quantitative variables.

For both qualitative and quantitative variables, it is useful to introduce some form of normalization. For quantitative variables, the scale on which variable x_j is measured clearly influences the dimension of d_j and therefore its contribution to the sum (6.1). This observation suggests that we can divide the squared distance by the variance of x_j. Similarly, for qualitative variables, we should keep in mind the number of levels of variables x_j, because the correspondence of observations between two subjects does not have the same significance if x_j has 2 or 20 possible alternatives. A simple way of considering this is to divide d_j by the number of levels of x_j. However, these indications are not followed systematically, because we can easily produce examples where the effect of these normalizations is more harmful than useful, leading to groups that are less easily distinguished than they were originally.

Once functions d_j are chosen, the problem remains of combining them to obtain dissimilarities $d(i, i')$. The simplest option is clearly to add them, by

$$d(i, i') = \sum_{j=1}^{p} d_j(x_{ij}, x_{i'j}). \tag{6.1}$$

Whatever combination is adopted, conditions

$$d(i, i) = 0, \qquad d(i, i') \geq 0, \qquad d(i, i') = d(i', i)$$

should be satisfied, and also $d(i, i') = 0$, if and only if all the d_j components are 0.

If all the variables are quantitative, we can also use the distances listed in table 6.1. In more common cases when variables are both quantitative and qualitative, it is reasonable to calculate the dissimilarities separately for the three sets of quantitative variables, qualitative variables and ordinal qualitative variables; obtaining respectively $d^{(1)}(i, i')$, $d^{(2)}(i, i')$, and $d^{(3)}(i, i')$; and last, combining them in the form

$$d(i, i') = \frac{w_1\, d^{(1)}(i, i') + w_2\, d^{(2)}(i, i') + w_3\, d^{(3)}(i, i')}{w_1 + w_2 + w_3}$$

where w_1, w_2, and w_3 are weights that may be chosen subjectively to make the three components of comparable size.

The values of $d(i, i')$ are arranged in a $n \times n$ *dissimilarity matrix* D, with zero diagonal and nonnegative elements. When the property of symmetry is valid for all functions $d_j(i, i')$, matrix D is symmetric. Because this symmetric property is required by most of the algorithms used, we can fulfill the requirement by redefining D as $(D + D^\top)/2$.

Once dissimilarity matrix D is constructed, it constitutes the basis for most of the grouping methods currently used. Each of these sets is determined to amalgamate subjects with low dissimilarity and separate those with high dissimilarity. These methods are usually grouped according to the following scheme:

$$\text{Clustering} \begin{cases} \text{nonhierarchical} \\ \text{hierarchical} \begin{cases} \text{agglomerative} \\ \text{divisive} \end{cases} \end{cases}$$

There are also other algorithms that do not fit into this scheme. They are not based on matrix D, and therefore are not treated here, where we confine ourselves to classical procedures.

Table 6.1. SOME COMMON TYPES OF DISSIMILARITY USING CLUSTERING METHODS WITH QUANTITATIVE VARIABLES

Name	$d(i, i')$
Euclidean distance simple: $w_j = 1$ weighted with variance $w_j = 1/s_j^2$ weighted with range: $w_j = 1/R_j^2$	$\left(\sum_{j=1}^{p} w_j (x_{ij} - x_{i'j})^2\right)^{1/2}$
Mahalanobis distance (where Σ is a positively defined matrix)	$\{(\tilde{x}_i - \tilde{x}_{i'})^\top \Sigma^{-1} (\tilde{x}_i - \tilde{x}_{i'})\}^{1/2}$
Minkowsky distance (for a parameter $\lambda \geq 1$)	$\left(\sum_{j=1}^{p} w_j (x_{ij} - y_{i'j})^\lambda\right)^{1/\lambda}$
Manhattan distance	$\sum_{j=1}^{p} w_j \lvert x_{ij} - x_{i'j}\rvert$
Canberra metric (one of several variants), where terms in which denominator is 0 are excluded	$\sum_{j=1}^{p} \frac{\lvert x_{ij} - x_{i'j}\rvert}{\lvert x_{ij}\rvert + \lvert x_{i'j}\rvert}$
L_∞ norm	$\max_j \lvert x_{ij} - x_{i'j}\rvert$

6.1.3 Nonhierarchical Methods

The best-known and by far the oldest nonhierarchical method is called K-means and was designed for continuous variables. The basic idea is that of identifying aggregating points, called *centroids,* around which to construct groups, attributing observations to the closest centroid. The centroids are not irrevocably fixed but are themselves subject to sequential updating as the algorithm proceeds.

Let us assume that we have subdivided observations into K groups, according to a certain criterion. We note that total dissimilarity, summing all the elements of D, can be decomposed as

$$\sum_{i,i'} d(i, i') = \sum_{k=1}^{K} \sum_{G(i)=k} \left(\sum_{G(i')=k} d(i, i') + \sum_{G(i')\neq k} d(i, i') \right) = D_{\text{within}} + D_{\text{between}}$$

where $G(i)$ indicates the group to which the ith individual is assigned, and

$$D_{\text{within}} = \sum_{k=1}^{K} \sum_{G(i)=k} \sum_{G(i')=k} d(i, i')$$

$$D_{\text{between}} = \sum_{k=1}^{K} \sum_{G(i)=k} \sum_{G(i')\neq k} d(i, i')$$

are the overall dissimilarity *within* groups and *between* groups, respectively. As we want to choose the groups in such a way as to minimize the dissimilarity within them, we try to minimize D_{within}. Because the total dissimilarity depends neither on K nor on the way in which the groups are created, this aim is equivalent to maximizing D_{between}.

Since the number of possible clusters which can be constructed for a fixed value of K is finite, in principle this minimization is achievable in a finite number of operations by scanning all possible choices. Clearly, this is not a viable option, as the number of possible groupings grows at impressive speed with n, and we must resort to suboptimal algorithms.

A classic algorithm of this type is that of K-means, which uses the Euclidean distance to construct dissimilarities between quantitative variables. For a known property of the sample mean, we can write

$$D_{\text{within}} = \sum_{k=1}^{K} \sum_{G(i)=k} \sum_{G(i')=k} \|\tilde{x}_i - \tilde{x}_{i'}\|^2 = 2\sum_{k=1}^{K} \sum_{G(i)=k} \|\tilde{x}_i - m_k\|^2 \qquad (6.2)$$

where m_k is the mean vector of the subjects of the kth group, that is, the vector form of the arithmetic mean of each variable.

The method aims at minimizing this expression of D_{within}, given group number K and the initial position of centroids m_k. The algorithm then proceeds iteratively, clustering individuals round the centroids, which are subject to iterative uploading, until convergence. This convergence is ensured but does not necessarily correspond to an absolute minimum of the objective function.

The procedure is presented in detail in algorithm 6.1. Step 2.a guarantees that deviance (6.2) is minimum once the centroids have been chosen, and step 2.b guarantees the deviance is minimum once the subjects have been allocated to their groups.

Figure 6.1 illustrates the outcome of the K-means method applied to two sets of simulated data, with very simple structure, so that in both cases three groups of points are evident in a basically nonambiguous way. For illustrative purposes, we use only $p = 2$ variables. The top panels illustrate the method with $K = 3$ and show the final outcome with two choices of initial centroid; the bottom-left panel

---

Algorithm 6.1 K-means

---

1. Choose K and initial arbitrary centroids $m_1, \ldots, m_K$.
2. Cycle for $r = 1, 2, \ldots$:

    a. for $i = 1, \ldots, n$, assign $\tilde{x}_i$ to group k, so that $\|\tilde{x}_i - m_k\|$ is minimum,
    b. for $k = 1, \ldots, K$, let m_k be equal to the arithmetic mean of the subjects belonging to group k,

    until centroids $m_1 \ldots, m_K$ stabilize.

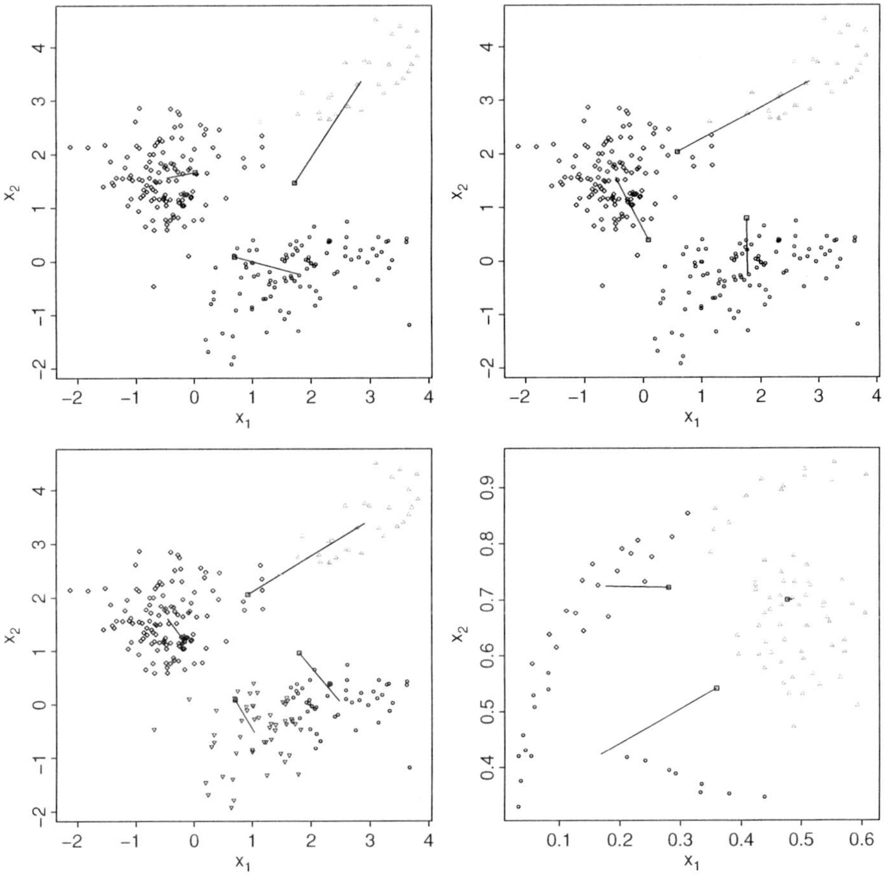

Figure 6.1 Simulated data with three groups each. Top and bottom-left panels: data set C1; bottom-right panel: data set C2. Groups are distinguished by different symbols; squares: initial position of centroids, chosen randomly; line segment: direction of final positions of centroids.

refers to the case when $K = 4$. For these data, the chosen groups, distinguished by type of symbol, correspond satisfactorily to those that are true, in the sense that "true" applies to the top panels whatever their initial configuration. The bottom-left panel obviously contains one group too many, but the union of two of the chosen groups corresponds substantially to one of the true groups. The same type of outcome is also maintained when the configuration of the initial centroids changes greatly.

However, the result of the method is very different in the bottom-right panel, which refers to other data, with a group structure of a more filiform type. In this case, the individual groups are obviously different from those that are true, this is also the case when starting with other choices of initial centroids. This different type of result is due to the choice of metric used, because the Euclidean distance allows for spherical structures.

This method has two limitations: (1) it requires the initial choice of various elements; (2) it can only be applied to quantitative variables. This second restriction can be overcome by substituting the Euclidean distance with another form of dissimilarity, adapting to the case of qualitative variables, and introducing the concept of *medoid*, that is, a unit representative of the group that minimizes within-group dissimilarity. Conversely, the requirement to specify a value for K is, in many cases, a problem, when we have no information on the structure of the data to guide us.

6.1.4 Hierarchical Methods

To overcome the foregoing inconvenience and to solve the abrupt specification of a value for K, methods that structure the data hierarchically and organize them into groups are often used. This is done by associating the set of points with a binary tree structure, so that the leaves of the tree correspond to the units and the nodes to subsets of the points. Due to the nature of a binary tree, this introduces a hierarchy in the subsets associated with the branches.

There are two large families of hierarchical methods: *agglomerative* and *divisive*. We start with those that are agglomerative, which are more highly developed and frequently used.

An *agglomerative* method starts from an initial state in which $K = n$, that is, a state in which each individual constitutes a separate group, and then proceeds by successive aggregations of previously formed groups having low dissimilarity. This sequence of aggregations continues until $K = 1$, that is, when all the individuals belong to the same group.

Clearly, this method of proceeding gives rise to a hierarchical structure in which the subdivision into K groups "contains" the subdivision in $K + 1$ groups, in the sense that the former is obtained from the latter by aggregating two groups. figure 6.2 shows an example of such a tree; in this context, this type of diagram is called a *dendrogram*. The reason for the different lengths of the vertical stems is given shortly.

To turn the general framework into an operational procedure, we need to introduce a measure of dissimilarity between the two groups. At the start of the agglomeration process, when all the groups are formed of a single unit, it is clear that $d(i, i')$ also constitutes the dissimilarity between the two degenerate groups formed by $\{i\}$ and $\{i'\}$. In later stages, we agglomerate groups formed of several units and, correspondingly, need a dissimilarity measure between groups composed of more than one unit. If G and G' represent two groups, the three most frequently used measures are:

$$d_S(G, G') = \min_{i \in G, i' \in G'} d(i, i'), \qquad d_C(G, G') = \max_{i \in G, i' \in G'} d(i, i'),$$

$$d_M(G, G') = \frac{1}{n_G \, n_{G'}} \sum_{i \in G} \sum_{i' \in G'} d(i, i'),$$

which are called *single link*, *complete link*, and *average link*, respectively. Obviously the grouping changes with the adopted measure.

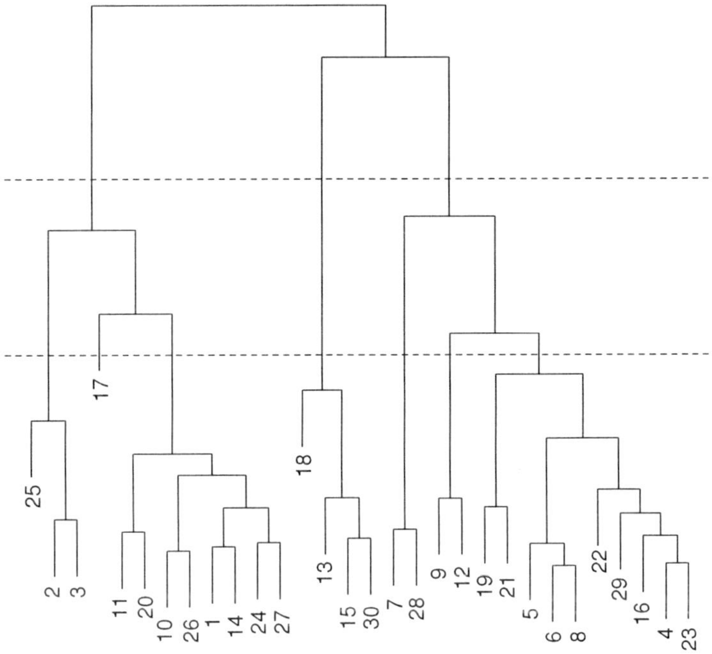

Figure 6.2 A dendrogram.

Information about the dissimilarity between two groups can be incorporated into a dendrogram by making the height of the vertical line connecting two successive ramifications on the same branch proportional to the fall in dissimilarity obtained by passing from K to $K + 1$ groups.

This fact can be used as a guide to use the dendrogram for the operative choice of number of groups, if this is not known a priori. For example, in figure 6.2, the two dashed lines identify $K = 3$ and $K = 7$ groups. We usually cut the tree horizontally at the level where the vertical stems are longer, and the number of intersecting stems represents the number of prechosen groups. "Objective rules" also exist, but which of them is preferable is not immediately obvious. Again, the analyst must make an evaluation.

To appreciate the effect of different types of link, we examine figure 6.3, which shows the same data as the first three panels in figure 6.1. From top to bottom, the first pair of panels refers to the single link, the second to the complete link, and the third to the average link. For each pair, the left panel presents the dendrogram and the right panel the clusters corresponding to $K = 3$, with the same symbols as in figure 6.1 to distinguish the groups.

In this example, the single link method clearly does not work nearly as well as the others. This negative result is due to the spheroidal form of the groups. In fact, in figure 6.4, in which the data from the last panel of figure 6.1 were used, the groups determined by the complete and average links do not correspond to the true groups, whereas the single link does allow them to be identified. This means

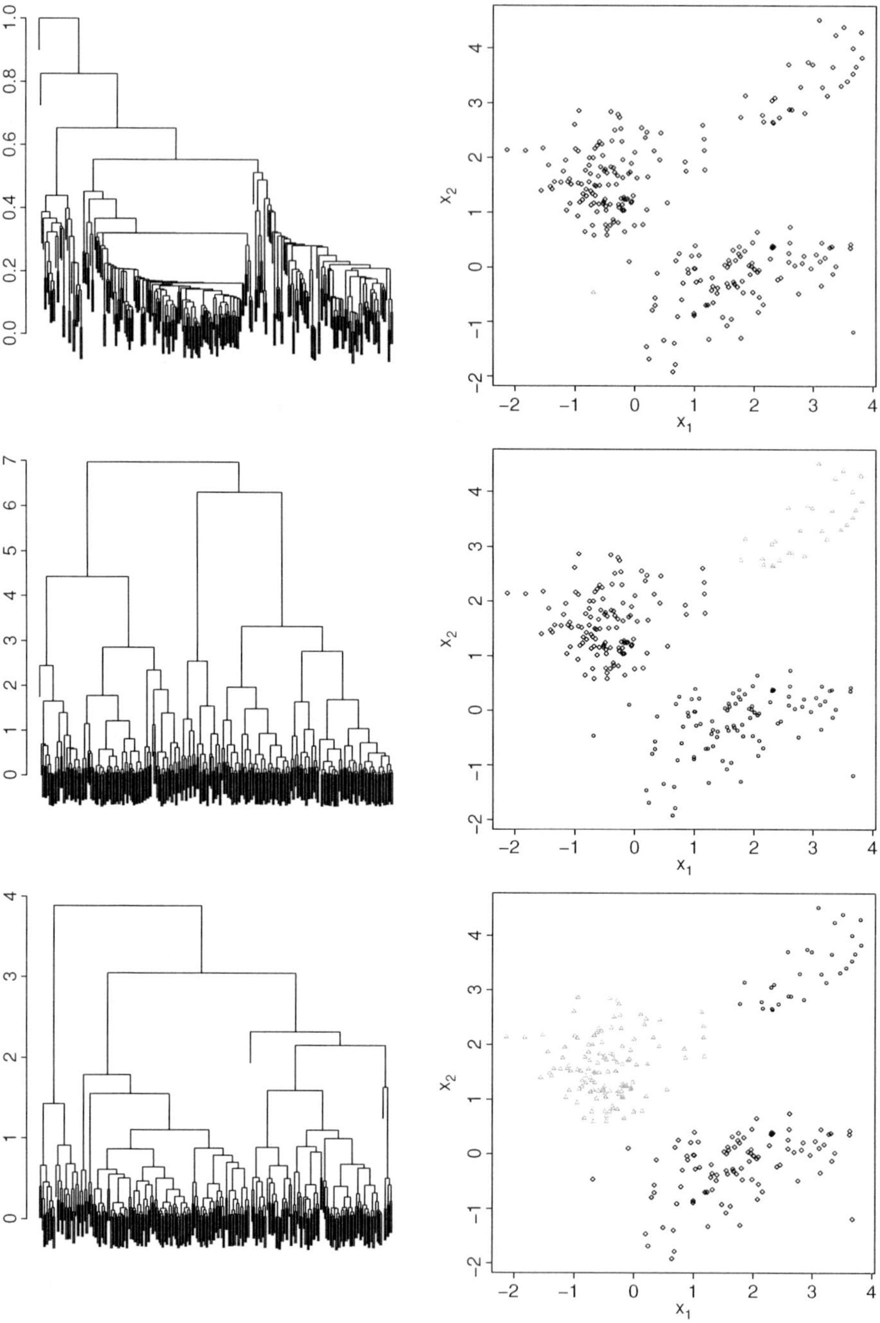

Figure 6.3 Simulated data C1: Groups made with agglomerative hierarchical method and three types of link (from top to bottom: single, complete, and average links). Left: dendrogram; right: clustering when $K = 3$.

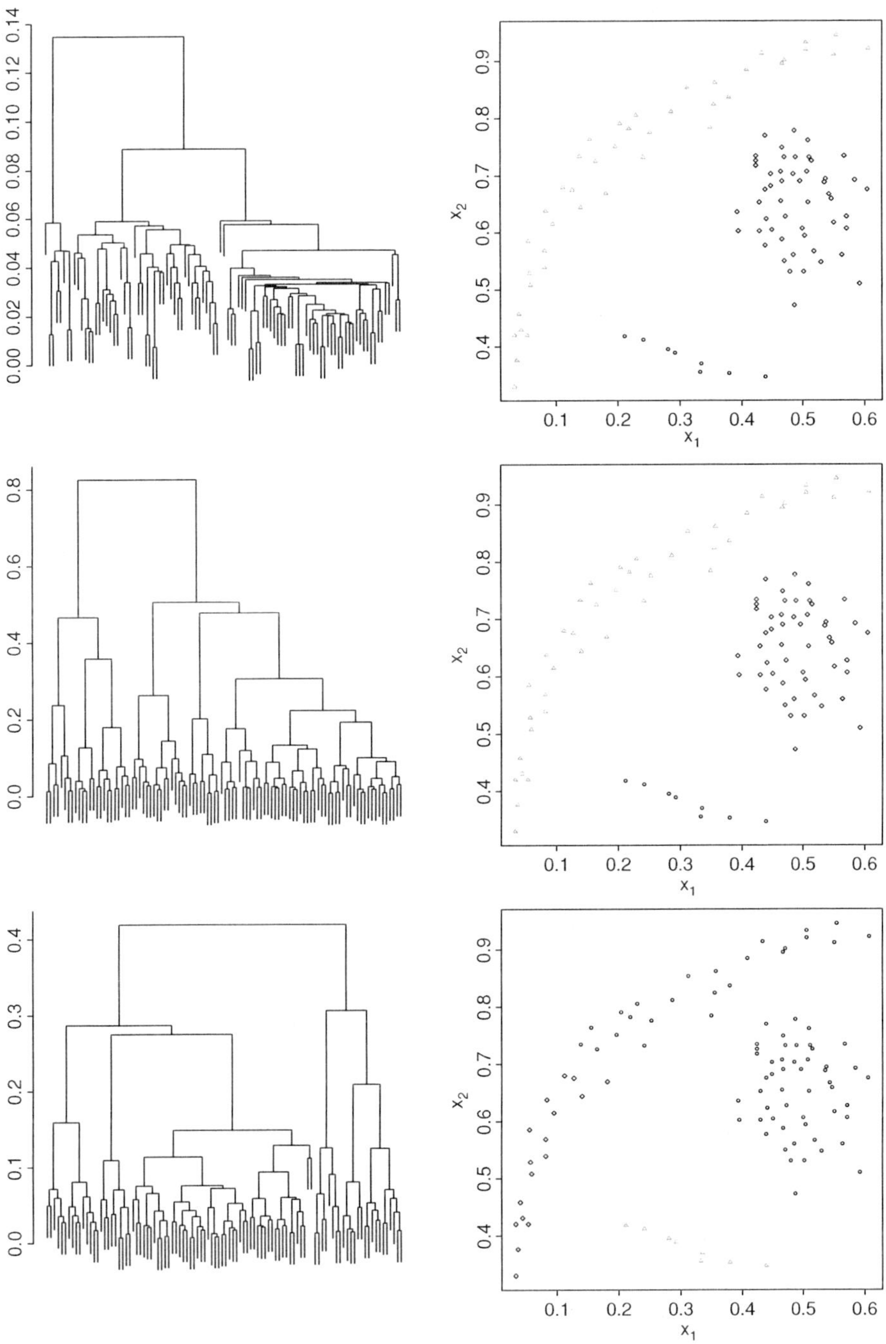

Figure 6.4 Simulated data C2: Groups made with agglomerative hierarchical method and three types of link (from top to bottom: single, complete, and average links). Left: dendrogram; right: clustering when $K = 3$.

that the single link tends to work better with filiform types of geometric structures and the complete link with spheroidal structures.

In one sense, divisive methods represent the dual approach to agglomerative methods. Here we follow a logic similar to the previous case but start from the opposite extreme—that is, first forming one group that includes all units, and then proceeding by successive subdivisions.

The division of a group is evaluated according to the dissimilarity between the various choices of two subgroups that can be formed starting from the original one. These dissimilarities between subgroups are evaluated with the same forms of links already seen for agglomerative methods. However, divisive methods have been less thoroughly explored and are less frequently used than agglomerative ones.

Bibliographical notes

A pioneering work on cluster methods is by Hartigan (1975). Another classic account is found in Mardia et al. (1979, ch. 13), which, although more concise, is still clearly described. A work that in its time significantly influenced the formulation of the concept of dissimilarity is that of Gower (1971). A relatively more recent treatment, with particular emphasis on computational aspects, is by Kaufman & Rousseeuw (2009).

6.2 Associations Among Variables

The previous section concerned the clustering of units and, in a more general sense, their forms of association. We now deal with the dual problem of relations among variables.

6.2.1 Elementary Notions of Graphical Models

A large proportion of statistical methodology is concerned with studying how variables are connected to each other. This broad problem has various forms, according to whether the variables are quantitative or qualitative, whether there is a natural distinction between explanatory and response variables, and so on. In fact, much of what we have seen in the previous chapters deals with the problem of relationships between variables in the asymmetric case, that is, one or more variables that are responses to explanatory ones. Here, we briefly deal with the symmetric case, in which all variables play the same role.

The best-known concept of dependence between two variables is probably that of correlation. If x_r and x_s are the vectors of observations on two quantitative variables, recorded from the same n units, the sample correlation between them can be written as

$$\mathrm{corr}\{x_r, x_s\} = \frac{\langle x'_r, x'_s\rangle}{\|x'_r\|\ \|x'_s\|} = \cos(\text{angle between } x'_r \text{ and } x'_s)$$

where x'_r and x'_s represent the deviations of x_r and x_s from their respective arithmetic means, and the notation $\langle x'_r, x'_s\rangle$ indicates the inner product.This is not the most common way the correlation is expressed algebraically, but it has the advantage of showing its geometric interpretation, and it highlights the fact that correlation measures the degree of alignment of the directions of x'_r and x'_s. If we have p numerical variables, say, $x_1, \ldots, x_p$, we calculate the correlation matrix formed by all the pairs of $\mathrm{corr}\{x_r, x_s\}$, for $r, s = 1, \ldots, p$.

The population version of the concept of correlation, referred to two random variables, X_r and X_s, is given by

$$\mathrm{corr}\{X_r, X_s\} = \frac{\mathbb{E}\{X'_r X'_s\}}{\|X'_r\| \, \|X'_s\|} = \frac{\mathrm{cov}\{X_r, X_s\}}{\sqrt{\mathrm{var}\{X_r\} \ \mathrm{var}\{X_s\}}}$$

where $X'_r = X_r - \mu_r, X'_s = X_s - \mu_s$ denote the centred variables after subtracting their mean values, μ_r and μ_s, and we have used the nonstandard notation

$$\|U\| = \mathbb{E}\{U^2\}^{1/2}$$

referred to a 0-mean random variable U. As for the sample version, a set of p random variables leads to the introduction of a correlation matrix formed by all pairs $\mathrm{corr}\{X_r, X_s\}$; see also section A.2.1.

Although the correlation matrix is a fundamental tool in studying dependence structures, it does have limitations. One is the fact that a correlation reflects exclusively the dependencies of *linear* type between variables, but as this is discussed in every introductory textbook on statistics, we do not discuss it now.

Another source of difficulty in interpreting the values of the correlation matrix is illustrated by the following simple numerical example, taken from Mardia et al. (1979, p. 170). In a sample of children, the variables are:

$$x = \texttt{intelligence}, \quad y = \texttt{weight}, \quad z = \texttt{age}$$

and a sample correlation matrix is

$$R = \begin{pmatrix} 1 & 0.6162 & 0.8267 \\ 0.6162 & 1 & 0.7321 \\ 0.8267 & 0.7321 & 1 \end{pmatrix}.$$

The high correlation between weight and intelligence, 0.6162, indicates a relationship between variables that is very surprising and unlikely on general grounds. The problem lies in the third variable, age, and how it interacts with the other two.

Therefore, for better indications, we must examine the dependence between x and y after the effect of z has been removed. This leads us to obtain the residual vectors

$$e^{(x)} = x - \left(\hat{\beta}_0^{(x)} + \hat{\beta}_1^{(x)} z\right), \qquad e^{(y)} = y - \left(\hat{\beta}_0^{(y)} + \hat{\beta}_1^{(y)} z\right),$$

after the linear dependence on z has been removed by fitting a simple regression model of type (2.1) on each of x and y and to consider the correlation between these residual vectors. We then arrive to the introduction the *partial correlation* between x and y given z, which is defined as

$$\text{corr}\{x, y\}^* = \text{corr}\left\{e^{(x)}, e^{(y)}\right\}.$$

Again, a population version of the partial correlation is introduced, replacing sample vectors by random variables and sample moments by population moments, as for the correlation.

In our numerical example, the partial correlation between `weight` and `intelligence`, once `age` is fixed, drops to a much more reasonable 0.0286; this means that they are essentially uncorrelated. Repeating the previous operation for all the variables—that is, considering all the possible pairs of variable—we obtain the partial correlation matrix

$$R^* = \begin{pmatrix} 1 & 0.0 & 0.7 \\ 0.0 & 1 & 0.5 \\ 0.7 & 0.5 & 1 \end{pmatrix}$$

where we round the values to only one decimal place; in particular, 0.0286 is rounded to 0. This is reasonable when we consider that the observed correlations are subjected to sampling variability. To proceed in a canonical way, we would have to test a statistical hypothesis formally, but that is not the point we wish to focus on now.

The matrix of partial correlations, R^*, is much easier to interpret than that of marginal correlations, R, particularly when we associate it with a graph like that shown in figure 6.5, which is made up of one node for every variable and one nondirected edge for every nonzero partial correlation. The graph shows that x and y are correlated only "through" z, and are uncorrelated conditionally on the value assumed by z.

Now assume joint normality of the three parent random variables, say, (X, Y, Z). Because independence and lack of correlation are equivalent conditions in the context of multivariate normal distributions, we have a situation of *conditional independence* between X and Y, conditionally on the value of Z: we write $X \perp\!\!\!\perp Y|Z$.

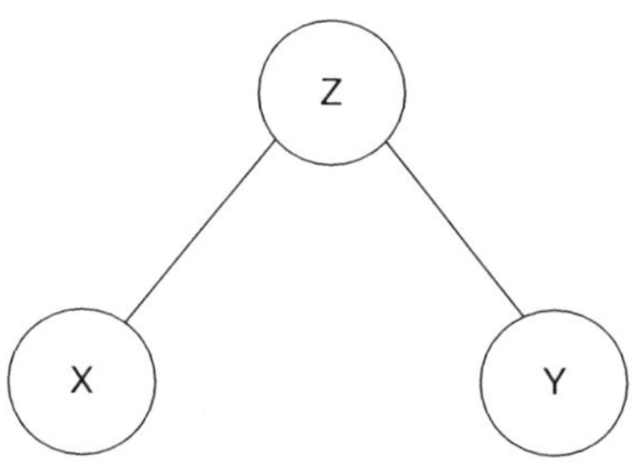

Figure 6.5 A simple graphical model.

The need to develop tools for examining and correctly interpreting complex dependence structures becomes more pressing as the number of available variables increases. The theoretical apparatus is often called a *graphical model* because it is linked to the idea of expressing the dependence structure by means of a graph. This theory is highly structured: it does not handle only continuous variables, nor does it refer only to analysis of association structures of a symmetric nature, but it also covers the asymmetric case, in which one or more variables play the role of the response variable with respect to the explanatory variables, as described in previous chapters. Here we merely mention the analogy of the previous case when using categorical variables.

Now move to the case where X and Y represent two categorical variables. Their joint distribution is identified by the set of probabilities

$$\pi_{jk} = \mathbb{P}\{X = x_j, Y = y_k\}$$

where x_j and y_k vary in the set of levels for variables X and Y, respectively. It is also useful to rewrite these probabilities in another form, based on the identity

$$\pi_{jk} = \pi_{j+}\,\pi_{+k}\,\frac{\pi_{jk}}{\pi_{j+}\,\pi_{+k}}$$

where symbol $+$ indicates the sum of the values over the corresponding index (e.g., $\pi_{j+} = \sum_k \pi_{jk}$). This yields

$$\log \pi_{jk} = \beta_j^X + \beta_k^Y + \beta_{jk}^{XY} \qquad (6.3)$$

where $\beta_j^X = \log \pi_{j+}$, and analogously for the other terms.

This factorization of probabilities allows a clearer interpretation of the ingredients. Because X and Y are categorical variables, the right-hand side is similar to the same type used in two-way analysis of variance. In our case, too, the various parameters are subject to constraints, such as

$$\sum_j \pi_{j+} = 1 = \sum_k \pi_{+k}.$$

Terms β_j^X and β_k^Y of (6.3) play the role of main effects and reflect the marginal distribution of X and Y. "Interaction" term β_{jk}^{XY}, which depends on the relationship between probabilities π_{jk} and their value in the independence case, $\pi_{j+}\,\pi_{+k}$, constitute an *association measure* between factors X and Y. Specifically, if $\beta_{jk}^{XY} = 0$ for all j and k, we have a situation of independence between X and Y; conversely, positive values indicate that the probability of event $\{X = x_j \cap Y = y_k\}$ is higher than in the independence hypothesis, and there is therefore a positive association between event components $\{X = x_j\}$ and $\{Y = y_k\}$. In reverse, negative values of the parameter indicate a "repulsion" situation, or negative association between events.

Now assume that, on the basis of a sample of n elements, a *frequency table* has been constructed, of which entry n_{jk} represents the observed frequency of events $\{X = x_j \cap Y = y_k\}$. Denote the expected value of n_{jk} by μ_{jk}. From (6.3), it immediately follows that

$$\log \mu_{jk} = \log n + \beta_j^X + \beta_k^Y + \beta_{jk}^{XY} \tag{6.4}$$

which is a special case of the generalized linear models (2.42). In this particular case, the link function is the logarithm and (6.4) is an example of the *log-linear model*.

We can also use theoretical apparatus and the iterative weighted least squares algorithm for GLMs. Starting from observed values n_{jk}, we can estimate the parameters of (6.4) and carry out other inferential operations. We can therefore verify that the available data allows the removal of component β_{jk}^{XY} from (6.4), inasmuch as it is not significant, and we conclude that X and Y are independent variables.

However, we often have to deal with more than two variables, sometimes many more. As in analysis of the correlation structure of a continuous multivariate variable, it is essential to use tools allowing for systematic examination of dependence structures that rapidly become complicated. The concept of conditional independence also plays an important role in the case of categorical variables.

If we consider three categorical variables, (X, Y, Z), and indicate by μ_{jkl} the number of obtained observations for the general cell of the corresponding three-way table, representation as the corresponding log-linear model, as in (6.4), is

$$\begin{aligned}\log \mu_{jkl} = \log n &+ \beta_j^X + \beta_k^Y + \beta_l^Z \\ &+ \beta_{jk}^{XY} + \beta_{kl}^{YZ} + \beta_{jl}^{XZ} + \beta_{jkl}^{XYZ}\end{aligned}$$

where the significance of the new symbols is similar to those already introduced. Note here that term β_{jkl}^{XYZ} is also introduced, whereas in the Gaussian case a term expressing an association between three components did not exist, because the particular nature of normal distribution allows us to express all associations among variables via correlations, and thus only involves pairs of variables.

The specification of the foregoing model for the independence of X and Y conditional on Z is given by

$$\log \mu_{jkl} = \log n + \beta_j^X + \beta_k^Y + \beta_l^Z + \beta_{kl}^{YZ} + \beta_{jl}^{XZ}$$

and figure 6.5 shows the relative graph (with different labeling of nodes).

In the applicational context on which we focus, log-linear and graphical models are used in studying associations between variables (often categorical) that variously represent aspects of customers' behavior. These associations, both positive and negative, give us useful suggestions for company commercial actions.

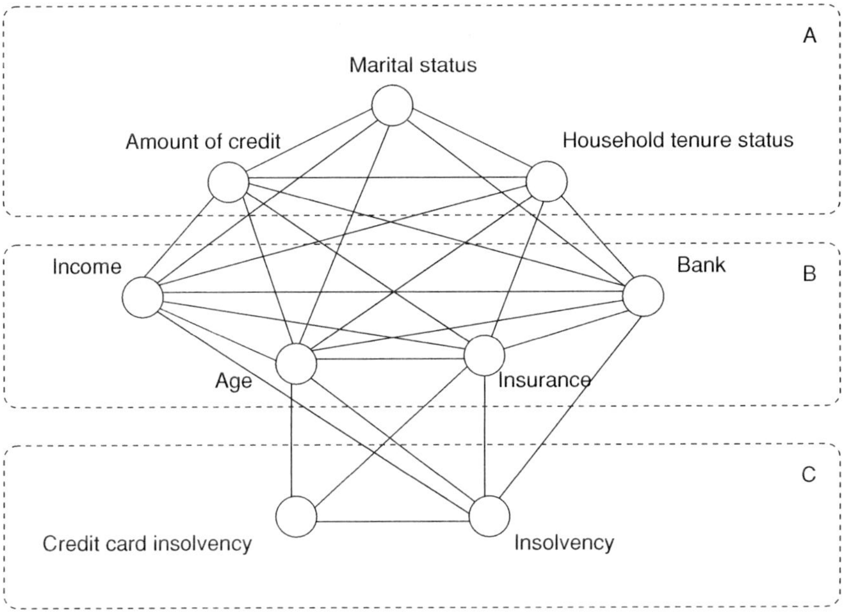

Figure 6.6 A graphical model for credit scoring.

In terms of computational cost, a high value of n has a small effect because determination of frequencies n_{jk} is fast and computing time increases linearly with n. Once the frequency table has been obtained, later processing has a computational cost that does not depend on n. However, difficulties may arise if the number of variables involved is high, and even more so if the number of possible levels of these variables is large, because this can lead to a huge frequency table. We discuss this aspect in the following subsection from a different point of view.

To illustrate the capacity of the representation of complex dependency structures, consider figure 6.6, taken from Hand et al. (1997). It shows a model to evaluate which variables influence the occurrence of insolvency in returning a bank loan. The study was based on a survey of about 23,000 holders of loans issued by a large U.K. bank. Financing not exceeding £10,000 was allowed, not covered by secure guarantees.

The variables in the model described customers according to their demographic characteristics, which are `Age` (categories: 17–30, 31–40, or over 40) and `Marital Status` (Married, Other), and socioeconomic ones, which were `Income` (up to £700, £700–£1500, over £1500) and an indicator variable of housing tenure status. Information derived from the credit history of the customer, encoded by some indicator variables, was also available: `Bank` indicates a current account with the loan company, `Credit card insolvency` indicates past difficulties with credit card payments. Last, there is information about finances: the loan `Amount` (up to £3000, over £3000), `Insolvency`, measured with an indicator variable of a certain number of missed payments, and an indicator variable for taking out loan protection insurance, because all customers were given the opportunity, by means of a small increase in the monthly payment, to buy insurance to protect themselves from some types of insolvency.

One node of the graph in figure 6.6 is associated with each variable. The edges denote associations among the variables; the absence of an edge means that the corresponding variables are conditionally independent, given the values of the other variables. The advantage of visualizing the model by means of the graph is the parallel between graphical separation and conditional independence, which means that if all paths from a node of set A to a node of set C pass through a node of set B, then set A is conditionally independent of set C, given B. This implies that, knowing the values of the variables in B, knowledge of the variables in C does not add any information about the nodes in A, and vice versa.

Interpreting that, if income, age, and the indicators of bank and insurance company are known, then the amount of credit, marital status, and housing tenure status do not provide extra information in predicting loan insolvency or financial insolvency. In other words, the set of nodes grouped in sets A, B, and C in figure 6.6 behave the same way as X, Z, and Y in figure 6.5. Therefore, for this type of financing, the variables for income, age, and the indicators of insurance, bank, and credit card insolvency contain all the information regarding insolvency. This causes a significant reduction in the size of the problem and enables us to identify customer profiles with an insolvency probability about three times the marginal probability.

6.2.2 Association Rules

Let us denote by $A_1, A_2, \ldots, A_p$ a set of binary variables, whose possible values are labeled 0 and 1. Although in an abbreviated fashion, the previous section showed how the basis of n observations of such variables can let us develop a model to represent the dependence structure of such variables compactly.

To develop the connected log-linear model, we first have to construct the p-ways frequency table. If all the variables are dichotomous, the number of cells in the table is 2^p cells, a number that "explodes" rapidly as p increases and is higher still if some of the variables have more than two levels. If $p = 20$, for example, the number of cells is $2^p = 1048576$. The potential number of parameters to be estimated for the connected log-linear model is slightly lower, but is gigantic in any case.

To explain the following, we refer to a background applied problem, where high values of p are easy to observe. In *market basket analysis*, variable A_j is the indicator variable, often called an *item*, that is, a customer has purchased the jth product from the company catalog ($j = 1, \ldots, p$). According to data on the purchases made by n customers, we identify the associations existing among variables $A_1, \ldots, A_p$, or at least pick up those that are considered interesting.

If the company in question has a catalog containing a small number of items, for example, a service company, then the methods discussed in the previous section are perfectly adequate. Instead, if the company is a supermarket, then p is easily of the order of thousands, a frequency table with 2^{1000} cells cannot even be stored in a computer, and it cannot be processed to develop a log-linear model. Besides computing complications, there is also a serious inferential problem with this table, which will inevitably be extremely sparse, that is, with very many zeroes, hence violating the standard assumptions for inferences about log-linear models.

In short, we must explore other routes. An alternative and currently very popular method comes from the field of machine learning and similar areas. It refers to the idea of *association rule*, intended as a proposition of the type

$$\text{condition} \Rightarrow \text{consequence}$$

as for example

$$\text{it is raining} \Rightarrow \text{the ground is wet}$$

The concept of rule constitutes a classic paradigm in the field of artificial intelligence as a way of representing knowledge. The variant of this concept of more direct interest to us is the *probabilistic (association) rule*, which assigns a probability to the previous "consequence," once the condition has been fulfilled. For example, the rule

$$\text{the customer purchases bread and jam} \Rightarrow \text{the customer purchases butter}$$

does not intend to be deterministic, and therefore a probability is typically associated with it.

On the basis of n purchases carried out by as many customers, our aim is thus to choose rules that in probability theory correspond to conditional probabilities of the type:

$$\mathbb{P}\{E_2|E_1\} = \frac{\mathbb{P}\{E_1 \cap E_2\}}{\mathbb{P}\{E_1\}} = \pi_{12} \tag{6.5}$$

where E_1 is an event related to a group of variables and E_2 an event determined by another set of variables; all the rules need not involve the same number of variables. Obviously, to evaluate these probabilities numerically, we make use of the relative frequencies of the same events, as observed in the data. For example, `jam` is the indicator variable of the purchase of jam, and so on for other variables. So a simple probabilistic rule is of the type

$$\mathbb{P}\{\texttt{butter} = 1|\texttt{bread} = 1, \texttt{jam} = 1\} = 0.71 \tag{6.6}$$

where E_1 involves two indicator variables and E_2 one. In this context, sets of events such as E_1 or E_2 are called *itemsets* and are called k-itemsets if the events specify the values of k indicator variables. For example, E_2 is a 1-itemset and E_1 a 2-itemset.

We can infer many rules from a data set, even for a limited number of variables, but to be useful, a rule must satisfy various conditions, as follows:

- Obviously, a rule must have a high level of *confidence*, that is, value π_{12} of (6.5) must be high, ideally, 1.

- The rule must also be capable of being applied to a suitable number of cases. For example, the rule of (6.6) has a good level of confidence, but if we later discover that hardly anyone buys both bread and jam, it is practically useless. The rule must therefore have a high *support*, given by $\mathbb{P}\{E_1\}$ in (6.5). The term *support* is also sometimes used to refer to $\mathbb{P}\{E_1 \cap E_2\}$. Itemsets with supports larger than a fixed threshold are called *frequent itemsets*.
- In a predictive approach, another characteristic for a good rule requires that knowledge that the "condition" is verified should produce a better prediction of the "consequence." A measure of this is given by the ratio between confidence and the support of the consequent event, which is a measure of expected confidence when the condition is not known. An estimate of this association measure, $P(E_2|E_1)/P(E_2) = P(E_2 \cap E_1)/P(E_1)P(E_2)$, is called *lift*. Although this term coincides with that of section 5.2.4, the two concepts are separate. Note that this lift is the exponential of term β^{XY} in (6.3).
- Last, the rule must be "interesting." The rule "if a person has a baby, then she is a woman" has a confidence level of 1, and support is not negligible, but the rule states nothing of interest. Identifying what is 'interesting' is not always easy, because it often involves specific aspects of the essential problem. However, there have been proposals to introduce quantitative criteria, such as the "*J* measure," which is fundamentally given by the Kullback-Leibler divergence between conditional distribution $(\mathbb{P}\{E_2|E_1\}, \mathbb{P}\{\bar{E}_2|E_1\})$ and unconditional distribution $(\mathbb{P}\{E_2\}, \mathbb{P}\{\bar{E}_2\})$, weighting the divergence with $\mathbb{P}\{E_1\}$.

The problem of operatively identifying the rules remains. Conceptually, the type of operations required is elementary: we first calculate the empirical frequencies of various subsets and then select those that are the most useful with respect to the criteria. Although the required operations are very simple, the size of the possible events to consider is mind-boggling, even when the number p of variables is not very high and the computational cost becomes unmanageable.

However, the APriori algorithm comes to our rescue. Developed specifically for this problem, APriori is highly efficient and can select a set of associated rules that are interesting in some way, even though a limited number of data readings is available. The APriori algorithm, presented in algorithm 6.2, uses a hierarchical "level-wise" search, where k-itemsets (i.e., containing k indicator variables) are used to explore $(k+1)$-itemsets to find frequent itemsets. This is done by following the a priori property: any $(k+1)$-itemset that is not frequent cannot be a subset of a frequent k-itemset and hence should be removed. Initially, the set of frequent 1-itemsets is found. This is used to find the set of frequent 2-itemsets, which in turn is used to find the set of frequent 3-itemsets, and so on until no more frequent k-itemsets can be found.

The results and conclusions are rather different from those discussed in previous sections for other models. At least two considerations must be made.

Algorithm 6.2 APriori algorithm for association rules

1. Assign a threshold t_s for support and one t_c for confidence.
2. Let $k = 1$; generate frequent 1-itemsets with support larger than the indicated threshold t_s.
3. Cycle for $k = 2, 3, \ldots$ until no new frequent itemsets are identified:

   a. generate candidate itemsets with length $(k + 1)$ from frequent k-itemsets;
   b. prune candidate *itemsets* containing infrequent k-itemsets (support lower than threshold t_s);
   c. obtain the support of each candidate $(k + 1)$-itemset by scanning the entire data set;
   d. eliminate infrequent candidates, keeping only those whose support is larger than threshold t_s.

4. For every nonempty subset of each *frequent itemset*, choose the rules that have confidence larger than threshold t_c.

- The final result is not a global model illustrating the complex behavior of the phenomenon but a selection of particular aspects that are considered of interest. The aim of the study is therefore part of the identification of interesting data *patterns*; see section 1.1.2.
- As the association rules selected in this way are not inserted in an inferential process, we have no information about their level of generalizability. It is not difficult to construct a statistical significance test for a fixed proposition, but the problem is that, in principle, we carry out a large number of such tests and select only those that are the most significant. We therefore process repeated hypothesis tests that completely change the real significance level, which in the end is very different from the nominal one.

A final remarks deals with the field of application of the association rules. As already noted, the most classical application is market basket analysis, but the same concepts are relevant for other uses. An example is text analysis, in which indicator variable A_j may indicate the presence or absence in a certain fragment of text of the jth term of a vocabulary list with p terms $(j = 1, \ldots, p)$.

Bibliographical notes

The theory of graphical models is excellently explained by Whittaker (1990) and is still very pertinent today. Two other classic texts are those by Lauritzen (1996) and Cox & Wermuth (1998), the former more mathematical in nature, the latter combining theoretical and applicative aspects. Association rules are discussed, among others, by Hand et al. (2001, ch. 13). The APriori algorithm was developed by Agrawal et al. (1996), combining previous works by the same authors.

6.3 Case Study: Web Usage Mining

We return to the real-life case analyzed in section 5.10.4. Here, we concentrate on the segmentation of visitors to the company website.

In this section, we follow two lines of analysis that complement what was already shown in section 5.10.4. First, we look for behavioral segments of visitors in terms of visited pages, in particular by classifying visitors into homogeneous groups according to visits to pages belonging to the eight areas already used in the previous analysis. We then analyze sequences of visited pages by identifying the most likely navigation paths in the website.

6.3.1 Profiling Website Visitors

Website managers are interested in behavioral segmentation of visitors for future marketing decisions, and cluster analysis (see section 6.1) is typically used.

Table 6.2 shows which visitors reached which areas in sessions with a single page. Clearly, as people visiting only one page are easily classified by the area including that page, we remove them from subsequent analysis.

A more specific method of analysis is needed for the 4,572 visitors going to two or more pages. Table 6.3 lists some descriptive indicators of the distributions of the number of hits for each area. All distributions are highly skewed to the right, and it seems reasonable to use some kind of data transformation to run a clustering procedure. Because the variables represent counts, all measured on the same scale as the number of hits, it also seems reasonable to consider the base 2 logarithm of each of them incremented by 1. This transformation, once rounded upwards, is the number of binary digits needed to write the counting number.

The choice of a complete linkage is natural in implementing hierarchical cluster analysis in this case, where we are looking for very homogeneous groups. Figure 6.7 shows the dendrogram for identifying the optimal number of groups, in which the dashed line shows the level at which we decide to cut the tree, obtaining

Table 6.2. Web Usage Mining: Percentage of the Single-Page Sessions for Each Area

Area	% Visits
Business area	23.21
Communications	1.45
Company	1.00
Consulting	8.70
Contacts	0.12
Events	1.42
Home	1.33
White papers	50.78

Table 6.3. Web Usage Mining: Means, Medians, and Percentiles for Number of Hits to Each Area

Area	Mean	Median	3rd quartile	90th percentile	99th percentile
Business area	2.275	2	3	6	9
Communications	0.2937	0	0	0	2
Company	0.8871	0	0	3	6
Consulting	0.7220	0	0	2	4
Contacts	0.1562	0	0	1	1
Events	0.1223	0	0	0	1
Home	0.6177	0	0	2	3
White papers	0.5693	0	0	2	3

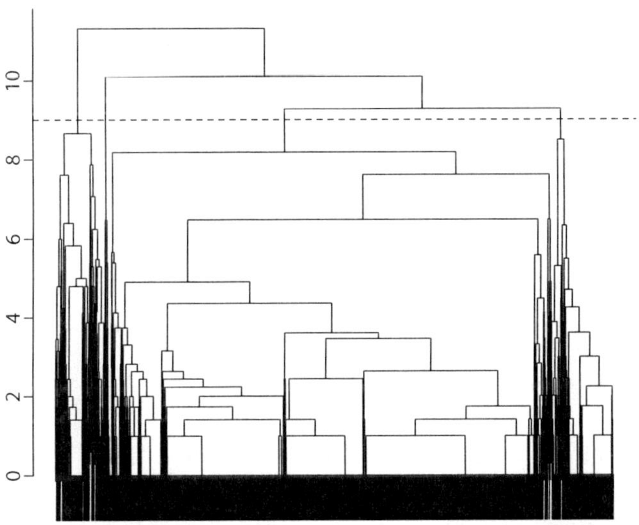

Figure 6.7 Web usage mining: Dendrogram for cluster procedure with complete linkage.

four groups. Cuts allowing the choice of three or four groups are all of similar height, so we decide to use the larger number of clusters (i.e., four).

Table 6.4 lists averages and standard deviations for each cluster of the eight variables used. These outcomes show the following.

- Cluster A is characterized by a large number of visits to the business area, home, company, consulting, and contacts pages. These visitors are probably the most interested ones, who may become customers of the consulting branch: they look at all the company information, the consulting area, and the business areas in which the company works.
- Cluster B is the smallest and has a very high number of visits to the white papers area. These customers are interested in the workings of the

Table 6.4. WEB USAGE MINING: MEANS AND STANDARD DEVIATIONS (IN BRACKETS) FOR NUMBER OF VISITS TO EACH AREA

Area	Cluster A	Cluster B	Cluster C	Cluster D	Overall mean
Business area	9.43 (5.83)	0.58 (2.02)	1.76 (1.92)	0.30 (0.73)	2.27 (3.36)
Communications	0.67 (1.59)	0.02 (0.14)	0.27 (1.24)	0.16 (0.62)	0.29 (1.22)
Company	3.15 (6.40)	0.07 (0.43)	0.69 (2.03)	0.53 (1.90)	0.89 (2.79)
Consulting	2.23 (3.80)	0.41 (1.81)	0.13 (0.46)	3.73 (3.76)	0.72 (2.13)
Contacts	0.37 (0.78)	0.02 (0.14)	0.12 (0.39)	0.21 (0.59)	0.15 (0.47)
Events	0.43 (0.79)	0.11 (0.37)	0.09 (0.35)	0.13 (0.58)	0.12 (0.45)
Home	1.67 (2.14)	0.62 (1.35)	0.43 (0.88)	1.10 (2.41)	0.62 (1.36)
White papers	0.40 (1.83)	14.75 (14.18)	0.46 (0.94)	0.02 (0.14)	0.57 (2.38)
Number of visitors	409	53	3603	507	4572

company, probably with the aim of learning more about what the company actually does and how, rather than doing business with it.

- Cluster C is the largest group, and shows a low level of interest in consulting and events. None of the areas is visited more than another, and this group may be considered as one of general surfers.
- Cluster D shows great interest in consulting, home page, and contacts. These visitors are less interested in white papers and business, so they are probably less interested in understanding in detail how the company works, but they still seem to be interested in the company's products. They are probably a group of potential customers, although less determined and perhaps less knowledgeable than those of Cluster A.

We also implement the nonhierarchical k-means algorithm. As discussed in section 6.1, we need to select the number of expected clusters. In practice, we find solutions for a range of values for the numbers of clusters and examine the value of the within-group sum of squares associated with each solution. As the number of groups increases, the within-group sum of squares decreases. However, we may find some sudden change indicating the best solution. The top panel of figure 6.8 plots this quantity for a range of number of clusters. Here, the centers of each cluster are randomly selected. The plot suggests looking at the four-cluster solution, where the "elbow" is slightly sharper. The means of the clusters are plotted in the bottom panel of figure 6.8 and show the following:

- Group 1, with 175 sessions, is characterized by a high number of visits to every area, and is the group with the most loyal visitors.
- Group 2, with 2752 sessions, includes visitors with a high average number of hits on the white paper area, but few to all other areas.
- Group 3, with 388 sessions, shows visitors' great interest in the company, contacts, and events, but little in white papers. Visitors are potential customers for events organized by the company.

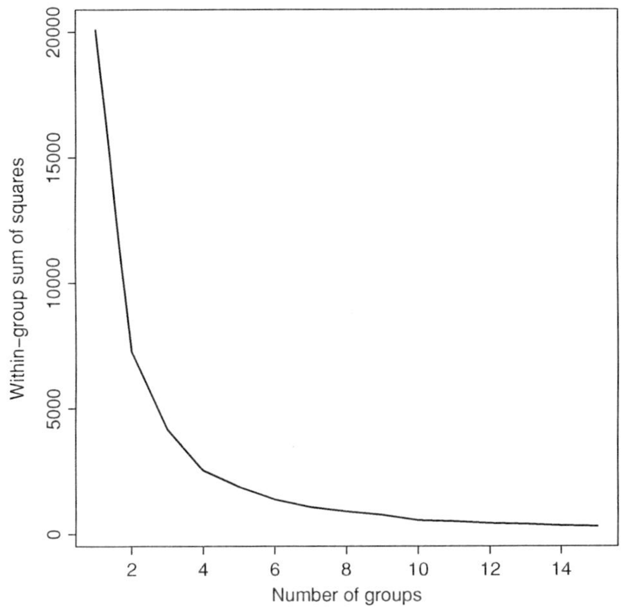

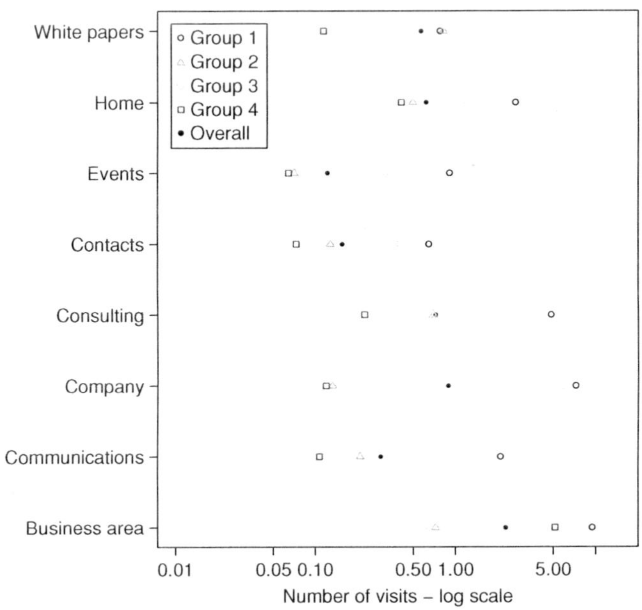

Figure 6.8 Web usage mining. Top: plot of within-group sum of squares against number of clusters. Bottom: means of the clusters for each area.

- Group 4, with 1257 sessions, comprises visitors who are only interested in the business area; all other areas are seldom visited.

These differences between the sets of clusters obtained by the two procedures are noticeable and represent a typical practical situation. Cross-tabulation of the two sets of groups is given in the table.

Hierarchical clusters	k-means clusters			
	1	2	3	4
A	139	1	19	250
B	2	49	0	2
C	27	2228	351	997
D	7	474	18	8

However, the aim of website managers, that is, to find homogeneous groups among visitors, did not require a single segmentation result. They may actually be interested in examining and using each of them for different marketing goals.

For example, the segmentation yielded by the hierarchical procedure can be used to plan actions for customers interested in consulting products. Cluster A is sufficiently small and well characterized to be viewed as a target of marketing action by direct proposals for consulting services on the part of the company. Cluster B comprises people interested in the contents and methods adopted by the company; it may include researchers or other consultants who useful as contacts to improve methodology and share know-how.

The set of groups obtained by *k*-means may also be used to segment potential customers of other products offered by the company, such as organization of events: group 3 is a typical target interested in events. Group 1 is mainly a subset of hierarchical Cluster A, including the most loyal and interested visitors of that cluster. Group 4 isolates visitors interested in the business area, who were not identified in the other sets.

Therefore, both segmentations may be used by website managers to decide on various marketing actions, directed to different targets and with different goals, and new visitors may be included in a specific cluster (one for each of the two segmentations proposed) depending on their surfing habits.

6.3.2 Sequence Rules and Usage Behavior

In section 5.10.4 we saw how visits to a single page can be predicted by analyzing data on the order in which web pages are visited. Here, a finer analysis is proposed to predict navigation paths and page sequences by considering every single page instead of areas and analyzing all observed paths, not only final hits on the contacts pages.

Association rules (see section 6.2.2) can be used to see the most probable navigation paths in the website and predict the pages that will be viewed according to the path the visitor has taken so far.

In consideration of the large number of pages visited, corresponding to the events we wish to associate, we use a simple modification of the APriori algorithm, called the *sequential pattern discovery using equivalence classes* (spade) algorithm, proposed by Zaki (2001) which identifies the navigation paths visited most often. In this case, the order of visits to pages is crucial to understanding the sequential path of the session. To take into account the order of sequences, we only consider rules in which events are naturally ordered and, to simplify computation, associate each sequence to the ordered lists of sessions in which it occurs. Frequent sequences can thus be found efficiently by means of intersections on these lists.

Figure 6.9 shows the frequency bar plot for inspecting the item distribution of pages visited. To reduce the number of items, we only plot item frequency for items with support greater than 2%.

The algorithm found a total of 186 sequences with support of at least 0.5%. By selecting only rules with at least 60% of confidence, we obtain the 15 sequences shown in table 6.5. Here, the support indicates that the percentage of users who visited the two pages were in sequence, and the confidence represents the probability that the second page of the sequence was seen by visitors interested in the first (group of) page(s).

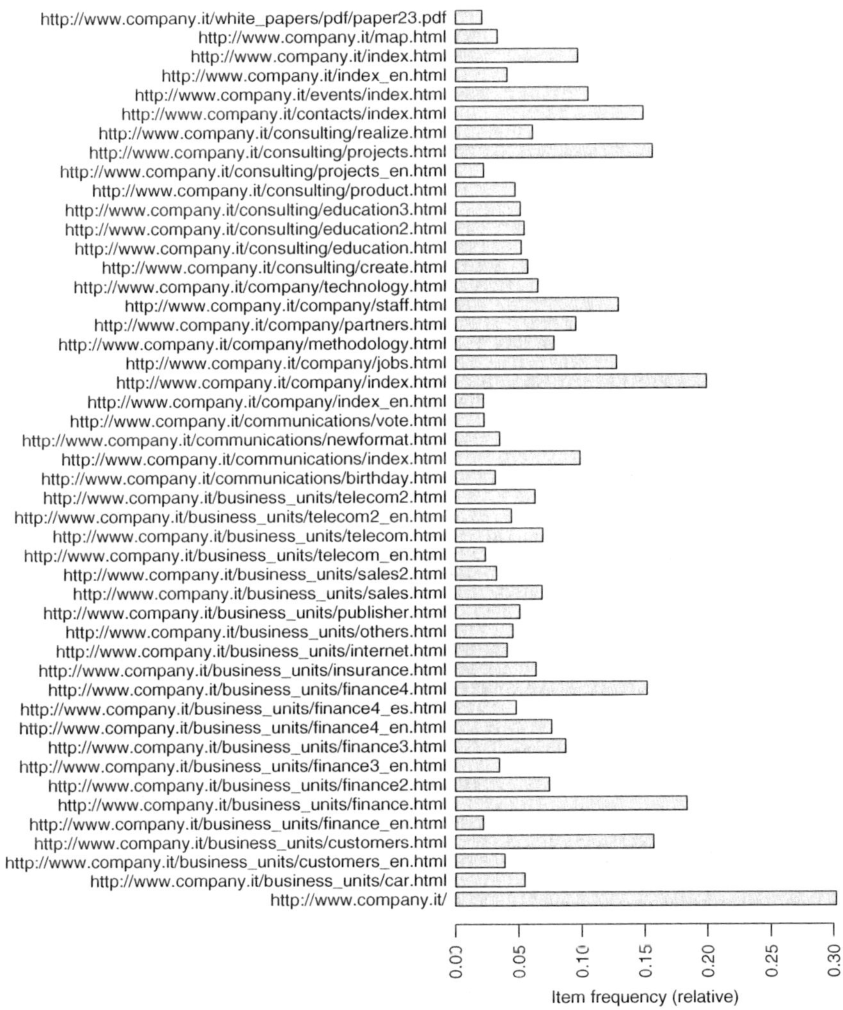

Figure 6.9 Web usage mining: Item frequencies of page views with support greater than 2%.

Table 6.5. Web Usage Mining: The Most Frequent Sequences Occupying More Than One Page

	Rule	Support	Confidence	Lift
1	<http://www.company.it/business_units/finance4.html, http://www.company.it/business_units/customers.html> => <http://www.company.it/business_units/finance4.html>	0.0056	0.7228	10.04
2	<http://www.company.it/company/index.html, http://www.company.it/company/staff.html, http://www.company.it/company/partners.html> => <http://www.company.it/company/jobs.html>	0.0069	0.7054	35.35
3	<http://www.company.it/company/staff.html, http://www.company.it/company/partners.html> => <http://www.company.it/company/jobs.html>	0.0073	0.6931	34.73
4	<http://www.company.it/, http://www.company.it/company/index.html, http://www.company.it/company/staff.html> => <http://www.company.it/company/partners.html>	0.0065	0.6552	49.53
5	<http://www.company.it/, http://www.company.it/company/index.html> => <http://www.company.it/company/staff.html>	0.0100	0.6525	36.86
6	<http://www.company.it/, http://www.company.it/company/staff.html> => <http://www.company.it/company/partners.html>	0.0068	0.6520	49.29
7	<http://www.company.it/company/index.html, http://www.company.it/company/technology.html> => <http://www.company.it/company/partners.html>	0.0054	0.6498	49.12
8	<http://www.company.it/company/index.html, http://www.company.it/company/partners.html> => <http://www.company.it/company/jobs.html>	0.0074	0.6467	32.40
9	<http://www.company.it/company/index.html, http://www.company.it/company/staff.html> => <http://www.company.it/company/partners.html>	0.0099	0.6434	48.64
10	<http://www.company.it/consulting/create.html> => <http://www.company.it/consulting/realize.html>	0.0052	0.6398	75.38
11	<http://www.company.it/company/index.html, http://www.company.it/company/technology.html> => <http://www.company.it/company/staff.html>	0.0053	0.6359	35.93
12	<http://www.company.it/company/partners.html> => <http://www.company.it/company/jobs.html>	0.0082	0.6185	30.99
13	<http://www.company.it/company/technology.html> => <http://www.company.it/company/partners.html>	0.0055	0.6128	46.32
14	<http://www.company.it/company/index.html, http://www.company.it/company/staff.html> => <http://www.company.it/company/jobs.html>	0.0093	0.6060	30.36
15	<http://www.company.it/company/technology.html> => <http://www.company.it/company/staff.html>	0.0054	0.6000	33.90

Among the page sequences visited by a moderately large number of people, the rule with the highest confidence describes a path from two pages presenting those business units including finance and customers, and then goes back to a finance page, with a conditional probability of 72%. With almost the same confidence, there are two sequences that, from the pages describing the company as a working environment (index, staff, partners, etc.), then go to the page listing job offers, including people looking for new jobs.

Two paths moving from generic company pages to information about staff and partners still have high confidence (65%). This is a typical sequence followed by people potentially interested in doing business with the company: first they look for generic information about it; when they see it appears to be interesting, they look for the personal characteristics of the people working there (this is mainly a consulting company, so the standards of its personnel are crucial for the services offered). A further passage may be direct contact with the aim of collaborating with the company.

Note that only a few of the identified rules involve pages from different areas of the website. The pages do not seem to be very well connected, or else people visiting the site are only interested in specific goals and go directly to the pages of interest.

Appendix A

Complements of Mathematics and Statistics

A.1 Concepts on Linear Algebra

We recall some standard facts in linear algebra and establish notation. A matrix is an array of elements, or *entries,* all taken from the same set, organized into rows and columns. These entries commonly belong to the set of real numbers, and this is the case we deal with here. Matrix A has dimension $m \times n$ if it has m rows and n columns; we can also say that A is an $m \times n$ matrix and write $A = (a_{ij})$, where the parentheses contain the generic element of A.

The transposed matrix of A, obtained by switching rows and columns, is denoted $A^\top$. A matrix v of dimension $n \times 1$ is called the (column) vector of dimension n or, equivalently, the $n \times 1$ vector, and we write $v \in \mathbb{R}^n$; analogously, a matrix of dimension $1 \times n$ is called a row vector.

The identity matrix of order n is indicated by I_n, 1_n is the $n \times 1$ vector having all elements equal to 1 and 0 is the zero-matrix whose dimension will be clear from the context.

If $A = (a_{ij})$ is a square matrix of order n, that is, an $n \times n$ matrix, we use the following notation and terminology:

i. A is symmetric if $A^\top = A$;
ii. $\det(A)$ is the determinant of A; the property $\det(A\,B) = \det(A)\ \det(B)$ holds;

iii. If $\det(A) \neq 0$, we say that A is nonsingular and there is an inverse matrix, A^{-1}, so that $A\,A^{-1} = A^{-1}\,A = I_n$; we can also write $(A^\top)^{-1} = (A^{-1})^\top$ and $(A\,B)^{-1} = B^{-1}\,A^{-1}$, if both the inverses exist;

iv. A symmetric matrix A is positive semi-definite if $u^\top Au \geq 0$ for every nonzero vector $u \in \mathbb{R}^n$; in this case, we can write $A \geq 0$; we can also write $A \geq B$ to indicate that $A - B \geq 0$;

v. A symmetric matrix A is positive definite if it is symmetric and $u^\top Au > 0$ for every nonzero vector $u \in \mathbb{R}^n$; in this case, we write $A > 0$; we also write $A > B$ to indicate that $A - B > 0$;

vi. A is orthogonal if its transpose and inverse are equal, that is, $A^\top = A^{-1}$; in this case, $\det(A) = \pm 1$;

vii. $\mathrm{tr}(A)$ is the trace of A, that is, the sum of the elements on its main diagonal; $\mathrm{tr}(AB) = \mathrm{tr}(BA)$ holds for two matrices A and B, which need not to square, presuming both products AB and BA are possible;

viii. A is idempotent if $A = A^2$; for an idempotent matrix, the rank is equal to the trace, that is, $\mathrm{rk}(A) = \mathrm{tr}(A)$;

ix. A is a diagonal matrix if all the elements outside the main diagonal $(a_{11}, a_{22}, \ldots, a_{nn})$ are 0; we can also write $A = \mathrm{diag}(a_{11}, \ldots, a_{nn})$;

x. The so-called *matrix inversion lemma*

$$(A + BCD)^{-1} = A^{-1} - A^{-1}B(C^{-1} + DA^{-1}B)^{-1}DA^{-1} \qquad (A.1)$$

holds when the matrices are of conformable dimensions and the required inverse matrices exist; in particular, if b and d are column vectors and $c = 1$ is a scalar, then (A.1) becomes

$$(A + b\,d^\top)^{-1} = A^{-1} - \frac{1}{1 + d^\top A^{-1} b} A^{-1} b\,d^\top A^{-1} \qquad (A.2)$$

which is called the *Sherman–Morrison formula*.

A.2 Concepts of Probability Theory

A.2.1 Multivariate Random Variables

If $X_1, \ldots, X_p$ are random variables defined on the same probability space, then the vector

$$X = \begin{pmatrix} X_1 \\ X_2 \\ \vdots \\ X_p \end{pmatrix}$$

is a *multivariate random variable*. The expectation or mean value $\mathbb{E}\{X\}$ of X is defined as the vector of the expectations of the components, if they all exist.

That is, we define

$$\mathbb{E}\{X\} = \begin{pmatrix} \mathbb{E}\{X_1\} \\ \mathbb{E}\{X_2\} \\ \vdots \\ \mathbb{E}\{X_p\} \end{pmatrix}$$

and the *variance matrix* (or dispersion matrix) is defined as

$$\text{var}\{X\} = \begin{pmatrix} \text{var}\{X_1\} & \text{cov}\{X_1, X_2\} & \dots & \text{cov}\{X_1, X_p\} \\ \text{cov}\{X_2, X_1\} & \text{var}\{X_2\} & \dots & \text{cov}\{X_2, X_p\} \\ \vdots & \vdots & \ddots & \vdots \\ \text{cov}\{X_p, X_1\} & \text{cov}\{X_p, X_2\} & \dots & \text{var}\{X_p\} \end{pmatrix}$$

presuming the existence of every element of the matrix. However, the existence of the elements on the main diagonal is sufficient to guarantee the existence of all the others, keeping in mind the Cauchy-Schwartz inequality. Note also that $\text{var}\{X\}$ is a symmetric matrix and that $\text{var}\{X_i\}$ is a notation equivalent to $\text{cov}\{X_i, X_i\}$.

The matrix obtained by dividing the generic term $\text{cov}\{X_i, X_j\}$ by the product of the respective standard deviations, $\sqrt{\text{var}\{X_i\}} \times \sqrt{\text{var}\{X_j\}}$, is called the *correlation matrix*.

If $\text{var}\{X\}$ is a diagonal matrix, we say that X has uncorrelated components.

A.2.2 Some General Properties

We state some simple properties of the expectation and variance matrix of multivariate random variables; for proofs, see, for example, Azzalini (1996, Appendix A.4). For this section, we assume that $X = (X_1, \dots, X_p)^\top$, with $\mathbb{E}\{X\} = \mu$, $\text{var}\{X\} = V$.

Lemma A.2.1
If A is a $q \times p$ matrix, b a $q \times 1$ vector, and

$$Y = AX + b,$$

then

i. $\mathbb{E}\{Y\} = A\mu + b$,
ii. $\text{var}\{Y\} = A\,VA^\top$.

Lemma A.2.2
Variance matrix $V = \text{var}\{X\}$ is positive semi-definite and is also positive definite if there are no zero vectors b for which $b^\top X$ has degenerate distribution.

Lemma A.2.3
If $\text{var}\{X\} = V > 0$, there is a square matrix C of order p, so that $Y = CX$ has uncorrelated components with unit variance, that is, $\text{var}\{Y\} = I_p$.

Lemma A.2.4
Let $A = (a_{ij})$ be a square matrix of order p. Then

$$\mathbb{E}\left\{X^\top AX\right\} = \mu^\top A\mu + \text{tr}(AV).$$

A.2.3 Multivariate Normal Distribution

We want to extend the concept of normal distribution from the scalar to the p-dimensional case. In the multidimensional case, the normal (or Gaussian) distribution plays a key role to an even greater extent than in the scalar case.

The following is a constructive definition equipped with certain properties. More details are given, for example, by Azzalini (1996, Appendix A.5); for a more detailed presentation, see Mardia et al. (1979, ch. 2 and 3).

Let $Z_1, \ldots, Z_p$ be independent random variables $N(0, 1)$, so vector $Z = (Z_1, \ldots, Z_p)^\top$ is a multivariate random variable that we can reasonably consider the first case of a multivariate normal variable. However, the distribution of Z is very specific, and we want to introduce a much wider class, keeping the properties of the simple distribution.

In the univariate case, the normal distribution class can be generated by transformations of the type $X_0 = \mu + \sigma Z_0$, if $Z_0 \sim N(0, 1)$ and $\sigma \neq 0$ (note that $\sigma < 0$ is not excluded). A similar operation in the p-dimensional case is of the type

$$X = \mu + \Sigma^{1/2} Z$$

where $\mu \in \mathbb{R}^p$ and $\Sigma^{1/2}$ is a $p \times p$ matrix of full rank.

The probability density function of Z is given by the product of p copies of density $N(0, 1)$. From this, applying known rules to calculate the distributions of transformed random variables, the density function of X is

$$p(x) = \frac{1}{(2\pi)^{p/2} \det(\Sigma)^{1/2}} \exp\left\{-\tfrac{1}{2}(x-\mu)^\top \Sigma^{-1}(x-\mu)\right\} \tag{A.3}$$

for $x \in \mathbb{R}^p$.

Therefore, let us decide *by definition* that a random variable X with a distribution of type (A.3) is said to have normal (or Gaussian) multivariate p-dimensional distribution with parameters μ and $\Sigma = \Sigma^{1/2}(\Sigma^{1/2})^\top$. We thus adopt the notation

$$X \sim N_p(\mu, \Sigma).$$

The case in which Σ does not have full rank is admissible, even though our construction and (A.3) are assumed in the case with full rank.

The family of multivariate normal distributions has many formal properties that make its use as a probabilistic model particularly advantageous. Some of the simplest, already implicit in what has been stated so far, are listed here.

a. The contour lines of $p(x)$ are *ellipses* of equation

$$(x - \mu)^\top \Sigma^{-1}(x - \mu) = \text{constant}.$$

b. If Σ is a diagonal matrix, the components of X are not only uncorrelated but also independent, as we can immediately see from the expression of $p(x)$.
c. Because $\mathbb{E}\{Z\} = 0$ and $\text{var}\{Z\} = I_p$, lemma A.2.1 immediately gives

$$\mathbb{E}\{X\} = \mu, \qquad \text{var}\{X\} = \Sigma.$$

To better perceive the nature of normal distribution, it is useful to examine figure A.1, which shows the case when $p = 2$. The left panel shows some contour lines of the probability density of Z, which are circumferences, since its variance matrix is the identity. The same panel also shows the 100 points randomly sampled from Z. The right panel refers to the transformed variable

$$X = \begin{pmatrix} 1 \\ -0.5 \end{pmatrix} + \begin{pmatrix} 0.9 & 0.4 \\ 0.4 & 1.1 \end{pmatrix} Z \sim N_2\left(\begin{pmatrix} 1 \\ 0.5 \end{pmatrix}, \begin{pmatrix} 0.97 & 0.80 \\ 0.80 & 1.37 \end{pmatrix}\right) \tag{A.4}$$

and shows the density contour lines corresponding to those in the left panel. This signifies that the ellipses on the right represent the deformation of the

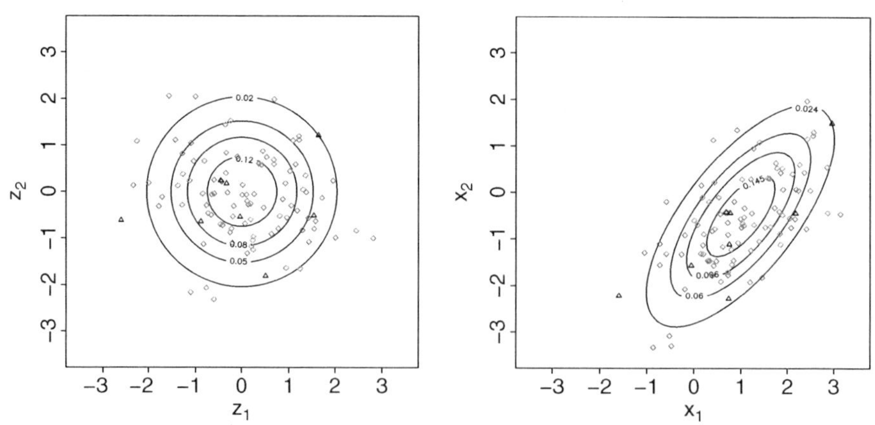

Figure A.1 Contour lines of density function and sample points from bivariate normal distributions. Left: variable Z with independent components $N(0, 1)$; right: those of its transformation (A.4).

circumferences on the left, according to transformation (A.4); the value of the density associated with these curves is modified according to factor $\det(\Sigma)^{1/2}$ in (A.3). The right panel also shows the previous sample points as modified by the adopted transformation. Some of the points are marked by symbols different from the majority of the sample, to facilitate matching of the corresponding points in the two panels. The inclination of the main axis of the ellipses denotes a correlation between the two components, which in this case is $0.694 = 0.80/\sqrt{0.97 \times 1.37}$.

One of the most important properties of the family of multivariate normal distributions is that they are closed to affine transformations, including those that reduce the dimension. More precisely, if $a \in \mathbb{R}^q$ and B is a $q \times p$ matrix, then

$$Y = a + BX \sim N_q(a + B\mu, B\Sigma B^\top). \tag{A.5}$$

This includes the special case in which the scalar linear combination of components having multivariate normal distribution has normal distribution.

As a particular case of the previous property, the class is closed with respect to the marginalization operation, in the following sense. We subdivide the components of X into two sets, the first of q and the second of $p - q$ components. For notational simplicity, we assume the first set corresponds to the first q components of X, although this is not essential. In other words, we introduce the partitions

$$X = \begin{pmatrix} X_1 \\ X_2 \end{pmatrix}, \quad \mu = \begin{pmatrix} \mu_1 \\ \mu_2 \end{pmatrix}, \quad \Sigma = \begin{pmatrix} \Sigma_{11} & \Sigma_{12} \\ \Sigma_{21} & \Sigma_{22} \end{pmatrix}$$

where μ and Σ are partitioned in the same way as X. And so, as a particular case of the general property (A.5), we obtain

$$X_1 \sim N_q(\mu_1, \Sigma_{11}).$$

The property of closure of the normal distribution class with respect to the conditioning operation also holds. Specifically, the distribution of X_1 conditional on $X_2 = x_2$ is

$$(X_1 | X_2 = x_2) \sim N_q(\mu_1 + \Sigma_{12}\Sigma_{22}^{-1}(x_2 - \mu_2), \Sigma_{11\cdot 2}) \tag{A.6}$$

where

$$\Sigma_{11\cdot 2} = \Sigma_{11} - \Sigma_{12}\Sigma_{22}^{-1}\Sigma_{21}.$$

As x_2 varies, the conditional mean value $\mu_1 + \Sigma_{12}\Sigma_{22}^{-1}(x_2 - \mu_2)$ corresponds to the equation of a plane, called the *regression hyperplane*. Conditional variance $\Sigma_{11\cdot 2}$ is "smaller" than marginal variance Σ_{11}, where "smaller" means inequality between matrices, with equality only when Σ_{12} is the zero matrix. Note that the conditional variance does not depend on x_2.

There are various connections between multivariate normal distribution and χ^2 distribution, of which the simplest is given by

$$(X-\mu)^\top \Sigma^{-1}(X-\mu) = Z^\top Z \sim \chi_p^2.$$

In addition, if $X \sim N_p(\mu, I_p)$ and A is a symmetric positive semi-definite $p \times p$ matrix of rank q, then

$$Q_1 = X^\top A X \sim \chi_q^2(\delta_1), \quad Q_2 = X^\top (I_p - A)X \sim \chi_{p-q}^2(\delta_2)$$

with noncentrality parameters $\delta_1 = \mu^\top A\mu$, $\delta_2 = \mu^\top (I_p - A)\mu$, respectively, and the two quadratic forms Q_1 and Q_2 are stochastically independent.

A.3 Concepts of Linear Models

A.3.1 Linear Models and the Least Squares Criterion

We assume that the relation between response variable y and explanatory variables $x_1, \ldots, x_p$ is of the type

$$y = \beta_1 x_1 + \cdots + \beta_p x_p + \varepsilon \tag{A.7}$$

where ε is a component, called *error*, that expresses the deviation between empirical observations and systematic component $\beta_1 x_1 + \cdots + \beta_p x_p$, also called *linear predictor*. Regression parameters $\beta_1, \ldots, \beta_p$ are real numbers; therefore, in the absence of constraints on the model, $\beta = (\beta_1, \ldots, \beta_p)^\top$ is any point in $\mathbb{R}^p$.

We make use of a set of n observations $(n \geq p)$ of variables $x_1, x_2, \ldots, x_p, y$, which therefore satisfy the relations

$$y_i = \beta_1 x_{i1} + \cdots + \beta_p x_{ip} + \varepsilon_i, \qquad (i = 1, \ldots, n) \tag{A.8}$$

On the basis of these n replicas, we estimate parameters β and carry out other inferential operations.

Assume that error component ε is a random variable that in successive observations from model (A.7), is such that

$$\mathbb{E}\{\varepsilon_i\} = 0, \qquad \mathrm{var}\{\varepsilon_i\} = \sigma^2, \qquad \mathrm{cov}\{\varepsilon_i, \varepsilon_j\} = 0 \quad \text{se } i \neq j, \tag{A.9}$$

where σ^2 is a positive constant value for all replications. Consequently,

$$\mu_i = \mathbb{E}\{Y_i\} = \beta_1 x_{i1} + \cdots + \beta_p x_{ip}, \qquad \mathrm{var}\{Y_i\} = \sigma^2$$

when Y_i represents the random variable that generated observation y_i.

Formulation (A.7) is said to be a linear model, and assumptions (A.9) are called second-order hypotheses, because they involve moments up to the second order.

To estimate parameters β on the basis of n sample observations, according to model (A.8), it is common to adopt the least squares criterion, which selects

β values that minimize the sum of square deviations between observed and interpolated values, which in turn minimizes $Q(\beta)$, given by

$$Q(\beta) = \sum_{i=1}^{n} \{y_i - (\beta_1 x_{i1} + \cdots + \beta_p x_{ip})\}^2$$

where the unknown β is now treated as a free variation quantity in $\mathbb{R}^p$.

The whole formulation lends itself to more compact notation by means of matrices. We therefore create vector y, of n observations of the response variable, and do the same for ε. Analogously, we form a matrix X with dimension $n \times p$, whose jth column is formed from n observations on variable x_j; we assume that matrix X has full rank p. Therefore, we can rewrite (A.8) in a compact matrix form as

$$y = X\beta + \varepsilon$$

with the second-order hypothesis given by

$$\mu = \mathbb{E}\{Y\} = X\beta, \quad \text{var}\{Y\} = \sigma^2 I_n.$$

The least squares criterion lies in the solution of the optimization problem

$$\min_{\beta \in \mathbb{R}^p} D(\beta) \qquad \text{where} \qquad D(\beta) = \|y - X\beta\|^2.$$

The following presentation is taken from Azzalini (1996, ch. 5), to which we refer for missing details.

A.3.2 The Geometry of Least Squares

We now analyze the various components in the game from the purely geometric point of view, leaving aside statistical and probabilistic aspects for the moment. We consider vectors $y, x_1, \ldots, x_p$ containing, respectively, the values of the response variable and p explanatory variables as elements of vector space $\mathbb{R}^n$.

As β varies in $\mathbb{R}^p$, expression $X\beta = \beta_1 x_1 + \cdots + \beta_p x_p$ can be seen as a linear combination of columns $x_1, \ldots, x_p$ of X with coefficients β — that is, the parametric equation of a subspace of $\mathbb{R}^n$ spanned *by the columns* of X. This subspace, which we call $\mathcal{C}(X)$, is a vector space on $\mathbb{R}$ with dimension p. The property that, if $X\beta \in \mathcal{C}(X)$ and $a \in \mathbb{R}$ then also $a(X\beta) = X(a\beta) \in \mathcal{C}(X)$ holds; moreover, if $X\beta$ and Xb are two elements of $\mathcal{C}(X)$, then also $X\beta + Xb = X(\beta + b) \in \mathcal{C}(X)$; clearly, the other properties of vector spaces also hold.

Model (A.7–A.9) then states that $\mu = \mathbb{E}\{Y\}$ lies in $\mathcal{C}(X)$, and the least squares criterion chooses which vector of $\mathcal{C}(X)$ minimizes the Euclidean distance between vector y and space $\mathcal{C}(X)$. We indicate by $\hat{\mu} = X\hat{\beta}$ this element of $\mathcal{C}(X)$, identified by coefficients $\hat{\beta} \in \mathbb{R}^p$. The situation is illustrated is figure A.2.

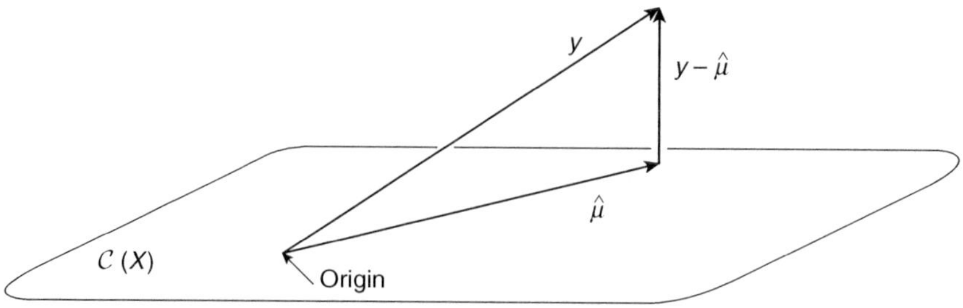

Figure A.2 Projection of y on $\mathcal{C}(X)$.

On the basis of known results of vector space geometry, we know that vector $\hat{\mu} \in \mathcal{C}(X)$, which minimizes the distance from y, is such that

$$(y - X\hat{\beta}) \perp \mathcal{C}(X)$$

and this requires $(y - \hat{\mu})$ to be orthogonal to the vectors that constitute the basis of $\mathcal{C}(X)$. Therefore, it is necessary that

$$(y - X\hat{\beta})^\top X = 0,$$

that is,

$$X^\top X\beta = X^\top y, \tag{A.10}$$

which are called *normal equations*.

The inversion of matrix $X^\top X$ is legitimate because the condition that X has rank p implies that $X^\top X$ is still of rank p. Therefore, the minimum of $D(\beta)$ is obtained for β and is

$$\hat{\beta} = (X^\top X)^{-1}X^\top y. \tag{A.11}$$

The same result can be obtained by minimizing $D(\beta)$ in an analytical instead of a geometrical way. The *projection* vector of y on $\mathcal{C}(X)$ is

$$\begin{aligned}\hat{\mu} &= X\hat{\beta}\\ &= X(X^\top X)^{-1}X^\top y\\ &= Py \end{aligned}\tag{A.12}$$

where $P = X(X^\top X)^{-1}X^\top$ is called the *projection matrix* on $\mathcal{C}(X)$. This identifies an operator, associated with matrix X, whose role is precisely that of projecting a vector $y \in \mathbb{R}^n$ by transforming it into $Py \in \mathcal{C}(X)$ with a minimum distance from y. We can immediately verify that P is symmetric and idempotent because $P^2 = P$;

this signifies that $Py = P(Py)$, so projecting a projection has no effect. We note that these observations imply that

$$\text{rk}(P) = \text{tr}(P) = \text{tr}((X^\top X)^{-1}X^\top X) = p.$$

We can therefore split y into two components: its projection $\hat{\mu}$ on $\mathcal{C}(X)$, and the component of the *residuals* given by the difference vector

$$y - \hat{\mu} = y - X(X^\top X)^{-1}X^\top y = (I_n - P)y. \tag{A.13}$$

These two components are orthogonal to each other; in fact, $y - \hat{\mu}$ is orthogonal to every element of $\mathcal{C}(X)$ and not only $\hat{\mu}$. For any vector $Xa \in \mathcal{C}(X)$, we have

$$\begin{aligned}(Xa)^\top(y - \hat{\mu}) &= (Xa)^\top(y - Py)\\ &= a^\top X^\top(y - X(X^\top X)^{-1}X^\top y)\\ &= 0.\end{aligned}$$

Matrix $I_n - P$ is also a projection matrix: it projects the elements of $\mathbb{R}^n$ in the space orthogonal to $\mathcal{C}(X)$. As calculated for the rank of P, we have $\text{rk}(I_n - P) = n - p$.

The orthogonality between the projection vector and one of the residuals has an immediate corollary: expanding the norm of $\hat{\mu} + (y - \hat{\mu})$, we obtain

$$\|y\|^2 = \|\hat{\mu}\|^2 + \|y - \hat{\mu}\|^2 \tag{A.14}$$

which is an instance of the Pythagorean theorem, in which y plays the role of the hypotenuse and $\hat{\mu}$ and $y - \hat{\mu}$ that of the sides.

A.3.3 The Statistics of Least Squares

We now examine, from a statistical point of view, the quantities introduced in the previous section. This naturally brings us to consider y observations and error components as determinations of random variables Y and ε, respectively. We have

$$\begin{aligned}\mathbb{E}\left\{\hat{\beta}\right\} &= \mathbb{E}\left\{(X^\top X)^{-1}X^\top Y\right\}\\ &= (X^\top X)^{-1}X^\top \mathbb{E}\{Y\}\\ &= PX\beta\\ &= \beta\end{aligned} \tag{A.15}$$

and therefore $\hat{\beta}$ is an unbiased estimate of β; in addition,

$$\mathbb{E}\{\hat{\mu}\} = \mu.$$

For the variance matrix of the estimates, we have

$$\begin{aligned}\operatorname{var}\left\{\hat{\beta}\right\} &= (X^\top X)^{-1}X^\top \operatorname{var}\{Y\}\left((X^\top X)^{-1}X^\top\right)^\top \\ &= (X^\top X)^{-1}X^\top(\sigma^2 I_n)X(X^\top X)^{-1} \\ &= \sigma^2(X^\top X)^{-1} \end{aligned} \tag{A.16}$$

and

$$\begin{aligned}\operatorname{var}\{\hat{\mu}\} &= X\operatorname{var}\left\{\hat{\beta}\right\}X^\top \\ &= \sigma^2 X(X^\top X)^{-1}X^\top \\ &= \sigma^2 P.\end{aligned}$$

Up to now, we have only looked at the estimation of β. Although to a lesser extent than β, we are also interested in estimating σ^2. The least squares criterion does not tell us how to proceed. Because we have $\mathbb{E}\{\varepsilon_i^2\} = \sigma^2$ for generic term ε_i, it is reasonable to estimate σ^2 with the arithmetic mean of the $\hat{\varepsilon}_i^2$, where $\hat{\varepsilon}_i$ is the general component of the residual vector

$$\hat{\varepsilon} = y - \hat{\mu}$$

and therefore we consider

$$\hat{\sigma}^2 = \frac{\sum_i \hat{\varepsilon}_i^2}{n} = \frac{\|\hat{\varepsilon}\|^2}{n} \tag{A.17}$$

as an estimate of σ^2. Note that this expression can be rewritten in various other forms, bearing in mind the relations

$$\begin{aligned}\|\hat{\varepsilon}\|^2 &= D(\hat{\beta}) \\ &= (y-\hat{\mu})^\top(y-\hat{\mu}) \\ &= y^\top(I_n - P)^\top(I_n - P)y \\ &= y^\top(I_n - P)y = \varepsilon^\top(I_n - P)\varepsilon \\ &= y^\top y - y^\top X\hat{\beta}.\end{aligned}$$

To calculate expectation of (A.17), we have

$$\begin{aligned}\mathbb{E}\{n\hat{\sigma}^2\} &= \mathbb{E}\left\{y^\top(I_n - P)y\right\} \\ &= \mu^\top(I_n - P)\mu + \operatorname{tr}((I_n - P)\sigma^2 I_n) \\ &= \sigma^2(n-p),\end{aligned} \tag{A.18}$$

using lemma A.2.4. The term $\mu^{\top}(I_n - P)\mu$ is 0, because $I_n - P$ projects onto the space orthogonal to $\mathcal{C}(X)$ where μ lies, and therefore

$$(I_n - P)\mu = (I_n - X(X^{\top}X)^{-1}X^{\top})X\beta = 0.$$

Thus, $\hat{\sigma}^2$ is subject to bias, which tends to 0 as $n \to \infty$. If we need an unbiased estimate for σ^2, this is given by

$$s^2 = \hat{\sigma}^2 \frac{n}{n-p} = \frac{\|y - \hat{\mu}\|^2}{n-p}. \tag{A.19}$$

A.3.4 Constrained Estimation

We now consider the problem of estimating β when linear constraints are present in the β coefficients, that is, β is such that

$$H\beta = 0 \tag{A.20}$$

where H is a $q \times p$ matrix (with $q \leq p$) with rank q formed of specified constants. The solution to this problem is particularly useful in the framework of hypothesis testing on the components of β, but it is also of independent interest.

First consider the geometric meaning of condition $H\beta = 0$. It requires μ to lie in the subset of $\mathcal{C}(X)$, which satisfies q conditions specified by $H\beta = 0$. This subset represents a *vector subspace*, here called $\mathcal{C}_0(X)$, of dimension $p - q$ of space $\mathcal{C}(X)$, as shown in figure A.3.

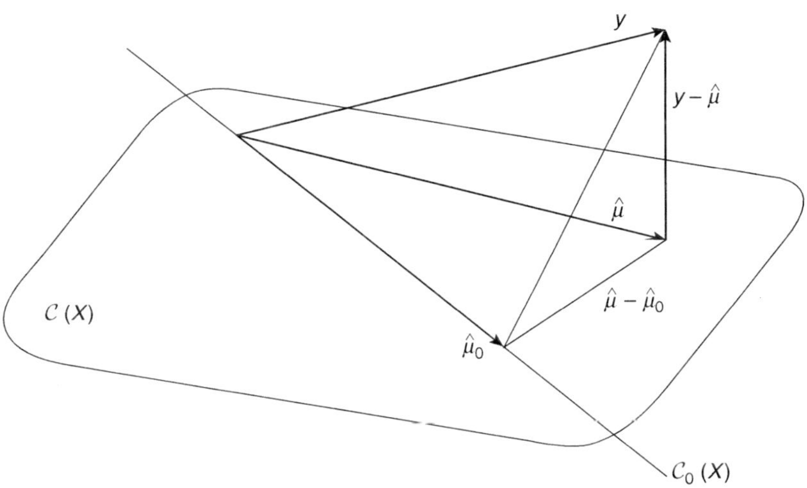

Figure A.3 Projection of y on $\mathcal{C}(X)$ and subspace $\mathcal{C}_0(X)$.

To obtain the constrained minimum of $D(\beta)$, we must minimize

$$D^*(\alpha, \beta) = (y - X\beta)^\top (y - X\beta) + 2(H\beta)^\top \alpha,$$

where α is a vector of Lagrange multipliers, with constraint (A.20). After some algebraic manipulation, we reach the solution

$$\hat{\beta}_0 = \hat{\beta} - (X^\top X)^{-1} H^\top K H \hat{\beta} \tag{A.21}$$

where

$$K = \{H(X^\top X)^{-1} H^\top\}^{-1}. \tag{A.22}$$

The corresponding projection of y on $C_0(X)$ is given by

$$\begin{aligned} \hat{\mu}_0 &= X\hat{\beta}_0 \\ &= \hat{\mu} - X(X^\top X)^{-1} H^\top K H \hat{\beta} \\ &= (P - P_H)y = P_0 y \end{aligned}$$

having set

$$\begin{aligned} P_H &= X(X^\top X)^{-1} H^\top K H (X^\top X)^{-1} X^\top, \\ P_0 &= P - P_H. \end{aligned} \tag{A.23}$$

Consider the conclusions we have reached so far. We have a new projection matrix, P_0, which projects any vector of $\mathbb{R}^n$ on space $C_0(X)$. If we apply this matrix to y, we obtain $\hat{\mu}_0$, which, by its very nature, is the element of $C_0(X)$ with the minimum distance from y. It is also easy to verify the following further properties:

- Vector $y - \hat{\mu}_0$ is orthogonal to every element of $C_0(X)$. If Xc is an element of $C(X)$ so that $Hc = 0$, we have $(y - \hat{\mu}_0)^\top Xc = 0$. In particular,

$$(y - \hat{\mu}_0) \perp \hat{\mu}_0.$$

- The projection of $y - \hat{\mu}_0$ on $C(X)$ is $P(y - \hat{\mu}_0) = \hat{\mu} - \hat{\mu}_0$, which is such that

$$\hat{\mu} - \hat{\mu}_0 \perp \hat{\mu}_0.$$

Last, we obtain the following decomposition:

$$y = \hat{\mu}_0 + (\hat{\mu} - \hat{\mu}_0) + (y - \hat{\mu})$$

where the three summands on the right-hand side are orthogonal to each other, and therefore allow us to write

$$\|y\|^2 = \|\hat{\mu}_0\|^2 + \|\hat{\mu} - \hat{\mu}_0\|^2 + \|y - \hat{\mu}\|^2 \tag{A.24}$$

which is an extension of (A.14).

A.3.5 Normality Assumptions

If we add to the second-order hypothesis (formulated in section A.3.1 around the distribution of the random variable of the error component) that of normality, for which

$$\varepsilon \sim N_n(0, \sigma^2 I_n),$$

we can obtain more stringent results for the distributive properties of the inferential quantities already seen. First, it immediately follows that

$$Y \sim N_n(X\beta, \sigma^2 I_n)$$

leading to

$$\hat{\beta} \sim N_p(\beta, \sigma^2 I_p).$$

The interpretation of $\hat{\beta}$ changes, in the sense that it can be seen as a maximum likelihood estimate, besides being descended from the least squares criterion. In fact, maximization of the likelihood function corresponds to maximization with respect to β of the term within $\exp(\cdot)$ of (A.3), if we assume that $\mu = X\beta$, and this coincides with the minimization of $D(\beta)$.

The components of projection and error of Y also have normal distribution, as

$$\begin{aligned} \hat{Y} = PY \quad &\sim \quad N_n(PX\beta,\ \sigma^2 P), \\ \hat{\varepsilon} = (I_n - P)\,Y \quad &\sim \quad N_n((I_n - P)X\beta,\ \sigma^2(I_n - P)) \end{aligned}$$

for which we can apply the results for quadratic forms of random normal variables noted in section A.2.3. It therefore follows that

$$\hat{Y}^\top \hat{Y} \sim \sigma^2 \chi^2_p(\delta), \qquad \|\hat{\varepsilon}\|^2 \sim \sigma^2 \chi^2_{n-p},$$

where the noncentrality parameter is $\delta = \beta^\top X^\top X \beta$ and the two quadratic forms are independent.

These facts thus establish the distribution of the decomposition of total variability $Y^\top Y$ into two components, that is, error component $\|\hat{\varepsilon}\|^2$ and regression component $\hat{Y}^\top \hat{Y}$. These properties yield the distribution of the F test connected with the analysis of variance table.

The sum of the squares of regression component $\hat{Y}^\top \hat{Y}$ is then further decomposed into individual components, one for each explanatory variable with corresponding decomposition of the degrees of freedom.

Appendix B
Data Sets

Appendix B describes the data used in this volume. They are also available at the website: `http://azzalini.stat.unipd.it/Book-DM/`.

B.1 Simulated Data

Some of the data used were obtained by means of simulation of pseudo-random numbers, as follows:

a. Yesterday's data and tomorrow's data. A table with 30 rows (other than those with variable names) and 3 columns, contains variables `x`, `y.yesterday`, `y.tomorrow`, with self-explanatory names. These data are used in chapter 3 and section 4.8.
b. Data for three classes, of sizes 120, 80, and 100, are used in chapter 5. The data table contains 300 rows (other than those with variable names) and 3 columns, for two explanatory variables, z_1 and z_2, and one class indicator. Some of the numerical examples in chapter 5 refer to data in the first two classes.
c. Two data collections, C1 and C2, are used in section 6.1, each with two variables, with 250 and 100 points.

B.2 Car Data

The car data, first used in section 2.1.1 and then in section 2.1 and chapter 4, were obtained by simple manipulation of original data that referred to the characteristics of 203 automobile models imported into the United States in 1985. The original data are available at: `ftp://ftp.ics.uci.edu/pub/machine-learning-databases/autos`. Their manipulation on our part simply consisted of converting one unit of measurement to another and eliminating some variables. The new variables are as follows:

Variable	Description
`make`	manufacturer (factor, 22 levels)
`fuel type`	type of engine fuel (factor, 2 levels: diesel, gasoline)
`aspiration`	type of engine aspiration (factor, 2 levels: standard, turbo)
`body style`	type of body style (factor, 5 levels: hardtop, wagon, sedan, hatchback, convertible)
`drive wheels`	type of drive wheels (factor, 3 levels: 4wd, fwd, rwd)
`engine location`	location of engine (factor, 2 levels: front, rear)
`wheel base`	distance between axes (cm)
`length`	length (cm)
`width`	width (cm)
`height`	height (cm)
`curb weight`	weight (kg)
`engine size`	engine size (L)
`compression rate`	compression rate
`hp`	horsepower
`peak-rpm`	number of peak revolutions per minute
`city distance`	city distance covered (km/L)
`highway distance`	highway distance (km/L)
`n.cylinders`	number of cylinders

B.3 Brazilian Bank Data

The data used in sections 2.3.3 and 2.4 were obtained by simple manipulation of original data referring to a customer satisfaction survey by a Brazilian bank. For 500 subjects, randomly selected from the bank's customers, some information from marketing research was obtained. Some characteristic variables of customers and their satisfaction are:

Variable	Description
`id`	customer identification
`satisfaction`	(factor: 4 levels)
`education`	(factor: 5 levels)
`age`	(years)
`gender`	gender
`car`	indicator of car ownership
`phone`	indicator of phone use
`fax`	indicator of fax use
`pc`	indicator of PC ownership
`pincome`	annual income (in Brazilian reais)
`ok`	satisfaction index (factor: 2 levels)

Customers were also asked which products of the bank they used and if they also used similar products supplied by other banks. The names of the variables regarding the products of the bank that commissioned the survey end with the number $n = 1$; the number of similar variables that refer to other banks end with the number $n = 2$. The following is the list of surveyed products, with self-explanatory names:

`savings`*n*	`installment.loan`*n*
`creditcard`*n*	`investment.fund`*n*
`bankcard`*n*	`commodities.fund`*n*
`cd`*n*	`annuities.fund`*n*
`specialchecking`*n*	`car.insurance`*n*
`auto.bill.payment`*n*	`home.insurance`*n*
`personal.loans`*n*	`life.insurance`*n*
`mortgage`*n*	

B.4 Data for Telephone Company Customers

The data for telephone customers, used in section 4.10.1 and later in chapter 5 in the two case studies in section 5.10, was obtained by simple manipulation of original data referring to the characteristics of 30,619 customers of a European telephone company with postpay contracts. To be part of the set, the customers had to be active in the 10 consecutive months to which the data refer, which are conventionally indicated by numbers from 1 to 10 ($nn = 01, \ldots, 10$).

The original data were processed simply by eliminating some of the original variables. For the customers, the variables are:

- characteristic variables of customer and of company contract

Variable	Description
`id`	customer identification
`tariff.plan`	customer tariff plan (factor, 5 levels)
`payment.method`	(factor, 3 levels: PO: postal order, CC: credit card, DD: direct debit)
`gender`	(factor, 3 levels: M: male, F: female, B: company)
`age`	(years)
`activ.zone`	geographical activation zone (factor, 4 levels)
`activ.chan`	channel of activation (factor, 8 levels)
`vas1`	presence of a first value-added service
`vas2`	presence of a second value-added service

- variables for traffic in the 10 available months. For each month, indicated by the first part of the name (q01, q02,. . . , q10), the following variables are available:

Variable	Description
q*nn*.out.ch.peak	total monthly number of outgoing calls at peak tariff times
q*nn*.out.dur.peak	duration of total monthly outgoing calls at peak tariff times
q*nn*.out.val.peak	total monthly outgoing call value at peak tariff times
q*nn*.out.ch.offpeak	total monthly number of outgoing calls at off-peak tariff times
q*nn*.out.dur.offpeak	duration of total monthly outgoing calls at off-peak tariff times
q*nn*.out.val.offpeak	total monthly outgoing call value at off-peak tariff times
q*nn*.in.ch.tot	total monthly number of incoming calls
q*nn*.in.dur.tot	duration of total monthly incoming calls
q*nn*.ch.sms	total monthly number of SMS sent
q*nn*.ch.cc	number of monthly calls to customer services

- the variable `status`, that is which is the indicator variable of possible deactivation in the thirteenth month, that is, two months after the final month for which traffic is available (factor, 2 levels: 0—active, 1—deactivated).

B.5 Insurance Data

The data on insurance customers, used in section 4.10.2, was obtained by simple manipulation of original data on the characteristics of a sample of 5,000 customers of a European insurance company. To be part of the set, the customers had to take out one policy in at least one of the company's lines of business.

Processing the original data consisted simply of eliminating some of the original variables. For these customers, the available variables are as follows.

- Customers' characteristic variables

Variable	Description
id	customer identification
gender	(factor, 3 levels: M: male, F: female, — missing)
age	(years)
occupation.1	occupational categories of employment 1 (factor, 11 levels)
occupation.2	occupational categories of employment 2 (factor, 17 levels)
zip	postcode (numeric)
area	geographical area of residence (factor, 33 levels)
region	geographical region of residence (factor, 10 levels)
city	indicator variable of residence in urban areas

- Variables regarding to canceled claims and policies:

Variable	Description
number.claims.last	number of claims in last year
number.claims.3	number of claims in last 3 years
amount.claims.last	amount of claims in last year
amount.claims.3	amount of claims in 3 years
number.cancel.last	number of policies canceled in last year
number.cancel.3	number of policies canceled in 3 years

- Variables relating to products. For each product, indicated by number n at the end of the name of the variable (for nonlife products $n = 1, \ldots 9$ and life products $n = 1a, 1b, 2a, 2b, 3a, 3b$), the following variables are available:

Variable	Description
n.nonlife.0	number of private car third-party liability policies
prem.nonlife.0	total amount of premiums for private car third-party liability policies
number.bank.1	number of bank products of type 1
number.bank.2	number of bank products of type 2
net.bank.2	net asset value funds
tot.bank.2	total amount of funds
ac.bank.2	total amount of funds acquired
number.non-life.*n*	number of nonlife policies of type *n*
prem.non-life.*n*	total amount of premiums for nonlife policies of type *n*
number.life.*n*	number of life policies of type *n*
prem.life.*n*	total amount of premiums for life policies of type *n* in last year
pre.payed.life.*n*	total amount of paid premiums for life policies of type *n*
i.cancel.last	policies canceled in last year
i.cancel.3	policies canceled in 3 years
i.bank.1	at least one bank product of type 1
i.bank.2	at least one bank product of type 2
i.non-life.*n*	at least one nonlife policy of type *n*
i.life.*n*	at least one life policy of type *n*

B.6 Choice of Fruit Juice Data

The data on fruit juice purchases were presented by Foster et al. (1998, ch. 11) and are available through the distribution system for statistical information `StatLib` at the website `http://lib.stat.cmu.edu/`.

The data refer to 1,070 fruit juice purchases of two different brands (MM and CH) in certain U.S. supermarkets, supplied with some contributory variables. The data used in chapter 5 were slightly processed in the sense that some characteristics of little importance were excluded. The variables used are as follows:

Variable	Description
`choice`	prechosen brand (factor, with 2 levels)
`id.cust`	customer identification
`week`	identifier of week of purchase
`priceCH`	reference price for brand CH (USD)
`priceMM`	reference price for brand MM (USD)
`discountCH`	discount applied to product CH (USD)
`discountMM`	discount applied to product MM (USD)
`loyaltyCH`	loyalty indicator for product CH
`loyaltyMM`	loyalty indicator for product MM
`store`	store identifier (factor, with 5 levels)

Variable `loyaltyMM` is constructed starting from the value 0.5 and updating with every purchase by the same customer, with a value that increases by 20% of the current difference between the current value and 1, if the customer chose MM, and falls by 20% of the difference between the current value and 0 if the customer chose CH. The corresponding variable `loyaltyCH` is given by 1 − `loyaltyMM`. There are five stores in question, numbered from 0 to 4.

B.7 Customer Satisfaction

The data on customer satisfaction, used in section 5.10.3, were obtained by simple manipulation of original data on responses to a questionnaire from a survey of 4,000 customers of a European IT company producing and selling software and offering consulting services. The survey was carried out by an independent marketing research company specializing in such surveys. Processing of original data consisted simply of eliminating some of the original variables. These were:

- products/services used
  Question: *Which products/services of the company do you use?*

Variable	Product/service
`V2`	1
`V3`	2
`V4`	3
`V5`	4
`V6`	5
`V7`	6
`V8`	7
`V9`	others

- satisfaction with staff and products (except `V11`, all variables are factors with 10 levels: 1: totally disagree,..., 10: totally agree)

Variable	Question	Answer
V11	In the last year, did you have contacts with company personnel for consultancy, information, or solutions to problems?	1: no 2: yes, once 3: yes, sometimes 4: yes, often
V24	The products are easy to use	
V25	The products can easily be adapted to customers' needs	
V26	The products are exactly what I need	
V27	Product results are reliable	
V28	Differing products are easily integrated	

- Question: *Please rate the importance of the following aspects in evaluating an IT company* (each variable is a factor with 10 levels: 1: not at all important,..., 10: very important):

Variable	Item
V29	expertise of personnel
V30	capacity to offer an efficient consulting service
V31	problem solving
V32	reliability of products/services
V33	flexibility of products/services
V34	efficiency of products/services
V35	working speed of products
V36	helpfulness of personnel
V37	efficiency in serving customers
V38	predisposition towards customers' needs
V39	capacity to respond to customer's needs
V40	flexibility in making changes
V41	capacity for technological innovation

- Question: *Please rate your satisfaction with the following aspects* (each variable is a factor with 10 levels: 1: not at all important,..., 10: very important):

Variable	Item
V42	expertise of personnel
V43	capacity to offer an efficient consulting service
V44	problem solving
V45	reliability of products/services
V46	flexibility of products/services
V47	efficiency of products/services
V48	working speed of products
V49	helpfulness of personnel
V50	efficiency in serving customers
V51	predisposition towards customers' needs
V52	capacity to respond to customer's needs
V53	flexibility in making changes
V54	capacity for technological innovation

- customers' overall satisfaction and characteristics:

Variable	Question	Answer
V56	Recalling all aspects analyzed in this questionnaire, how satisfied are you with the company, overall?	1: extremely satisfied 2: very satisfied 3: quite satisfied 4: quite dissatisfied 5: very dissatisfied 6: extremely dissatisfied
V58	occupational category of employment	12 categories
V59	employment status	1: employer 2: manager
V60	age	
V61	length of service in company	
V62	education	1: university degree 2: high school diploma 3: middle school diploma oth: other
V63	gender	

B.8 Web Usage Data

The data on web usage, used in sections 5.10.4 and 6.3, refer to hits made by 26,157 anonymous visitors to the website of a consulting company. Data were collected from web log files, collecting all relevant information about hits on every page of the website.

A *user session* describes the sequence of web pages viewed consecutively by a visitor, without leaving the website or the connection. We call these sequences of pages "visits." The website to which the data refer does not have a cookie system or other way of identifying the same visitor in different sessions, so we consider each session in the analysis as a new visitor, and we call the same event "session" or "visit" indifferently.

All pages visited in a year are included in the data set. Sessions are labeled with a identification number, and no personal information is available. The website has 215 pages and the total number of page views (hits) on the entire site was 47,387. Some of the pages have similar contents and were aggregated in eight categories (home, contacts, communications, events, company, white papers, business units, consulting). The day and time of all visits to every single page are also recorded. For each single event (visit to a page) the available variables are:

`ID`	identification number of single event (page visited)
`sessionID`	identification number of session
`screen`	screen resolution used by customer (if available)
`url`	address of visited page
`dt`	day and month of event
`yr`	indicator of year of event: data refer to two consecutive years, called 1 and 2.
`hr`	time of event

Appendix C

Symbols and Acronyms

AIC	Akaike information criterion
CART	classification and regression trees
CRM	customer relationship management
d.f.	degree or degrees of freedom
DWH	data warehouse
GAM	generalized additive model
GCV	generalized cross validation
GLM	generalized linear model
KDD	knowledge discovery in databases
LDA	linear discriminant analysis
MARS	multivariate adaptive regression splines
OLAP	online analytical processing
OLTP	online Transaction Processing
PCA	principal component analysis
QDA	quadratic discriminant analysis
ROC	receiver operating characteristic
SQL	structured query language
SVM	support vector machine

$\det(\cdot)$	determinant of a matrix
$\mathrm{tr}(\cdot)$	trace of a matrix

$\mathrm{rk}(\cdot)$	rank of a matrix
D	deviance
L	likelihood function
$\ell(x)$	logistic function $\mathrm{e}^x/(1+\mathrm{e}^x)$
$\mathbb{E}\{\cdot\}$	expectation of a random variable
$\mathrm{var}\{\cdot\}$	variance (or matrix of variance) of a random variable
$\|\cdot\|$	Euclidean norm
$\mathbb{R}, \mathbb{R}^p$	set of real numbers, p-dimensional Euclidean space
$I(x)$	indicator function 0–1 of logical predicate x
I_A	set of indicator variables of factor A
I_n	identity matrix of order n
1_n	$n \times 1$ vector of elements, all 1

REFERENCES

Afifi, A. A. & Clark, V. (1990). *Computer-Aided Multivariate Analysis*, 2nd ed. New York: Van Nostrand Reinhold.

Agrawal, R., Mannila, H., Srikant, R., Toivonen, H., & Verkamo, A. I. (1996). Fast discovery of association rules. In U. M. Fayyad, G. Piatetsky-Shapiro, P. Smyth, & R. Uthurusamy (eds.), *Advances in Knowledge Discovery and Data Mining* (pp. 307–328). Cambridge, Mass.: AAAI/MIT Press.

Agresti, A. (2002). *Categorical Data Analysis*, 2nd ed. Hoboken, N.J.: Wiley.

Agresti, A. (2010). *Analysis of Ordinal Categorical Data*, 2nd ed. Hoboken, N.J.: Wiley.

Akaike, H. (1973). Information theory as an extension of the maximum likelihood principle. In B. N. Petrov & F. Csaki (eds.), *Second International Symposium on Information Theory* (pp. 267–281). Budapest: Akademiai Kiado.

Atkinson, K. E. (1989). *An Introduction to Numerical Analysis*, 2nd ed. New York: Wiley.

Azzalini, A. (1996). *Statistical Inference Based on the Likelihood*. London: Chapman & Hall.

Azzalini, A. & Scarpa, B. (2004). *Analisi dei dati e data mining*. Milan: Springer-Verlag Italia.

Bellman, R. E. (1961). *Adaptive Control Processes*. Princeton, N.J.: Princeton University Press.

Berry, M. J. A. & Linoff, G. (1997). *Data Mining Techniques: For Marketing, Sales, and Customer Support*. New York: Wiley.

Bishop, Y. M. M., Fienberg, S. E., & Holland, P. W. (1975). *Discrete Multivariate Analysis: Theory and Practice*. Cambridge: Cambridge University Press.

Bowman, A. W. & Azzalini, A. (1997). *Applied Smoothing Techniques for Data Analysis: The Kernel Approach with S-Plus Illustrations*. Oxford: Oxford University Press.

Breiman, L. (1996). Bagging predictors. *Machine Learning*, 24, 123–140.

Breiman, L. (2001a). Random forests. *Machine Learning*, 45, 5–32.

Breiman, L. (2001b). Statistical modeling: The two cultures. *Statistical Science*, 16(3), 199–215.

Breiman, L., Friedman, J. H., Olshen, R. A., & Stone, C. J. (1984). *Classification and Regression Trees*. Monterey: Wadsworth.

Burnham, K. P. & Anderson, D. R. (2002). *Model Selection and Multimodel Inference: A Practical Information-Theoretic Approach*, 2nd ed. New York: Springer Verlag.

Casella, G. & Berger, R. L. (2002). *Statistical Inference*, 2nd ed. Pacific Grove: Duxbury Press.

Claeskens, G. & Hjort, N. L. (2008). *Model Selection and Model Averaging*. Cambridge: Cambridge University Press.

Cleveland, W. (1979). Robust locally weighted regression and smoothing scatterplots. *Journal of the American Statistical Association*, 74, 829–836.

Cleveland, W. & Devlin, S. (1988). Locally-weighted regression: An approach to regression analysis by local fitting. *Journal of the American Statistical Association*, 83, 596–610.

Cleveland, W. S., Grosse, E., & Shyu, M.-J. (1992). Local regression models. In J. M. Chambers & T. Hastie (eds.), *Statistical Models in S* (pp. 309–376). Pacific Grove: Duxbury Press.

Cook, R. D. & Weisberg, S. (1999). *Applied Regression Including Computing and Graphics.* New York: Wiley.

Cox, D. R. (1997). The current position of statistics: A personal view. *International Statistical Review*, 65, 261–276.

Cox, D. R. & Hinkley, D. V. (1979). *Theoretical Statistics*, 2nd ed. London: Chapman and Hall.

Cox, D. R. & Wermuth, N. (1998). *Multivariate Dependencies: Models, Analysis, and Interpretation*. London: Chapman and Hall.

Craven, P. & Wahba, G. (1978). Smoothing noisy data with spline functions: Estimating the correct degree of smoothing by the method of generalized cross-validation. *Numerische Mathematik*, 31, 377–403.

Cristianini, N. & Shawe-Taylor, J. (2000). *An Introduction to Support Vector Machines and other Kernel-Based Learning Method*. Cambridge: Cambridge University Press.

Davison, A. C. & Hinkley, D. V. (1997). *Bootstrap Methods and Their Application.* Cambridge: Cambridge University Press.

Dawid, A. P. (2006). Probability forecasting. In S. Kotz, C. B. Read, N. Balakrishnan, & B. Vidakovic (eds.), *Encyclopedia of Statistical Sciences*, 2nd ed., vol. 10 (pp. 6445–6452). New York: Wiley.

de Boor, C. (1978). *A Practical Guide to Splines*. New York: Springer Verlag.

Efron, B., Hastie, T., Johnstone, I., & Tibshirani, R. (2004). Least angle regression (with discussion). *Annals of Statistics*, 32, 407–499.

Efron, B. & Tibshirani, R. (1993). *An Introduction to the Bootstrap*. New York: Chapman and Hall.

Fahrmeir, L. & Tutz, G. (2001). *Multivariate Statistical Modelling Based on Generalized Linear Models*, 2nd ed. New York: Springer Verlag.

Fan, J. & Gijbels, I. (1996). *Local Polynomial Modelling and its Applications.* London: Chapman and Hall.

Fine, T. L. (1999). *Feedforward Neural Network Methodology*. New York: Springer Verlag.

Fisher, R. A. (1936). The use of multiple measurements in taxonomic problems. *Annals of Eugenics*, 7, 179–188.

Foster, D. P., Stine, R. A., & Waterman, R. P. (1998). *Business Analysis Using Regression: A Casebook*. New York: Springer Verlag.

Freund, Y. & Schapire, R. (1996). Experiments with a new boosting algorithm. In L. Saitta (ed.), *Machine Learning: Proceedings of the Thirteenth International Conference*, vol. 35 (pp. 148–156). San Francisco: Morgan Kaufmann.

Friedman, J. (1991). Multivariate adaptive regression splines (with discussion). *Annals of Statistics*, 19(1), 1–141.

Friedman, J., Hastie, T., & Tibshirani, R. (2000). Additive logistic regression: A statistical view of boosting (with discussion). *Annals of Statistics*, 28(2), 337–407.

Friedman, J. & Tukey, J. (1974). A projection pursuit algorithm for exploratory data analysis. *IEEE Transactions on Computers, Series C*, 23, 881–889.

Golub, G. H. & Van Loan, C. F. (1983). *Matrix Computations*. Baltimore, Md.: Johns Hopkins University Press.

Gower, J. C. (1971). A general coefficient of similarity and some of its properties. *Biometrics*, 27, 857–871.

Green, P. J. & Silverman, B. W. (1994). *Nonparametric Regression and Generalized Linear Models: A Roughness Penalty Approach*. London: Chapman and Hall.

Hand, D. J. (1981). *Discrimination and Classification*. Chichester: Wiley.

Hand, D. J. (1982). *Kernel Discriminant Analysis*. Chichester: Wiley.

Hand, D. J., Mannila, H., & Smyth, P. (2001). *Principles of Data Mining*. Cambridge, Mass.: MIT Press.

Hand, D. J., McConway, K. J., & Stanghellini, E. (1997). Graphical models of applicants for credit. *IMA Journal of Mathematics Applied in Business and Industry*, 8, 143–155.

Hartigan, J. A. (1975). *Clustering Algorithms*. New York: Wiley.

Hastie, T., Tibshirani, R., & Friedman, J. (2009). *The Elements of Statistical Learning: Data Mining, Inference, and Prediction*, 2nd ed. New York: Springer Verlag.

Hastie, T. J. & Tibshirani, R. J. (1990). *Generalized Additive Models*. London: Chapman and Hall. Reprint 1999.

Hoerl, A. & Kennard, R. (1970). Ridge regression: Biased estimation for non-orthogonal problems. *Technometrics*, 12, 55–67.

Hosmer, D. W. & Lemeshow, S. (1989). *Applied Logistic Regression*. New York: Wiley.

Hotelling, H. (1933). Analysis of a complex of statistical variables into principal components. *Journal of Educational Psychology*, 24, 417–441, 498–520.

Huber, P. (1985). Projection pursuit. *Annals of Statistics*, 13, 435–475.

Hurvich, C. M., Simonoff, J. S., & Tsai, C.-L. (1998). Smoothing parameter selection in nonparametric regression using an improved Akaike information criterion. *Journal of the Royal Statistical Society, Series B*, 60, 271–293.

Izenman, A. J. (2008). *Modern Multivariate Statistical Techniques*. New York: Springer Verlag.

Johnson, R. & Wichern, D. (1998). *Applied Multivariate Statistical Analysis*, 4th ed. Upper Saddle River, N.J.: Prentice Hall.

Jolliffe, I. (2002). *Principal Component Analysis*. New York: Springer Verlag.

Jones, L. (1992). A simple lemma on greedy approximation in Hilbert space and convergence rates for projection pursuit regression and neural networks. *Annals of Statistics*, 20, 608–613.

Kaufman, L. & Rousseeuw, P. J. (2009). *Finding Groups in Data: An Introduction to Cluster Analysis*. Hoboken, N.J.: Wiley.

Kendall, M. G. & Stuart, A. (1969). *The Advanced Theory of Statistics, 3rd ed., vol. 1: Distribution Theory*. London: Charles Griffin.

Kolmogorov, A. (1957). On the representation of continuous functions by superposition of continuous functions of one variable and addition. *Doklady Akademiia Nauk SSSR*, 114, 953–956.

Lauritzen, S. L. (1996). *Graphical Models*. Oxford: Oxford University Press.

Loader, C. (1999). *Local Regression and Likelihood*. New York: Springer Verlag.

Mardia, K. V., Kent, J. T., & Bibby, J. M. (1979). *Multivariate Analysis*. London: Academic Press.

McCullagh, P. & Nelder, J. A. (1989). *Generalized Linear Models*. London: Chapman and Hall.

McLachlan, G. J. (1992). *Discriminant Analysis and Statistical Pattern Recognition.* New York: Wiley.

Miller, A. J. (2002). *Subset Selection in Regression.* Boca Raton, Fla.: Chapman and Hall/CRC.

Ohlsson, E. & Johansson, B. (2010). *Non-Life Insurance Pricing with Generalized Linear Models.* Heidelberg: Springer Verlag.

Pearson, K. (1901). On lines and planes of closest fit to systems of points in space. *Philosophical Magazine,* 2(6), 559–572.

Plackett, R. L. (1950). Some theorems in least squares. *Biometrika,* 37(1–2), 149–157.

Quinlan, J. R. (1993). *C4.5: Programs for Machine Learning.* San Mateo, Calif.: Morgan Kaufmann.

Rao, C. R. (1973). *Linear Statistical Inference and its Applications,* 2nd ed. New York: Wiley.

Ripley, B. D. (1996). *Pattern Recognition and Neural Networks.* Cambridge: Cambridge University Press.

Stone, C. J., Hansen, M. H., Kooperberg, C., & Truong, Y. K. (1997). Polynomial splines and their tensor products in extended linear modeling (with discussion). *Annals of Statistics,* 25, 1371–1470.

Stone, M. (1974). Cross-validatory choice and assessment of statistical predictions (with discussion). *Journal of the Royal Statistical Society, Series B,* 36, 111–147. (Corr: 1976, vol. 38, p. 102).

Tibshirani, R. (1996). Regression shrinkage and selection via the lasso. *Journal of the Royal Statistical Society, Series B,* 58, 267–288.

Trefethen, L. N. & Bau, D. (1997). *Numerical Linear Algebra.* Philadelphia: Society for Industrial and Applied Mathematics.

Tse, Y.-K. (2009). *Nonlife Actuarial Models. Theory, Methods and Evaluation.* Cambridge: Cambridge University Press.

Vapnik, V. (1998). *Statistical Learning Theory.* New York: Wiley.

Venables, W. N. & Ripley, B. D. (2002). *Modern Applied Statistics with S,* 4th ed. New York: Springer Verlag.

Wand, M. P. & Jones, M. C. (1995). *Kernel Smoothing.* London: Chapman and Hall.

Wasserman, L. (2004). *All of Statistics: A Concise Course in Statistical Inference.* New York: Springer Verlag.

Weisberg, S. (2005). *Applied Linear Regression,* 3rd ed. New York: Wiley.

Whittaker, J. (1990). *Graphical Models in Applied Multivariate Statistics.* Chichester: Wiley.

Wolpert, D. H. & MacReady, W. G. (1999). An efficient method to estimate bagging's generalization error. *Machine Learning,* 35(1), 41–55.

Zaki, M. J. (2001). Spade: An efficient algorithm for mining frequent sequences. *Machine Learning,* 42, 31–60.

AUTHOR INDEX

SUBJECT INDEX